AF327643

Metallurgical and Ceramic Protective Coatings

Metallurgical and Ceramic Protective Coatings

Edited by

Kurt H. Stern

Chemistry Division
Naval Research Laboratory
Washington D.C.

CHAPMAN & HALL
London · Weinheim · New York · Tokyo · Melbourne · Madras

Published by Chapman & Hall, 2–6 Boundary Row, London SE1 8HN, UK

Chapman & Hall, 2–6 Boundary Row, London SE1 8HN, UK

Chapman & Hall GmbH, Pappelallee 3, 69469 Weinheim, Germany

Chapman & Hall USA, 115 Fifth Avenue, New York NY 10003, USA

Chapman & Hall Japan, ITP-Japan, Kyowa Building, 3F, 2-2-1 Hirakawacho, Chiyoda-ku, Tokyo 102, Japan

Chapman & Hall Australia, 102 Dodds Street, South Melbourne, Victoria 3205, Australia

Chapman & Hall India, R. Seshadri, 32 Second Main Road, CIT East, Madras 600 035, India

First edition 1996

© 1996 Chapman & Hall

Typeset in 10/12pt Times by AFS Image Setters Ltd, Glasgow

Printed in Great Britain by St Edmundsbury Press, Bury St Edmunds, Suffolk

ISBN 0 412 54440 7

Apart from any fair dealing for the purposes of research or private study, or criticism or review, as permitted under the UK Copyright Designs and Patents Act, 1988, this publication may not be reproduced, stored, or transmitted, in any form or by any means, without the prior permission in writing of the publishers, or in the case of reprographic reproduction only in accordance with the terms of the licences issued by the Copyright Licensing Agency in the UK, or in accordance with the terms of licences issued by the appropriate Reproduction Rights Organization outside the UK. Enquiries concerning reproduction outside the terms stated here should be sent to the publishers at the Glasgow address printed on this page.

The publisher makes no representation, express or implied, with regard to the accuracy of the information contained in this book and cannot accept any legal responsibility or liability for any errors or omissions that may be made.

A catalogue record for this book is available from the British Library

♾ Printed on permanent acid-free text paper, manufactured in accordance with ANSI/NISO Z39.48-1992 and ANSI/NISO Z39.48-1984 (Permanence of Paper).

Contents

Contributors

Carol S. Ashley
Sandia National Laboratories
Albuquerque NM 87185
United States

Robert Bianco
Department of Materials Science and Engineering
Ohio State University
Columbus OH 43210-1179
United States

N. Birks
Metallurgical and Materials Engineering Department
University of Pittsburgh
Pittsburgh PA 15261
United States

C. Jeffrey Brinker
Center for Microengineered Ceramics
University of New Mexico
Albuquerque NM 87131
United States

Richard A. Cairncross
Sandia National Laboratories
Albuquerque NM 87185
United States

Ken S. Chen
Sandia National Laboratories
Albuquerque NM 87185
United States

S.B. Dunkerton
The Welding Institute
Abington
Cambridge CB1 6AL
United Kingdom

H. Herman
Thermal Spray Laboratory
Department of Materials Science and Engineering
University of New York
Stony Brook NY 11794-2275
United States

Alan J. Hurd
Sandia National Laboratories
Albuquerque NM 87185
United States

R.L. Jones
Chemistry Division
Naval Research Laboratory
Washington D.C. 20375-5342
United States

J. Mazumder
Center for Laser Aided Materials Processing
University of Illinois
Urbana IL 61801
United States

G.H. Meier
Metallurgical and Materials Engineering Department
University of Pittsburgh
Pittsburgh PA 15261
United States

F.S. Pettit
Metallurgical and Materials Engineering Department
University of Pittsburgh
Pittsburgh PA 15261
United States

Robert A. Rapp
Department of Materials Science and Engineering
Ohio State University
Columbus OH 43210-1179
United States

Scott T. Reed
Sandia National Laboratories
Albuquerque NM 87185
United States

David Rickerby
Rolls-Royce Surface and Technology Group
Derby DE2 8BJ
United Kingdom

S. Sampath
Thermal Spray Laboratory
Department of Materials Science and Engineering
University of New York
Stony Brook NY 11794-2275
United States

Joshua Samuel
Sandia National Laboratories
Albuquerque NM 87185
United States

P. Randall Schunk
Sandia National Laboratories
Albuquerque NM 87185
United States

Robert W. Schwartz
Sandia National Laboratories
Albuquerque NM 87185
United States

Cathy S. Scotto
Sandia National Laboratories
Albuquerque NM 87185
United States

Kurt H. Stern
Chemistry Division
Naval Research Laboratory
Washington D.C. 20375-5342
United States

Acknowledgment

I would like to thank my wife Faith for extensive editorial assistance in the preparation of this book and for literature searches at the Library of Congress.

1

Introduction

Kurt H. Stern

1.1 PROTECTIVE COATINGS

Modern technology has placed increasing stresses on the material used for a variety of technological uses. Both the uses and the stresses have probably grown at a greater rate than the number of materials that can be used to meet them. This is particularly true for structural materials which are almost entirely metallic. Although much has been done by producing new alloys with improved properties, there is a limit to the protection that can be afforded by this means alone. For this reason coatings have played an increasing role in protecting the structural metals from stress.

1.2 STRESSES

Stresses can conveniently be classified into a few types.

1.2.1 Wear

Wear occurs when material is removed from a surface by abrasion, i.e. the moving of one surface over another which, in severe cases, may result in detachment of the coating, and by erosion, the loss of material from the impact of high velocity particles.

Metallurgical and Ceramic Protective Coatings. Edited by Kurt H. Stern.
Published in 1996 by Chapman & Hall, London. ISBN 0 412 54440 7

1.2.2 Chemical attack

This may occur both in the gaseous and liquid phase, particularly at high temperatures, and includes attack by ambient oxygen. Corrosive liquids include acids and bases, molten salts and slags, i.e. molten oxides and silicates. Chemical attack may include a combination of these, such as occurs in marine gas turbines when ambient oxygen, droplets of sea salt and sulfur impurities in the fuel combine to form molten sodium sulfate which attacks the metallic turbine blade coating.

1.2.3 High temperatures

Structural materials may undergo undesirable changes when their temperature is raised, such as mechanical weakening or increased susceptibility to other kinds of attack. One approach to this problem is the application of thermal barrier coatings which reduce the substrate temperature. Another is the use of ablation coatings which reduce the surface temperature of space vehicles reentering the earth's atmosphere by removing excess heat as the heat of vaporization of the coating material.

1.3 SELECTION OF COATING MATERIALS

The types of materials used for coatings are relatively few – metals and a few classes of inorganic (ceramic) compounds (oxides, carbides, borides and nitrides) singly or in combination. Note that many useful coating materials, such as paints and plastics, are not mentioned here because this book deals with stresses which are too severe for them.

In selecting a coating material it is necessary to consider the uses to which it will be put. Factors to be considered include:

(a) the properties of the coating material itself – its melting point, hardness, vapor pressure, density and thermal expansion coefficient;
(b) the resistance of the coating material to the attack expected;
(c) the compatibility of the coating and substrate over the temperature range of the expected application. This includes the minimizing of thermal stresses, by matching thermal expansion coefficients, and the provision of good coating–substrate adhesion. Some interdiffusion may be desirable, but an excessive amount, such as may occur with silicon plating at high temperatures, may only lead to bulk diffusion and alloy formation; and
(d) the cost. Ultimately whether or not a particular coating will be used depends on the trade-off between the benefits to be gained and the additional cost to be incurred. When the application is critical, and the consequences of failure disastrous, higher costs are usually justified, particularly when there exists no reasonable alternative. However, cost is always a 'moving target' since a great need frequently leads to improvements in existing methods and the development of new ones.

It has recently been argued [1] that the rapid introduction of tribological coatings will require closer cooperation between coating experts and mechanical engineers if a new coating is to be successfully introduced. Coating experts use the following parameters to identify a new coating; coating method, composition, thickness, hardness, coating–substrate adhesion and friction and wear data, e.g. pin-on-disk. The following information is required by the mechanical engineer to design a coated machine element: Young's modulus, Poisson's ratio, thermal expansion coefficient, thermal conductivity, density, specific heat of both coating and substrate and information on residual stresses to assess the overall stress level exhibited by the coated body.

The authors point out that the information produced by the coating expert is largely useless to the designer. Coatings will only be introduced if the elements needed in design analysis are furnished together with the coating.

1.4   METHODS OF APPLICATION

New methods for applying coatings, improvements in existing methods, and new applications have proliferated in recent years, driven by technological need. As Bunshah [2] has pointed out, there is no unique way to classify these methods, which may, for example, be classified as vapor vs. condensed phase, chemical vs. physical, or atomistic vs. bulk.

Most of the coating methods are described in journal articles and conference proceedings which leave the nonspecialist potential user with a bewildering number of choices with which he is probably not equipped to deal. Books, which might be expected to take a broader view, are in short supply. Prominent among current books is the volume edited by Bunshah [2] which deals largely with vapor phase methods and the recently published book on ceramic films and coatings edited by Wachtman and Haber [3], which includes both protective coatings and coatings for electronic and optical applications. Note that both of these books are multiauthor, in recognition of the fact that no single individual can be intimately familiar with all the curently used methods for applying coatings.

1.5 ELECTROPLATING – AN EXAMPLE OF A TECHNOLOGY

The electroplating of metals from solutions, nearly always aqueous and at ambient temperature, has its roots in the work of Faraday, who enunciated the fundamental laws bearing his name, laws which relate the quantity of electrical charge passed between the electrodes in solution and the quantity and valence of material deposited. Long before more modern methods of depositing coatings were developed, many of the principles relevant to electroplating, such as the thermodynamics of electrode potentials and the kinetics of diffusion and the movement of ions under an applied field in solution, as well as their discharge at the electrode surface, had been worked out.

There have also been more recent developments. Many of these developments are scattered in the journal literature and have not yet found their way into books for the general technical reader.

A recently introduced method for improving the properties of electroplated coatings is the pulsed reversed current (RC) technique [4]. The cathodic curent is periodically interrupted by an anodic pulse which briefly redissolves some of the just deposited coating. This introduces two more parameters into the plating process: the pulse frequency v and the duty cycle $T/(T + T')$, where T is the (cathodic) deposition time and T' is the (anodic) dissolution time. Obviously $T > T'$. Optimizing RC conditions can lead to improvements in coating smoothness, and texture, somewhat comparable to the use of organic brighteners.

Another technique which has recently received increased attention is the electrodeposition of layered alloy coatings, usually called cylic multilayered (or compositionally modulated) alloy coatings (CMA). Although the technique goes back to early work by Brenner [5], it was more recently revived and extended by Cohen, Koch, and Sardi [6] who showed that layered coatings of Ag–Pd could be plated from a single bath by periodic alteration of current or potential. Spacings of less than 100 nm were achieved. Subsequently, spacings of Cu–Ni alloys as small as 0.8 nm were achieved by Yahalom and Zadok [7]. As poined out by Lashmore and Dariel [8], electrodeposition presents, in principle, several advantages over vapor phase methods for the production of layered alloys. Among them is the strong tendency for the coating to grow epitaxially and thus form materials with a texture determined by the substrate; the process is inexpensive and can easily be scaled up for plating large parts; and it is carried out at room temperature, avoiding coating–substrate diffusion.

These alloys have been of interest for their mechanical, electrical and magnetic properties [9]. In the context of this book the interest is in mechanical properties. It has been shown [10] that CMAs exhibit increased fracture strength, as much as three times that of the constituent metals, and enhanced wear performance in both dry sliding and in lubricating conditions.

Much remains to be learned about CMAs – which alloys can be plated, optimization of the plating parameters and relation of the coating characteristics, such as layer spacing, to the properties of interest, but it is already clear that work in this field is worth pursuing.

Although the editor wished to include a chapter on electrodeposition in the present volume, he was unable to find anyone willing to write it. It is interesting that the most recent book on electroplating written by a single author is the monumental two-volume *Electroplating of Alloys* by Brenner [5], which includes historical background, basic principles and a thorough discussion of all the alloys that could be plated from aqueous solution. Brenner was chief of the Electrodeposition Section at the National Bureau of Standards (now NIST, the National Institute of Standards and Technology) who devoted his entire career to this field. There are very few recent books. Frederick Lowenheim's *Modern Electroplating* [11] is an excellent multiauthor text, published in 1974, which covers both general principles and includes chapters on all the important metals and many alloys. The recent book by Dini [12], although it has 'electroplating' in its title, deals primarily with the properties of the plated coatings, such as porosity, stress, corrosion and wear, rather than with methods for plating metals and alloys.

The current situation, that of rapidly expanding scientific and technical fields that no single person can master, as well as the increasing demands placed on scientists, particularly in technology, has probably mitigated against broad, single-author books. The trend seems to be toward multiauthor books, but even they require an editor to organize them. The present book is a good example. I was recently asked to write an entire book on coatings, but found this well beyond my competence, though I agreed to write some chapters related to molten salt electroplating. Urged to serve as editor, I spent six months learning about coatings and trying to decide what should be in the book. Thus I must take whatever responsibility attaches to the selection of topics.

1.6 PLAN OF THIS BOOK

The purpose of the present volume is to explore in detail several methods, primarily carried out in the condensed state, which have not been reviewed recently. Although the emphasis is on the condensed state, some of the methods cannot be classified unambiguously. For example, although pack cementation is carried out by packing articles to be coated into a container, surrounded by a powdered mixture, the actual formation of a corrosion-resistant layer occurs by gas phase transport from the powder to the metal.

The methods also differ widely in the degree to which they have become technologically established. Thus, pack cementation is a currently used industrial method, with relatively little basic science still to be done. On the other hand, the electrodeposition of refractory metals and compounds is almost entirely in the basic research phase, and is included because it has the potential for giving rise to a useful technology.

In addition to the chapters on coating methods, chapters on coating corrosion and the measurement of coating adhesion have also been included. Coating corrosion is a serious problem at high temperatures in corrosive atmospheres, and much effort has gone into reducing it. Coating adhesion is of great importance in preventing the removal of a coating from a substrate by mechanical stresses.

The book begins, because of the editor's interest, with three chapters on electrodeposition from molten salts. Refractory metals, because of their desirable properties, had been the object of interest for a long time until many futile efforts showed that hydrogen overvoltage problems prevented them from being plated from aqueous solutions. It was not until the 1960s that Senderoff and Mellors showed that adherent, smooth deposits of all the refractory metals could be plated from alkali metal fluoride melts. However, commercialization of the process has lagged because of the stringent conditions required for melt purification and atmospheric control, and the corrosive nature of fluoride melts. Although some metals, e.g. tantalum, have been plated commercially, further spread of the method may depend on current research in finding other melts, such as chlorides, and lowering the temperature from the 700–800 °C range required for fluorides.

Metalliding differs from electroplating primarily because it employs higher temperatures rather than an applied voltage. These higher temperatures cause diffusion of the anode material to produce alloys within the cathode material, rather than coatings on the surface. The process was invented by Cook at the General Electric Company during the 1960s. He found nearly four hundred combinations to be feasible, with the diffusing element being the more electromechanically active. In spite of this success, General Electric chose not to continue this effort and turned over its patents to the Metalliding Institute of Gannon University in Erie, Pennsylvania, which provides consulting services to companies inerested in using the methods. However, there have been virtually no publications on the method in the scientific literature, and the information in Chapter 3 is based on the original patent literature. The subject is included because it produces highly useful coatings (surface alloys) and deserves to be more widely known.

There are only a few classes of refractory compounds which can be electroplated – borides, carbides and silicides – all of them from molten salts. Titanium boride is plated commercially (from borate melts) because of its excellent corrosion resistance to hot, concentrated aqueous brines. But work on electroplating other compounds is still in the research phase; these coatings are applied by chemical vapor deposition (CVD) or alternative methods. The high temperatures and long times required for silicide coatings tend to promote silicon diffusion into the substrate (siliciding) and require a barrier coating to produce a coating of uniform composition. Nevertheless, for some applications, such as large objects, electroplating may be superior to other methods, and further research seems warranted. Use of melts less corrosive than fluorides and use of lower temperatures would, as for metals, make electrodeposition of compounds more useful.

Unusual among all the methods for applying ceramic coatings, in that it begins with colloidal dispersion, is sol–gel coating. Of interest because of its versatility, low processing temperature and low cost, sol–gel coating has been used to prepare moisture and diffusion barriers, to improve the fracture resistance of ceramics and to make plastics wear and scratch resistant. Brincker *et al.* describe the scientific aspects of film formation in detail. They begin with its application by dipping the substrate into the sol or by spin-casting the sol on the substrate, and they end with its consolidation as a ceramic coating. The physics required to describe the entire process is formidable. As an example of an application the authors describe the protection of optical devices on space vehicles from attack by atomic oxygen in the upper atmosphere.

Lasers provide a unique tool for high quality surface modification. They are used to form surface alloys by melting a thin layer at the surface of a metal while simultaneously adding the desired surface element. The rapid solidification, resulting from the passage of the laser over the surface, often leads to novel metastable phases and superior properties. If the dilution of the surface region by the substrate metal is minimized, the process is called laser cladding.

A rather different application of lasers is chemical vapor deposition (LCVD) in which deposition is achieved by photodissociation or by thermal dissociation. It differs from the usual CVD in that deposition is localized. Various aspects of this technique, and the use of lasers in surface modification generally, are described in detail by Mazumder in Chapter 5.

In contrast to laser cladding, thick coatings may also be applied by mechanical means, usually referred to as solid phase cladding. Dunkerton provides a comprehensive review of the methods used to form coatings in the 0.5–5.0 mm range, including explosive cladding, the high velocity impact of plates and isostatic pressing. Emphasis is on the newer methods, such as friction surfacing, in which a rotating consumable is brought into contact with a stationary substrate. The bar is heated by friction, resulting in a plasticized layer which spreads over the substrate. Advantages include no melting of the coating or the substrate, excellent adhesion and the wide variety of possible combinations. So far, the method has been applied to steels and aluminum alloys, and has found use in reclaiming worn parts and forming the edges of cutting tools.

Thermal barrier coatings, generally stabilized zirconia, are used primarily to protect gas turbines and their metallic coatings, such as the MCrAlY alloys (M = Ni, Co, Ni, Al, Y). Their application permits higher operating temperatures and results in significant increases in efficiency.

Jones describes work in this field: the optimizing of coating composition, such as Y_2O_3 or HfO_2, which stabilize ZrO_2 in its high temperature tetragonal or cubic form; methods used to produce the coatings; and their failure modes, such as oxidation, hot corrosion and cracking. This is a technologically important field in which science and technology interact.

Pack cementation has been used for several decades to produce wear- and corrosion-resistant coatings on inexpensive or otherwise inadequate substrates. The method is a quasi-CVD batch process in which the part to be coated is packed in a sealed container, surrounded by a mixture consisting of the metal to be coated, frequently chromium or aluminum, a halide salt activator, e.g. NaCl, NaF or NH_4Cl, and an inert filler material. Heating to 800–1000 °C forms volatile halides which deposit on the substrate and cause the coating metal to diffuse into the substrate. Bianco and Rapp describe the process in detail and discuss recent work on the formation of alloy coatings.

Coatings may also be applied by spraying metal or ceramic powders, melted by oxyfuel combustion or electric arc plasma, onto metallic substrates. Herman and Sampath provide a comprehensive overview of the method, including the relation between the feed material characteristics and the properties of the deposit. These coatings provide an important means for improving wear and corrosion protection.

All metals are subject to corrosion by gases, such as oxygen, sulfur and CO_2, as well as by molten salts (Na_2SO_4, $NaVO_3$) at high temperatures. Coatings confer oxidation resistance by forming protective reaction barriers, such as Al_2O_3, SiO_2 and Cr_2O_3. Birks, Meyer and Pettit discuss the types of degradation suffered by metallic and ceramic coatings, and methods for alleviating this problem. Engineers are constantly raising the operating temperatures of engines to improve thermodynamic efficiency; coatings which were adequate at lower temperatures may no longer be adequate, and the corrosion problem will persist. The contents of this chapter are always likely to be relevant.

Finally, we consider one of the most important requirements of a coating; its adhesion to the substrate. If a coating is to be used successfully, it must not detach under conditions of use. Thus the tests which measure adhesion must themselves be carefully evaluated to see that they meet this need. Rickerby describes the various adhesion tests, giving a fundamental viewpoint, e.g. the forces involved, and the practical tests – from those that measure the breaking of coating–substrate bonds to those that pull off the coating mechanically.

REFERENCES

1. Godet, M., Berthier, Y., Vincent, L. and Flamand, L. (1991) *Surf. Coat. Technol.*, **45**, 1–8.
2. Bunshah, R.F. *et al.* (1982) *Deposition Technologies for Films and Coatings*, Noyes, Park Ridge, NJ.
3. Wachtman, J.B. and Haber, R.A. (1993) *Ceramic Films and Coatings*, Noyes, Park Ridge, NJ.
4. Kollia, C., Loizos, Z. and Spyrellis, N. (1991) *Surf. Coat. Technol.*, **45**, 155–160.
5. Brenner, A. (1963) *Electroplating of Alloys*, 2 vols, Academic Press, New York.
6. Cohen, U., Koch, F.B. and Sardi, R. (1983) *J. Electrochem. Soc.*, **130**, 1987–95.
7. Yahalom, J. and Zadok, O. (1987) *J. Mater. Sci.*, **22**, 499–503.
8. Lashmore, D.S. and Dariel, M.P. (1988) *J. Electrochem. Soc.*, **135**, 1218–21.
9. Lashmore, D.S. *et al.* (1991) *Proc. Electrochem. Soc.*, **92**(10), 381–6.
10. Focke, T. and Lashmore, D.S. (1992) *Scripta Metall.*, **27**, 651–6.
11. Lowenheim, F. (1974) *Modern Electroplating*, 3rd ed., Wiley, New York.
12. Dini, J.W. (1993) *Electrodeposition: The Materials Science of Coatings and Substrates*, Noyes, Park Ridge, NJ.

2

Electrodeposition of refractory metals from molten salts

Kurt H. Stern

2.1 INTRODUCTION

The refractory metals are generally considered to consist of the nine elements from Group IV to VI of the periodic table, as shown in Table 2.1. The term *refractory* refers to their high melting points, and indeed the melting points range from 1675 °C for titanium to 3410 °C for tungsten. The melting points

Table 2.1 Selected properties of the refractory metals

Refractory metal	Atomic number	Melting point (°C)	Boiling point (°C)	Density (g/cm^3)
Titanium	22	1670	3280	4.51
Vanadium	23	1895	3400	6.1
Chromium	24	1867	2670	7.14
Zirconium	40	1853	4375	6.44
Niobium	41	2468	4700	8.66
Molybdenum	42	2615	4700	9.01
Hafnium	72	2220	5400	13.2 + 0.1
Tantalum	73	2996	5427	16.63
Tungsten	74	3410	5600	19.3

Metallurgical and Ceramic Protective Coatings. Edited by Kurt H. Stern.
Published in 1996 by Chapman & Hall, London. ISBN 0 412 54440 7

increase left to right and top to bottom of the periodic table. However, these melting points are by no means exceptional; several of the platinum group metals have similar high values (osmium 3000 °C, platinum 1769 °C), but these metals are too expensive and too soft to be useful in structural applications. In contrast, the refractory metals are as hard as most structural metals, e.g. iron and nickel, but are also, with some exceptions, fairly readily available. Some of them are used in alloys, such as vanadium to improve the properties of steel. However, the major attraction is their corrosion resistance; in a wide variety of gaseous and aqeous environments [1–4] it is generally superior to that of the structural metals. Since refractory metals cannot compete with the structural metals in either price or availability, it has proved useful to combine the strength, machinability and low price of the structural metals with the corrosion resistance of the refractory metals by using them as coatings on the structural metals. In addition, refractory metals have good thermal and electrical properties and may well be useful in joining dissimilar materials, such as metals and ceramics, through the formation of thin, reactive intermediate layers [1]. For example, the use of chromium to protect steel from aqueous corrosion is well known. Additionally, some of the refractory metals are in short supply, and considerable savings would result from using them as coatings.

Methods for producing metallic coatings can be divided into two major categories: vapor phase and condensed phase. As has been pointed out by Bunshah [5], beyond this general classification there is no unique way to categorize the many methods which have been developed. In general, vapor phase methods have been described in the book edited by Bunshah, whereas this book deals primarily with condensed phase methods.

The major condensed phase technique for producing refractory metal coatings is electrodeposition. Most electrodeposition, or electroplating, is carried out in aqueous or nonaqueous solutions at ambient temperature. This topic has been discussed in a previous chapter. However, the only refractory metal which can be plated in this way is chromium. It is not surprising that the first atempts to plate the refractory metals were made in these media. The early attempts to plate tungsten in this way, all of which ended in failure, have been recounted by Davis and Gentry [6]. Similar failures occurred with other refractory metals. Although the precise reasons are open to argument, it is an experimental observation that any attempt to plate *pure* refractory metals from aqueous solution always produces hydrogen first. However, alloys of refractory metals with ferrous metals can be plated[6]. Similarly, attempts to electrodeposit refractory metals from nonaqueous solutions at ambient temperatures were also unsuccessful [6]. In addition to more fundamental reasons, these solutions exhibit low electrical conductivity and low solubility of potential metal precursors.

By the 1950s it had become apparent that molten salts offered the only potential route to the electrodeposition of the refractory metals. Before proceeding to a discussion of experimental methods and the individual metals, we define several techniques, all of which can loosely be called 'electrodeposition'.

2.1.1 Electrowinning

The term *electrowinning* can be applied to any electrochemical process in which the metal is produced, in any form. In practice, it is commonly applied to the formation of the metal in nonadherent form, e.g. as a powder. This is useful if further processing is to take place.

2.1.2 Metalliding

Metalliding is a quasi-electrochemical process which results in the formation of a diffusion coating. The element to be diffused is made the anode of a molten salt electrochemical cell (usually fluoride), with the substrate as cathode. If the anode material is more active electrochemically than the cathode, the anode will diffuse into the cathode when the electrodes are electrically connected. Because the process is basically driven by diffusion into the substrate, there will be a diffusion gradient in the substrate, and the time dependence is diffusion limited. In principle, there is no sharp coating–substrate interface. Metalliding is discussed in Chapter 3.

2.1.3 Electrodeposition (electroplating)

Electrodeposition results in the formation of a coating of uniform composition. It is driven by a voltage applied between the anode, consisting of the metal to be plated, and the cathode, which is the substrate to be plated. The molten salt medium, in which the anode and cathode are immersed, consists of a soluble form of the metal to be plated dissolved in a molten salt solvent, such as an alkali metal halide, which does not participate in the plating process. Coating thickness is determined by the electrical charge passed (current × time), and there is no theoretical upper limit to the thickness which may be achieved. In general, there is a sharp interface between coating and substrate. With the elevated temperatures used in molten salt plating, there may be some overlap between this technique and metalliding, with some coating–substrate interdiffusion occurring at the high temperature end of the plating range. Although this is rarely the case for the refractory metals, it readily occurs with more active elements such as silicon [7].

2.1.4 Electroforming

Electroforming is, in principle, electroplating. It results when the coating is made so thick that the substrate can be removed, e.g. from a mandrel, or dissolved away to produce a freestanding article of the coating. For refractory metals it was first demonstrated by Mellors and Senderoff [8], who showed that the ductility of the metals produced electrochemically allowed the formation of complicated shapes, e.g. condenser tubing, in contrast to the brittle metals produced by thermal techniques.

2.2 EXPERIMENTAL TECHNIQUES

2.2.1 General

Several authors have discussed the factors which need to be considered for the successful electrodeposition of the refractory metals. These discussions include the more fundamenal aspects – melt composition, temperature and the detailed chemistry and electrochemistry of the deposition process – as well as the engineering aspects of cell design for maximum yield. Inman and White [2] addressed both of these in a major review published in 1978. A very useful review, covering both electroplating and metalliding was prepared by Sethi [3] at about the same time. More recently, Kipouros and Sadoway [9] have updated this material in a brief review. I have drawn on these reviews without listing all the individual papers cited in them, since these are readily available in the literature.

In what follows, emphasis is placed on factors important for electroplating, rather than engineering aspects, for which the review by Inman and White [2] should be consulted. The major part of this chapter is divided into two parts: the first deals with experimental techniques which are familiar to molten salt electrochemists, but not necessarily to electrochemists familiar only with aqueous techniques; the second covers the molten salt electrochemistry of the nine refractory metals in groups IVA, VA, and VIA of the periodic table, with particular emphasis on the reduction processes which lead to metal deposition.

2.2.2 Elements of molten salt electroplating

The basic elements of electrochemical cell construction are (a) the electrodes, anode and cathode, and (b) the container. The cell contains the melt (solvent and solute) and the electrodes.

The operation of the cell requires selection of (a) temperature, (b) electrochemical parameters, such as voltage and current density, and (c) atmosphere – generally an inert gas.

Electrodes

In molten salt plating, the anode is usually the metal to be plated. Most of the refractory metals are now available commercially in a form convenient for cell construction, generally rods or plates. If only irregularly shaped pieces can be obtained, they can be hung in a wire-mesh basket made of a metal which will not react electrochemically under plating conditions, e.g. platinum.

The cathode is the substrate metal to be plated. For research investigations, cathodes are typically small coupons or wires. For voltammetric studies, small surface areas, such as the cross section of a thin wire imbedded in an inert material, are advantageous, but for corrosive melts, such as fluorides, hardly any nonconductor is truly inert, and the melt is likely to creep into the annular space between the wire and its containing tube. Consequently, the effective cathode area is then unknown. For good coating adherence, a cleaning pretreatment of the cathode surface to remove surface impurities is usually required. This treatment can be mechanical – grinding and polishing – or chemical – acid etching and washing by water and/or organic solvents. Some molten salts, such as fluorides, are themselves good cleaning agents for removing oxide films, but in any case, successful plating requires clean and dry electrodes.

To obtain a uniform coating the anode–cathode geometry must be considered. It is advantageous for the current density (CD) to be uniform, so as to avoid edge effects, i.e. thicker or thinner coatings at the edges than in the middle. If the inside of a hole is to be plated, the anode should be a rod concentric with the hole. Since refractory metals are difficult to machine, the plated coating should be as close to final dimensions as possible.

Athough plating can be carried out between anode and cathode, as long as the decomposition potential of the melt is not exceeded, electrochemical studies and some plating processes require a reference electrode. The fundamentals of electrochemical studies are beyond the scope of this chapter, but some discussion of reference electrodes seems appropriate.

A reference electrode has no current passing through it during the plating process, so that the voltage between the working electrode (the cathode for metal plating) and the reference can be set independently of the anode–cathode voltage. Ideally, the reference electrode is thermodynamically reversible. For example, if the reference is silver, some silver ions must be in the solution in contact with the silver metal, but this solution must not be in contact with the plating solution, to avoid mixing. This is usually accomplished by enclosing the reference in an envelope of a nonconductor, such as glass or ceramic. Electrochemical contact with the plating solution is made either via a liquid junction (pinhole), or by using the mebrane properties of the envelope such as glass [10–12] or porcelain [13]. Most of the envelope materials are suitable for use in nonfluoride halide and sulfate melts, but molten fluorides attack glass and many ceramics. However, a reference electrode based on Ni/Ni (II) in a boron nitride envelope for use in fluorides has been developed [14]. If a reversible electrode is not readily available, an inert metal, such as gold or platinum, can be used as a quasi-reference electrode. Such electrodes are supposed to maintain a constant potential during the electrochemical process. However, this may not always be the case. For example, platinum electrodes were found to shift their potential in fluoride melts when the oxide activity, a generally uncontrolled variable, changed [15].

The melt

The molten salt acts as the medium in which electrolysis takes place. The solvent is rarely a single salt, but a mixture designed to keep the eutectic temperature as low as possible. For example, the melting points of the individual alkali metal fluorides are near 800 °C, but the eutectic temperature of the frequently used ternary eutectic LiF–NaF–KF (FLINAK) is 456 °C. In general, it is desirable to keep the plating temperature as low as possible, consistent with good coating properties. This requirement sometimes necessitates operating considerably above the eutectic temperature; for example, in fluoride melts no good coatings are obtained below 700 °C. Much of the rationale for this is not well understood, but probably involves shifting equilibria among various melt species and/or changes in the electrochemical parameters. Such phenomena have been studied for a few cases (see below).

The solute, a compound containing the refractory metal, must be appreciably soluble in the melt, generally 5–10%, and be reducible within the 'electrochemical window' of the solvent, i.e. the solute must be reducible in a potential range where the solvent is not. For example, Mellors and Senderoff [8] plated all the refractory metals from FLINAK, using the fluorides of the refractory metals, e.g. K_2TaF_7, as solutes. They are both soluble and reducible. On the other hand, the oxyanions of these metals exhibit very different behaviors. Thus the tungstates (WO_4^{2-}) and molybdates (MoO_4^{2-}) are both soluble and reducible; chromates (CrO_4^{2-}) are soluble but not reducible; and tantalates (TaO_3^-) are not soluble. Details will be discussed under the individual metals and/or their refractory compounds.

For good quality coatings the melt must be free of impurities. Since commercially available chemicals, even of reagent grade, are hardly ever satisfactory, they must be purified before use. Many recipes have been used to accomplish this [16, 17], but the procedure selected depends on the nature of the melt components and the requirements of the particular process. It always includes removal of water by vacuum baking at gradually increasing temperature (with as much water as possible removed at the lowest temperature to avoid hydrolysis), possibly followed by vacuum melting and bubbling an inert gas through the melt. This procedure removes water but not oxides and hydroxides, which must be removed chemically. Chemical steps include sparging with the appropriate dry hydrogen halide gas (HCl for chlorides, HF for fluorides) followed by an inert gas; for fluorides NH_4F has been suggested [17]. In order to remove metallic impurities, such as iron, a preelectrolysis of the melt may be carried out as a final step.

Voltammetry may be used to detect possible interfering impurities in the operating electrochemical window. These remarks apply primarily to the solvent. For solutes, vacuum drying at elevated temperatures is usually sufficient. Obviously, purification should be carried out on solvent and solute, separately, before they are mixed.

Containers

The container material should not react with the melt. Thus the material selected depends on the melt and the temperature. For nonfluoride melts, fused quartz may be adequate, particularly if the melt is dry. For more aggresive melts, such as fluorides, metal containers may be required. Nickel is adequate up to 1000 °C, but some of its alloys, e.g. Inconel, may be attacked. Vitreous (glassy) carbon and platinum are also useful, but are too expensive for large-scale plating.

In any case, the entire corrosion system must be protected from atmospheric corrosion. This is accomplished by surrounding it with an inert gas, generally argon, which itself should be oxygen-free. For laboratory-size studies an inert-atmosphere glove box offers the advantages of visual access to the melt, easy manipulation of additions to the melt and easy handling of the electrodes. If this is not available, and for all larger operations, a system must be built which accomplishes the same objectives. Many designs have been published, but they must all provide for:

1. Immersion of the apparatus in a furnace to keep the melt at the desired temperature.
2. Insertion and removal of electrodes without exposure to oxygen at elevated temperatures; possible additions of solute under the same conditions.
3. No electrical shorting between electrodes.
4. Insertion of a thermocouple or other temperature-measuring device, preferably into the melt, with protection from corrosion.

For two-electrode plating these requirements are probably sufficient. If plating is done with a three-electrode system, i.e. with a reference electrode, both the electrodes and the (metallic) melt container must be electrically isolated from any outer conducting wall, such as a metal envelope. A recent design proposal for large-scale operation [18] accomplishes isolation and also prevents a buildup of conducting salt film on the inside of the container.

Cell operation

In order to plate a coating, a current must pass between the anode and cathode through the melt. A power source to accomplish this may be operated either at constant current or constant voltage. Most modern power sources permit placing limits on the variable which is not controlled. If only a single substance is reducible over a wide voltage range, such an arrangement is probably quite adequate; the main requirement is the total current the instrument can handle. Since the quality of the coating, i.e. its physical characteristics, is dependent on the electrochemical parameters, particularly the CD, some experimentation will be required to optimize the process.

For three-electrode plating, and for more sophisticated electrochemical studies, a potentiostat is required. Potentiostats may be operated in either constant voltage or constant current mode. In constant voltage mode, the potential between the cathode and the reference is fixed, and the anode–cathode current is allowed to flow to maintain this voltage. Which method is preferable depends on the particular system. Discussion of the details of potentiostatic operation is beyond the scope of this chapter, but this information is generally available in electrochemistry texts and instrument manuals. A distincton can be made between research potentiostats, which allow for very sophisticated kinetic studies but are generally limited to currents less than 1A, and potentiostats used for plating, which are less sophisticated but have the higher current capabilities required for plating larger surface areas.

2.3 THE REFRACTORY METALS

2.3.1 General

The refractory metals consist of the nine elements in Groups IVA, VA, and VIA of the periodic table.

> IVA: titanium, zirconium, hafnium
> VA: vanadium, niobium, tantalum
> VIA: chromium, molybdenum, tungsten

As a group they are characterized by high melting points, generally sufficient hardness (comparable to those of the structural metals) and good corrosion resistance to acids and to oxygen at moderately elevated temperatures. Some of these properties are shown in Table 2.1. Hardness depends on the method of preparation and is therefore not included. Partly because some refractory metals are in short supply and some have high densities, coating them on structural metals has been found to be more desirable than using them in bulk.

Electroplating of the refractory metals has reached different stages for different metals. For some, efforts still consist primarily of studies to elucidate the electrochemical mechanism for the reduction process; others have already been successfully plated.

The stages of progress toward successful plating can be described as follows:

(a) Selection of system to be studied: solvent, solute, substrate, and temperature. The choices may be based on what is already known, the interests of the investigator and practical considerations such as the particular substrate required and the properties desired of the coating.

(b) Thermodynamics and kinetics of the reduction–deposition process. An excellent review of the electrochemistry of all the elements in various molten salt systems has been prepared by Plambeck [19]. This book covers not only reduction processes, but all molten salt electrochemistry and should be consulted before proceeding in this field.

(c) Selection of the best variables for plating, based on results obtained in (a) and (b).

(d) Design of a practical plating system. This generally involves a scale-up from laboratory size to pilot plant and larger. Compromises may have to be made at this stage between conditions obtainable in the laboratory and in the plant.

Connections between the various stages are not always clear-cut. For example, some investigators describe the details of various electrochemical processes because of their intrinsic interest, without necessarily pointing out their relevance to plating, whereas others study plating in a more Edisonian fashion, without necessarily taking advantage of available electrochemical knowledge.

A successful coating can be judged by meeting various criteria:

(a) The coating is adherent: it resists mechanical stresses, and is sufficiently ductile to follow mechanical deformation of the substrate. A cross section of the coating frequently exhibits columnar structure, and is firmly anchored to the substrate, possibly by interdiffusion at the coating–substrate boundary.

(b) There is a good thermal expansion match between coating and substrate to allow use over a wide temperature range. This factor has to be considered when selecting a candidate coating metal for a particular substrate.

(c) The coating covers the substrate uniformly, with respect to thickness and the absence of edge effects. The surface is smooth and requires minimal machining.

(d) The current efficiency (CE) of the process is high, preferably 100%, i.e. all the charge passed results in coating formation.

From an examination of the literature it is clear that the details of the plating process, and therefore successful plating, are highly dependent on the plating parameters:

(a) molten salt solvent
(b) solute, and its concentration
(c) temperature
(d) electrochemical parameters: voltage, CD
(e) substrate

In general, (a) and (b) determine the solute species in the melt; (c) affects their interconversion and its kinetics as well as the eletrochemical kinetics;

(e) frequently affects the physical properties of the deposit as well as possible interdiffusion zones and alloying.

Before proceeding to a discussion of the individual metals, the seminal work of Senderoff and Mellors in this field must be mentioned. During the 1960s these workers at the Union Carbide Corporation discovered that all of the refractory metals could be successfully plated from molten fluorides. Their electrochemical studies in these melts for all metals except titanium, hafnium and vanadium were published in the open literature and will be discussed in the appropriate section. Detailed 'recipes' are described in their patent [8].

Most subsequent work has been concerned with substituting less corrosive melts from the fluorides, and with lowering the temperature from the 700–800 °C range.

In what follows, the individual metals will be discussed in order of increasing atomic weight in groups IVA, VA and VIA,

This chapter is not intended to be an exhaustive review of the known molten salt electrochemistry of the refractory metals. Rather, the focus is on those aspects related to electroplating, i.e. reduction processes. Inevitably, the selection of the included material is somewhat affected by personal bias.

2.3.2 The refractory metals

Titanium

Although its excellent corrosion resistance and ready availability make titanium a highly desirable coating material, more than 40 years' effort have not yet resulted in a well-established commercial process. This is not for lack of effort, as the published literature attests, but more likely results from the complex chemistry and electrochemistry of titanium in melts. Depending on the melt, Ti(IV), Ti(III), and Ti(II) are all species involved in the plating process, and some of them, particularly Ti(III), may react with the coating already formed to produce Ti(IV) and Ti(II) in an auto-oxidation–reduction reaction.

Already in 1955, Sibert and Steinberg [20] reported a fairly successful process for plating titanium on steel from $NaCl-K_2TiF_6$ melts near 900 °C, but smooth deposits, less than 0.1 mm, were obtained only for fairly short periods, after which the coating became porous and dendritic.

Wurm and coworkers [21–23] examined the chemical aspects of the plating process in $NaCl-KCl$ in the 600–700 °C range, using Na_2TiF_6 and K_2TiF_6 as solutes. They observed the formation of violet compounds on top of the metallic titanium formed on the cathode, and were able to identify these as salts of Ti(III), such as K_2NaTiF_6. Based on the analysis of frozen melts and on voltages observed in their two-electrode cell, they concluded that metallic sodium is the primary reaction product, i.e.

$$NaCl = Na + \tfrac{1}{2}Cl_2$$

and that this metal then reduces Ti(IV) to Ti(III) and subsequently to titanium metal. No mention is made of Ti(II). In an extensive series of plating studies [23] it was found that smooth deposits were obtained in KI–KF melts (note the presence of fluoride), but not in NaI–KI. Similar, but more detailed, conclusions about the mechanism were reached than in previous work [21], this time based on extensive current–voltage curves in their two-electrode cell.

A careful, exensive study of titanium electrochemistry in FLINAK and in $LiF-BeF_2-ZrF_4$ melts was reported by Clayton, Mamantov and Manning [24, 25] who also give references to earlier work. Linear sweep voltammetry, chronopotentiometry and chronoamperometry were used to measure electrochemical parameters for the processes

$$Ti(IV) + e = Ti(III)$$

and

$$Ti(III) + 3e = Ti(0)$$

Potentials are given versus the reversible Ni/Ni (II) reference electrode. Since all their studies were carried out at $500\,°C$, there is some question whether their results are directly applicable to plating, since this generally requires higher temperatures. A similar study in $NaBF_4$ at $420\,°C$ [25] gave similar results. The absence of Ti(II) in this work may be noted. In an extensive series of studies, a group of French workers [26–29] studied the reduction of $TiCl_4$ in several chloride melts. Since $TiCl_4$ is a vapor (m.p. $136\,°C$) even at $550\,°C$, the temperature of the experiment, it was introduced near the cathode by transpiration. Reduction waves were observed for $Ti(IV) \rightarrow Ti(III)$, Ti(II) and $Ti(II) \rightarrow Ti(0)$. The reduction process is strongly dependent on the melt cation. For example, in the presence of 'complexing' ions, such as K^+, Rb^+, or Cs^+, the anionic complex $TiCl_6^{2-}$ is formed. In $BaCl_2-KCl-LiCl$, Ti(III) is strongly complexed and the reduction proceeds through a single step: $Ti(III) \rightarrow Ti(0)$. At $700\,°C$ the situation is similar [28, 29]. The authors have summarized their results in a chart which relates the melt composition to the titanium reduction reactions. It is interesting that the anodic oxidation of titanium results in the formation of Ti(II) [30, 31] in LiCl–KCl, since this implies that it is subsequent steps which result in the formation of other oxidation states of Ti. For example, if Ti is added to the melt as $TiCl_4$ or K_2TiCl_6, and a titanium anode is used which generates Ti(II) on dissolution, complex processes, including those forming Ti(III) must be occurring. This complexity of titanium electrochemistry in chloride melts is also apparent from other recent reports. In a recent investigation, not readily available in the open literature [32], extensive electrochemical studies were carried out in the LiCl–KCl eutectic at $450\,°C$. Potentials were measured for the Ti/Ti(II) system versus the Pt/Pt(II) reference. The behavior of the electrode is

Nernstian, and thermodynamic data were calculated from its temperature dependence. The difficulties of plating titanium from a chloride melt are at least partly related to corrosion reactions such as

$$2Ti(III) + Ti = 3Ti(II)$$

From cyclic voltammetry and chronopotentiometry the reduction of Ti(II) was found to be a diffusion–controlled process. Thus the rate of reduction of Ti(II) is controlled by the redox reaction which is slower than the electron transfer at the cathode.

In a recent study, oriented toward industrial production, Rolland, Steden and Thonstad [33] studied the reduction of $TiCl_4$ in various combinations of LiCl, NaCl and KCl over a wide temperature range (450–850 °C) and CD (20–150 mA/cm^2). Between 20 and 50 mA/cm^2 no deposits at all were obtained at any temperature. Increased CD produced titanium deposits over the entire temperature range, but they were generally dendritic, poorly adherent and plated at low CE.

Based on all the above studies it appears unlikely that titanium can be plated in useful form from pure alkali metal chloride melts.

In contrast to the complicated electrochemistry observed in chloride melts, the situation in molten fluorides seems to be simpler, although not yet studied in great detail. On the positive side, Ti(II) seems to be absent, and the reduction proceeds through the reversible steps [34, 35].

$$Ti(IV) + e = Ti(III)$$

and

$$Ti(III) + 3e = Ti(0)$$

and the titanium salt K_2TiF_6 is easily handled, quite soluble and ionic. Indeed, fairly good titanium coatings of 20–30 µm on nickel were obtained at 700 °C [36]. On the negative side, temperatures at least this high are required for good plating, and fluoride melts are corrosive, necessitating the use of metallic containers. Vitreous carbon, frequently used for research, is probably too expensive and fragile for industrial operations. Removal of oxygenated species by chemical means and/or preelectrolysis is required to avoid formation of insoluble titanium oxide on the cathode. On the other hand, the introduction of TiO_2 into the melt has been reported [37] to improve the hardness of titanium by introducing oxygen along the *c* axis of the hexagonal structure.

Based on the results reported, further study of plating from fluoride melts might be useful, since the difficulties associated with chloride melts seem to arise from the basic electrochemistry in these media.

Zirconium

Although the literature on this metal is not as extensive as that on titanium, much of it exhibits parallels. Thus factors which affect the reduction of zirconium species include:

(a) melt composition, both anionic and cationic
(b) temperature

One major difference is the much lower stability of Zr(III), compared to Ti(III), and the only important melt species appear to be Zr(IV) and Zr(II) [38, 39]. The effect of temperature is clearly apparent in a polarographic study [40] in LiCl–KCl which indicates Zr(IV) as the only species at 450 °C, but finds Zr(II) at 550 °C. In a similar study by Inman *et al.* [41] in the same solvent, Zr(II) appeared to be the major product at the cathode. A more recent study of Zr(IV) reduction [42] in a pure chloride melt, but with varying cationic composition, indicates complexities in the reduction chemistry. For example in NaCl at 820 °C, four very closely spaced reductions were found. The authors attribute these to:

$$Zr(IV) + 2e = Zr(II)$$
$$Zr(II) + 2e = Zr(0)$$
$$Zr(IV) + 4e = Zr(0)$$

and a process involving the complex ion $ZrCl_6^{2-}$. In addition, the auto-oxidation–reduction.

$$Zr(IV) + Zr = 2Zr(II)$$

also occurs. In LiCl–KCl at 520 °C the two-step reduction $IV \rightarrow II \rightarrow 0$ occurs. In CsCl the reduction steps are similar, but the complex Cs_2ZrCl_6 is more stable.

In a study more directly related to plating, Mellors and Senderoff [43] found that only dendrites of zirconium metal were obbtained by reduction of $ZrCl_4$ in LiCl–KCl at 700 °C, and no deposit at all was produced in $NaBr–KBr–ZrBr_4$, again showing the importance of the melt anion. However, as with titanium, the most dramatic anion effect is the introduction of fluoride. Even in a $NaCl–ZrF_4$ melt, the only reduction is $Zr(IV) \rightarrow Zr(0)$ [38], and the same is true in FLINAK, even at 500 °C.

From a plating perspective, however, in addition to the importance of fluoride, the cationic composition of the melt and the temperature are also important. For example, only zirconium powder was produced from an $LiF–NaF–K_2ZrF_6$ melt, whereas when the melt contained potassium (LiF–KF, FLINAK) coherent dense plates of zirconium metal were obtained on the cathode at 675–800 °C, but not at lower temperatures [43].

Thus, although some interesting electrochemistry has been explored in chloride melts, and indeed zirconium powder can be produced, the formation of coherent zirconium deposits requires at least the presence of (a) fluoride, (b) potassium and (c) a temperature in excess of 700 °C. In order to ameliorate the corrosive effects of fluoride, it may be possible to work in mixed chloride–fluoride melts, but this has not yet been explored in detail.

Hafnium

There has recently been some interest in the electrochemical reduction of hafnium in molten salts [44–46]. In earlier work [47] the only hafnium species detected in LiCl–KCl at 450 °C was Hf(IV), consistent with a trend in which the lower-valent species in group IV decrease in stability with increasing atomic weight. The more recent studies have employed the NaCl–KCl eutectic in the 700–900 °C range, with or without the addition of fluoride. The question of whether Hf(II) exists in this melt is not entirely settled. Kuznetsov *et al.* [44] report this species to be an intermediate in the reduction of Hf(IV), with the process Hf(IV)→Hf(II) being reversible, whereas Poinso *et al.* [46] report Hf(IV) as the only species. Poinso *et al.* carried out their study with tungsten electrodes, whereas Kuznetsov *et al.* used platinum, which may alloy with hafnium. It is not clear whether this accounts for the different results.

There is general agreement that hafnium fluoride complexes, e.g. HfF_6^{2-}, are more stable than chloride complexes, and that in a mixed chloride–fluoride melt the metal is largely complexed with fluoride.

The metal can be produced by reduction from either chloride or chloride–fluoride melts [46], but in powdery form. Similar results had already been reported [48] for both $LiCl–RbCl–HfCl_4$ and $KCl–HfCl_4$ melts at 700–800 °C with 90% CE.

So far the only coherent hafnium plate has been reported for a $FLINAK–HfCl_4$ melt at 700 °C [49] with 90% CE [45].

Vanadium

In spite of the good corrosion resistance of vanadium in acid media [50], very little work has been done to explore the electrochemistry related to producing coatings of this metal. Some studies of vanadate electrochemistry in LiCl–KCl have been reported [51, 52]. The reduction chemistry in this solvent is complicated by the formation of insoluble compounds and oxide concentration-dependent disproportionation reactions. Recently the electrochemical reduction of vanadates (VO_3^-, VO_4^{3-}) in FLINAK was attempted [53], but it did not lead to formation of the metal.

The most interesting plating experiments in chloride and bromide melts are related to a program by the U.S. Bureau of Mines to electrowin and electrofine vanadium from molten salt baths. In one process [54] high purity, ductile vanadium was obtained from an $LiCl–NaCl–VCl_2$ melt in which commercially available vanadium carbide served as the starting anode material. The carbide is initially V_2C, but as the electrolysis proceeds it changes to VC, and finally to carbon. The authors give the electrode reactions as

$$\text{Anode:} \quad VC + 2X^- = VX_2 + C + 2e$$

$$\text{Cathode:} \quad VX + 2e = V + 2X^-$$

where $X = Cl$ or Br. The process is two-step because vanadium is first produced in the above process then electrorefined in a $KCl-LiCl-Vl_2$ melt. However, the deposited metal is dendritic. Other Bureau of Mines studies have primarily focused on the electrorefining of vanadium [55–57]. The main advantage of this process is the high purity obtained, particularly with the absence of N and O, which adversely affect the metal properties. But even when the deposit is dense, it is dendritic with frozen salt trapped in the interstices. None of this work is encouraging for the plating of good coatings and bromide melts.

The only report of successful vanadium plating is in the patent by Mellors and Senderoff [8] in which dense, structurally coherent metal was plated from K_3VF_6 or VF_3 in FLINAK at 770°C. No details of the relevant electrochemistry seem to have been published.

Niobium (Columbium)

Niobium, frequently called columbium in the older literature, has many properties which make it highly desirable as a coating material. Among them are its high melting point, excellent corrosion resistance in highly acid media, good weldability, and low neutron cross section.

In contrast to several other refractory metals for which plating was first attempted in chloride melts, the first and also the most successful work was carried out in pure fluoride melts. In their first paper [58], published after their general patent from fluoride melts had been issued [8], Mellors and Senderoff reported the first successful plating of coherent and adherent niobium plates from pure fluoride melts. Many of the required conditions, such as melt purity and absence of oxygen, have already been described earlier in this chapter. In this paper they also attribute the success of the method to the strong complexing power of the fluoride anion for the metal ion, as well as the importance of the solvent cation. For example, potassium enhances the stability of the fluoro complex and is a necessary melt constituent. Thus, plating is much more successful in $NaF-KF$ and FLINAK melts than in $NaF-LiF$. In a later paper [59] Senderoff and Mellors addressed the electrode reactions involved in the reduction of Nb(V). Using primarily chronopotentiometry as a tool, they found that the reduction proceeds in three steps:

(1) $Nb(V) + e = Nb(IV)$, reversible and diffusion controlled

(2) $Nb(IV) + 3e = Nb(I)$, reversible and diffusion controlled

(3) $Nb(I) + e = Nb(0)$, irreversible

A major problem is the rapid reaction of Nb(V) and Nb(I) to regenerate Nb(IV):

$$Nb(I) + 3Nb(V) = 4Nb(IV)$$

This reaction is so rapid that at low CD the CE falls to near zero in the presence of appreciable Nb(V), and at high CD only dendritic deposits are obtained. However, if the mean valence is reduced to 4.2 and the CD is sufficiently high, plating is successful. The desired valence is achieved by the presence of metallic niobium which lowers the valence by the reaction

$$Nb + 4Nb(V) = 5Nb(IV)$$

Niobium plating from fluoride melts has also been studied by Decroly *et al.* [60], but they were primarily concerned with a comparison of niobium and tantalum plating, and their work did not result in improving the plating process. The work of Senderoff and Mellors [58, 59] has led to further improvements in the plating procedure. Cohen [61] has applied periodic reversal (PR) techniques to improve both the coating structure and plating speed of niobium from molten fluorides. In this method a cathodic deposition step is followed by a briefer anodic dissolution which removes the uneven parts of the surface just deposited. The cycling takes place at low frequencies, typically a few cycles per minute. Cycling patterns and their effects on coating structure were studied in detail and the procedure was optimized to give the best grain size. An additional benefit is the increase in the plating rate by a factor of ten over direct current methods.

Further studies aimed at optimizing niobium plating from molten fluorides were reported by Capsimalis *et al.* [62]. Coatings from both KF–NaF and FLINAK were compared with respect to the detailed microstructure of the deposit. In both melts, temperatures above 725 °C were required to obtain coherent deposits. One interesting result is that the CE varies with CD, but the direction of change depends on the temperature.

Recently Taxil and Mahenc [63] have produced surface compounds, such as $NbNi_3$ by depositing niobium on nickel at 1000 °C. The surface compounds are created by diffusion, a process akin to metalliding (Chapter 3). These alloys may be useful as corrosion-resistant, insoluble anodes. On molybdenum and steel no such diffusion occurred, and anodizing in phosphoric acid produced Nb_2O_5 surfaces useful for electrolytic capacitors. Pulsed electrolysis was required to avoid dendrite formation.

In spite of the success obtained with fluoride melts, their drawbacks – high temperature, corrosivity and sensitivity to contamination – have led to several attempts to utilize chloride melts. These fall into two categories: pure chlorides, and melts consisting of chloride solvents and fluoride solutes. The fluoride solute is only present at low concentrations, but the high stability of the metal fluoride complex may be maintained even in a chloride melt. Thus the effective species might be a fluoride, even though the predominantly chloride melt would be less corrosive.

Workers who have studied chloride melts have generally used temperatures below 700 °C. One of the issues addressed is the predominant valence state of niobium. Saeki and Suzuki [64] reported that the anodic dissoluton of niobium between 500 and 650 °C resulted in a valence of 3.1, whereas the equilibrium composition was Nb_3Cl_8, i.e. a valence of 2.7. No mechanism for this reduction was suggested, but it might be due to the presence of the niobium anode. In a more detailed study [65] in the same melt between 380 and 600 °C two reversible redox reactions were detected:

$$Nb(V) + e = Nb(IV)$$

$$Nb(IV) + e = Nb(III)$$

Reduction of Nb(III) led to the formation of metallic niobium, but only at the higher temperature end of the range. At lower temperatures, insoluble nonstoichiometric niobium subhalides were formed. However, even when metal was formed, it was only as a powder. At higher temperatures the potential of the Nb(V)/Nb(IV) couple shifts close to the oxidation potential of Cl and the equilibrium pressure above the melt becomes so great that a chlorine atmosphere is required to avoid decomposition. Thus chloride melts containing niobium probably cannot easily be operated above 600 °C.

Somewhat similar results in the same melt were reported by Kuznetsov *et al.* [66], except that the reduction sequence was given as

$$Nb(V) \rightarrow Nb(IV) \rightarrow Nb(II) \rightarrow Nb(0)$$

At 630–650 °C the divalent state tends to disappear and reduction occurs direcly from Nb(IV). However, niobium was only obtained as a powder.

The shift from LiCl–KCl to NaCl–KCl appears to permit work at higher temperatures without melt decomposition. The reduction of K_2NbF_7 in this melt has been studied [67, 68] in the 700–800 °C range. The first step is the reduction of Nb(V) to Nb(IV). This species is directly reduced to the metal (note the absence of lower-valent species observed in pure chloride melts at lower temperatures), but the character of the deposit is highly potential dependent. Best results were obtained from the reduction of Nb(IV), but even they do not compare in quality with pure fluoride melts in the same temperature range. Thus, even if NbF_7^{2-} is a relatively stable species in the chloride sea of NaCl–KCl, good niobium plating seems to require a total fluoride environment.

Tantalum

In contrast to niobium, tantalum exhibits fewer stable valences between +5 and 0 in both chloride and fluoride melts. In contrast to niobium also, the earliest studies of tantalum reduction were carried out in molten chlorides.

In one of the earliest studies, Drossbach and Petrick [69] obtained metallic tantalum by the electrolysis of K_2TaF_7 in 1:1 NaCl–KCl at 850 °C, but the deposit was largely dendritic. The electrolysis seems to have been operated too near the cathodic limit of the melt, since Cl_2 was evolved. The use of an open cell introduced impurities of oxygen and water. Similar results were obtained by Efros and Lantranov [70] who confirmed Ta(III) as the only intermediate species, and also found the addition of Ta_2O_5 to be without effect.

The first, and thus far the only successful, tantalum plating was carried out by Senderoff, Mellors and Reinhart [71]. Using primarily chronopotentiometry as a diagnostic tool, they found a diffusion-controlled three-electron reduction to Ta(II) in FLINAK. The divalent species is the insoluble TaF_2 which forms on the cathode surface and is subsequently irreversibly reduced to the metal:

$$TaF_2(s) + 2e = Ta + 2F^-$$

More experimental details are given in the patent by Mellors and Senderoff. In general, a CD of 5–50 mA/cm was found suitable for the deposition of coherent plates.

This work has been extended to the plating of harder coatings by codepositing chromium with the tantalum from the same FLINAK melt [72]. The solute concentration was 10 wt% K_2TaF_7 and 0.1% chromium (as K_3CrF_6). Tantalum and chromium were used as anodes and the cathodes were copper and type 304 stainless steel. Most notable is the increase in hardness of the alloy coating, which was twice that of pure tantalum. The structure of the alloy was columnar, just as for tantalum, and the CE generally exceeded 90%. Another way to increase the hardness of the tantalum coating over the usually plated body-centered cubic (BCC) tantalum is to plate β-tantalum, which has a hardness ten times that of the BCC phase. In a preliminary study [73] β-tantalum was detected, but so far it is not clear what conditions are crucial for the deposition of this phase.

A recent study of tantalum deposition from a K_2TaF_7–FLINAK melt by chronopotentiometry [74] has essentially confirmed the mechanism proposed by Senderoff, Mellors and Reinhart [71] except that a chemical dissociation step

$$TaF_7^{2-} = Ta(V) + 7F^-$$

precedes the electrochemical step, and the intermediate is more likely to be Ta(III), rather than Ta(II).

There has been very little work done in pure chloride melts. Suzuki [75] has reported the reduction of $TaCl_4$ in LiCl–KCl between 500 and 675 °C. Tantalum was introduced into the melt by anodic dissolution, and its valence inferred from the weight loss of the anode and the amount of charge passed. Chronopotentiometric results were interpreted in terms of a diffusion-controlled, reversible reduction of Ta(IV) to Ta(III), followed by an irreversible reduction to the metal.

Other studies have used mixed chloride–fluoride melts. In contrast to previously cited studies [69–75], Konstantinov *et al.* [76] found that in the KCl–KF eutectic the reduction proceeds in a single reversible five-electron step at 700 °C. Also in contrast to previous authors, who found Ta_2O_5 to be electrochemically inactive, they observed additional voltammetric peaks on the addition of this material to the melt. They argued that K_3TaF_8 is formed in a K_2TaF_7–KF–KCl melt and that this reacts with Ta_2O_5:

$$3K_3TaF_8 + Ta_2O_5 + 6KF = 5K_3TaOF_6$$

Deposits of metallic tantalum were obtained by the reduction of these melts.

Baimakov *et al.* [77] tried specifically to determine both the effect of temperature on the reaction in a pure chloride melt, $LiCl$–KCl–$TaCl_5$, and the effect of added fluoride. In the chloride melt below $550\,^\circ C$ Ta(V) was first reduced to Ta(III), and then to the metal. Above $550\,^\circ C$ the first and second voltammetric waves merged to a single five-electron wave, presumably because Ta(III) becomes increasingly unstable at higher temperatures. As fluoride is added to this melt, TaF_7^{2-} is increasingly formed because fluoride complexes are stronger than chloride complexes. The equivalence point of the reaction is found by titration:

$$TaCl_n^{(n-5)^-} + nF^- = TaF_n^{(n-5)^-} + nCl^-$$

was found at $n = 6.6$–6.8. The process is similar for TaCl i.e. $TaCl_4^-$, except that Ta(III) disproportionates to Ta(V) and metallic tantalum.

From the above works, it seems likely that (a) the lower-valent species become less stable at higher temperatures and (b) the addition of fluoride to a chloride melt has a similar effect. It is thus not surprising that reduction in a KCl–$NaCl$–K_2TaF_7 melt at $720\,^\circ C$ would occur in a single five-electron step [78]. However, the character of the metallic tantalum was highly potential dependent, with the metal (in powdery form) only being plated over a narrow potential range.

Chromium

Chromium is the only one of the refractory metals that can be plated from aqueous solution. The decorative and protective coatings of this metal, e.g. on automobile trim, are well known. Nevertheless, closer examination shows they are frequently under stress and exhibit fine cracks. Although there have recently been improvements in aqueous chromium plating [79], molten salt electrorefining and electroplating justify their higher costs for more demanding applications and for high purity.

In principle, potential precursor species are the oxyanion CrO_4^{2-} or an oxidized form of chromium, such as Cr(II) or Cr(III). However, although CrO_4^{2-} is soluble in several melts, it either reacts with them, is not reducible or is reduced to species other than metallic chromium. For example, in $LiCl$–KCl at $450\,^\circ C$, CrO_4^{2-} is reduced to CrO_4^{5-}, which decomposes to yield CrO_3^{3-} and O^{2-} [80]. At higher CD, Li_5CrO_4 precipitates on the electrode. CrO_4^{2-} is stable in FLINAK up to $750\,^\circ C$, but it cannot be reduced below the reduction potential of the melt [81]. Consequently, only Cr(II) or Cr(III) are potential precursors for molten salt chromium plating; both of them have been studied.

Exceptionally high purity chromium has been obtained by molten salt electrorefining metal which had been produced by aqueous electrolysis. Several mixtures of alkali metal and alkali earth halides with $CrCl_2$ were investigated in a program by the U.S. Bureau of Mines [82]. Nitrogen and oxygen impurities generally did not exceed 20 ppm. However, the deposit was always dendritic.

Levy and Reinhardt [83] studied the reduction of Cr(III) in LiCl–KCl at 500 °C by voltammetry. The first step is the reduction to Cr(II). This species may react in two ways:

$$Cr(II) + 2Cl = CrCl_2(s)$$

$$Cr(II) + 2e = Cr(0)$$

Thus the metal can be formed preferentially by the proper choice of potential. The physical form of the metal was not reported.

The most rational system for plating chromium from LiCl–KCl has been developed by Inman and coworkers. Using voltammetry and chronopotentiometry, they found [84] that Cr(III) is first reduced to Cr(II), with the Cr(II) subsequently reduced to the metal. In a more detailed study of the plating process itself, Vargas and Inman [85] concentrated on the details of the electrocrystallization process. Beginning with $CrCl_2$ in LiCl–KCl, they noted that coatings produced at constant potential are invariably poor, e.g. dendritic or poorly adherent, and they attribute this to the fact that three-dimensional nuclei begin to grow before the surface is completely covered. Their solution to this problem is the 'initial pulse' method in which plating begins with a large nucleation pulse (or pulses) to produce a high nucleation density. The potential is then reduced to encourage the nuclei to coalesce into a coherent deposit. Although this scheme is certainly beneficial in producing more uniform coatings, considerable adjustment of plating parameters – pulse height, number of pulses, time intervals – is still required for optimizing the coating structure. Nevertheless, the procedure represents a real advance, not only for chromium plating, but for plating other refractory metals from chloride melts, which offer a lower temperature and easier handling than molten fluorides. The behavior of chromium coatings produced by the above method has already proven beneficial in protecting steel in nuclear, gas-cooled reactors from carbon pickup due to CO and CH_4 [86].

The electrodeposition of chromium from molten fluorides has also been investigated. Yoko and Bailey [87] studied the reduction of Cr(III) in FLINAK over an unusually wide temperature range, 612–983 °C. Using cyclic voltammetry, they found the reduction to proceed in two steps: (a) a fairly slow reduction to Cr(II) to yield a product which is soluble at 983 °C and an insoluble, but still electroactive, product below 893 °C; (b) a slow, quasi-reversible reduction of Cr(II) to Cr(0). There is some indication that the dendrite formation observed at lower temperatures changes to smoother plates at higher temperatures. Since CrF_2 is readily available, it seems unnecessary to begin with a fluoride of Cr(III), such as CrF_3 and K_3CrF_6, although excellent plates of chromium metal with these precursors in FLINAK were obtained by Mellors and Senderoff [8].

At any rate, given that chromium may be one of the few refractory metals platable from LiCl–KCl below 500 °C, and that improved chromium plating has also been achieved in aqueous solution [79], it is not surprising that less effort has been expended on fluoride melts.

Molybdenum

Molybdenum continues a trend observed earlier with chromium. Whereas metals in IVA and VA are easily plated from fluoride melts, but only with difficulty from chloride melts, those in VIA are relatively easily plated from chlorides (although they can be plated from fluorides). Moreover, in contrast to IVA and VA metals, the metal-containing species in the case of molybdenum and tungsten can be a molybdate (MoO_4^{2-}) or tungstate (WO_4^{2-}). Chromates are probably not reducible in either chlorides or fluorides.

The first reported electroplating of molybdenum was reported by Senderoff and Brenner [88]. Using K_3MoCl_6 as the solute, poor quality coatings – powders or dendrites – were obtained in NaCl–KCl at 900 °C, and much more coherent deposits in LiCl–KCl at 600–900 °C. But even in LiCl–KCl the deposits were brittle. Senderoff and Mellors [89] studied the deposition mechanism in LiCl–KCl. Their work, based on chronopotentiometry, is an early indication of the complexities of molybdenum chemistry and electrochemistry. They noted that melts of K_2MoCl_6 in LiCl–KCl are unstable on dilution even at 600 °C and postulated the existence of polynuclear ions such as $[Mo_2Cl_9]^{3-}$ which can dissociate on dilution to form volatile species:

$$[Mo_2Cl_9]^{3-} + 3Cl^- = 2[MoCl_6]^{2-}$$
$$= \tfrac{6}{5}MoCl_5(g) + \tfrac{4}{5}Mo\ (s) + 6Cl^-$$

The electrochemical results are interpreted in terms of a single, irreversible three-electron reduction at 600 °C, which becomes more reversible as the temperature is raised. More recently, White and Twardoch [90] have reexamined this problem in the same melt and found that MoO_2Cl_2 was evolved from the melt due to the presence of oxyanion and/or MoO impurities, since melts were stable after this evolution. Another aspect of the complicated behavior of molybdenum electroplating is provided by the striking effect of CsCl on the voltammograms. Similar results were obtained by Selis [91], who found that the kinetic parameters of the Mo/Mo(III) electrode in various alkali metal chlorides and bromides were subject to strong solvent effects.

These effects were attributed to differences in the degree of covalent bonding and electrostatic interactions.

In addition to mechanistic studies, successful molybdenum plating from chloride melts has been reported. This includes the work of White and Twardoch [90], who obtained good coatings in LiCl–KCl and LiCl–CsCl, and recent work by Kipouros and Sadoway [92], who found that the addition of activated alumina improved the suface characteristics of the metal plated from $KCl–K_2MoCl_6$ at 800 °C. The alumina apparently acts as a leveling agent, although it is not incorporated into the coating, and its precise mode of action remains a mystery.

Melts other than chlorides have also been studied for molybdenum electrodeposition. Senderoff and Mellors [93] found that K_3MoF_6, and gaseous $MoF_6 + Mo$ (s), can equally serve to plate molybdenum from molten FLINAK [8]. In all cases the electroactive species appears to be Mo(III).

In an exhaustive series of studies, Koyama *et al.* [94–96] reported the plating of smooth, adherent molybdenum coatings from molybdate melts. K_2MoO_4 served as the source of molybdenum and the following solvents were used: $KF–B_2O_3$, $KF–Na_2B_4O_7$ and $KF–Li_2B_4O_7$. Many compositions in each ternary system were studied at 750–900 °C, and the successful compositions were plotted on a ternary phase diagram. Excellent coatings were obtained by plating at various constant current densities. Most successful coatings were obtained from melts which were more than 50 mol% KF.

Baraboshkin and coworkers have used molybdenum plated from molten salts to examine theories accounting for the crystalline structures of the deposits. In one of their earlier studies [97] they examined two theories used to account for textural growth of electroplates: geometric selection and two-dimensional nucleation. In geometric selection, texture emerges from a chaotic orientation of grains on the substrate only through survival of favorable grain orientation during deposition, with no growth of new grains. In two-dimensional nucleation, textural growth is represented as alternate grain surface processes, comprising passivation and formation of two-dimensional nuclei with a specific orientation. In order to distinguish between the theories, plated layers were gradually thinned electrochemically and discrete layers examined in cross section. The results showed that the number of grains on cross section decreased with distance from the substrate, i.e. the crystals grow larger further away from the substrate because selection leaves only those grains whose crystallographic direction of growth is nearly normal to the substrate. In further work [98] factors determining the direction of maximum growth were examined. This rate is affected both by the lattice structure and the electrolysis conditions. Adjustments of the electrolysis conditions provide some control over the structure of the deposit. The appropriate parameters were used to plate single crystal layers of molybdenum from an $Na_2WO_4–K_2WO_4–WO_3–MoO_3$ melt [99].

It thus appears that molybdenum coatings of good quality can be plated from several melts: LiCl–KCl, FLINAK and KF–molybdate. Adherent and smooth coatings can be obtained, and the work of Baraboshkin *et al.* suggests that the coating structure can be controlled by a suitable choice of the plating parameters. Surprisingly, although the deposition mechanism has been studied in halide melts, virtually nothing about this is known in molybdate melts.

Tungsten

Tungsten has long been recognized as an attractive material for many uses. In particular, its high melting point, great hardness and oxidation resistance at high temperatures makes it an attractive candidate for applications both in bulk and as a coating on various metal substrates. Because its minerals are fairly readily available, there have been many attempts to produce the pure metal in both forms.

Electrolysis has been used for more than a hundred years to produce metallic tungsten [100]. However, in contrast to the electrowinning and electroplating of other metals, the situation with respect to both practical methods and understanding the electrochemical reduction mechanism is not yet clear. The chief reason seems to be the extreme dependence of the character of the deposit on the current efficiency, melt composition, temperature and current density.

Early work on tungsten plating has been described by Davis and Gentry [6], who also repeated much of it for confirmation. Their conclusions can be summarized as follows:

(a) Aqueous solutions–based on ten studies employing different solutions, all containing hexavalent tungsten, Davis and Gentry conclude that for thermodynamic reasons, rather than because of high overvoltage, hydrogen is always reduced preferentially to tungsten, and therefore the pure metal cannot be plated from aqueous solutions. However, alloys of tungsten with the ferrous metals can be plated with high CE up to 70% tungsten. For example, Brenner and coworkers [101] plated cobalt, nickel and iron alloys from alkaline solutions of hydrocarboxylic acids containing chlorides or sulfates of the ferrous metals and Na_2WO_4. Davis and Gentry [6] suggest that the deposition potentials for the solid solution alloys are much less than for pure tungsten. This may arise from a high affinity of the ferrous metals for tungsten, as evidenced by a smaller expansion of the ferrous lattice than would be expected if solid solution properties were additive.

(b) Organic solvents – a variety of such solvents were tried, but Na_2WO_4 is insoluble in most of them. From the few in which the solvent was somewhat soluble, no deposit was obtained.

(c) Molten salts – by far the largest amount of tungsten plating and studies of tungsten reduction have been carried out in molten salts. The number of salts and salt mixtures that have been used is not large:

    (1) LiCl–KCl
    (2) $ZnBr_2$-NaBr
    (3) alkali metal fluorides
    (4) borate–halide mixtures
    (5) tungstates

Sources of tungsten have been:

    (1) halides with various valence states of tungsten
    (2) tungstates
    (3) WO_3

It is clear that even this short list of solvents and solutes yields a rather large number of solute–solvent combinations. If, in addition to composition, one considers the variables over which an investigator exercises control – concentration, temperature, current density, substrate metal – it is not surprising that a rationalization of extant results is a daunting task. One should also consider whether the aim of a particular study was electrowinning or electroplating.

The earliest substantial study of tungsten electroplating is probably the work of van Liempt [102] who used molten alkali metal tungstates with and without added WO_3. Their use involves the concept of acid–base reactions in these melts, and the competition between electroplated tungsten and tungsten bronze. Van Liempt obtained powdered tungsten by the electrolysis of the ternary (Li, Na, K)WO_4 in the 700–1000 °C range with 60–80% CE. He proposed that Na is formed by electrolysis and subsequently reduces WO_4^{2-}, resulting in the formation of W and a basic tungstate:

$$6Na + 5Na_2WO_4 = W + 4(Na_2O \cdot WO_3)$$

In 'acid' melts, i.e. $Na_2O \cdot nWO_3$, with $n > 1$, tungsten bronzes, Na_xWO_3, are formed, e.g.

$$Na + W_2O_7^{2-} = WO_4^{2-} + NaWO_3$$

In addition, van Liempt found the effect of temperature to be very important. Only above 900 °C were good tungsten coatings obtained.

The electrodeposition of tungsten from tungstate–WO_3 melts has been pursued by several research groups. Baraboshkin and coworkers [103, 104] have plated tungsten from an 80–20 mol% Na_2WO_4–WO_3 melt in the 850–950 °C range on molybdenum, copper and nickel, although their main interest was the formation of epitaxial layers on single crystals. These melts have also been used to electrodeposit tungsten bronzes (Na_xWO_3) [105, 106]. The relation between melt composition, temperature and the composition of the product was clarified by Randin [107]. There is a dividing line between metal and bronze which is expressed by a linear dependence of temperature on WO concentration. Above the line the metal is plated, below it the bronze.

Another approach was taken by Fink and Ma [108], who found that borate melts were excellent solvents for WO^3. Their main purpose was electrowinning the metal in powder form. Their results were extended and adapted to tungsten plating by Davis and Gentry [6]. They obtained their best results – adherent coatings of columnar structure and a CE of 76% – from a melt of composition

	moles
$NaLiB_2O_4$	6
$NaLiWO_4$	2
WO_3	1

The best temperature range was 850–900 °C, and the best CD 10–100 mA/cm^2. This study served as the basis for much more detailed work by the U.S. Bureau of Mines [109]. Their electrolyte contained sodium and lithium metaborates, $NaBO_2$ and $LiBO_2$, carefully dehydrated, vacuum dried and fused at 900 °C. Na_2WO_4, Li_2WO_4 and WO_3 were similarly treated. The electrolyte had the following composition:

	w%	moles
$NaBO_2$	26.7	6
$LiBO_2$	20.2	6
Na_2WO_4	19.8	1
Li_2WO_4	17.7	1
WO_3	15.6	1

This is basically the melt used by Davis and Gentry [6]. The most favorable plating conditions, resulting in thick, dense and adherent tungsten coatings, were 50mA/cm^2 at 900 °C, with rotation of the cathode. CE was 96–100% on a variety of substrates. On the basis of the low hardness reported for this plated tungsten, Senderoff [110] surmised that the metal might be badly contaminated with oxygen because it was plated from an oxyanion melt.

In order to improve the properties of the electroplated tungsten, Senderoff and Mellors [111] applied the process which they had developed for electroplating the refractory metals in ductile form from fluoride melts (see previous sections). However, in contrast to the plating of other refractory metals from which the solute was a solid fluoride, tungsten plating required a tungsten valence near 4.5, which was achieved by bubbling gaseous WF^6 into a FLINAK melt in which finely divided tungsten metal was suspended.

More recent work has been concerned with (1) lowering the rather high temperature required in both fluoride and borate melts, and (2) finding a solid tungsten compound which would give results as good as the $WF_6 + W$ method. For solutes the list of candidates is rather small – solid chlorides and bromides, perhaps in valence states lower than $+6$. Alkali chlorides and bromides, and halide–borate solvents have been used with some success. In order to guide plating experiments, some mechanistic studies have also been carried out.

White and Twardoch [90] have tabulated many of the tungsten halide studies. They point out that these halides form clusters whose structure changes markedly with the tungsten valence. Quite recently it has become apparent that acid–base chemistry plays an important role in the morphology of the deposit. For example, in $ZnBr_2$–$NaBr$ melts [112] the acidity is changed by changing the ratio of these two components, and different deposit morphologies are observed when WO^6 is reduced at 350–400 °C, one of the lowest temperatures reported for tungsten plating. Ito and coworkers have reported the reduction of WO_4^{2-} in both LiCl–KCl and LiF–KF eutectics at 700 °C [113, 114]. Using O-specific ion electrodes they determined the equilibrium constant for the reaction

$$WO_4^{2-} = W(VI) + 4O^{2-}$$

where W(VI) stands for a fluoride complex. The equilibrium constant is $\sim 5 \times 10^{-12}$, i.e. WO_4^{2-} is very stable in these melts. However, the electrodeposited tungsten from both these melts was quite mossy. In a more detailed study [115], they found that the structure of the deposit in the fluoride eutectic was extremely sensitive to the basicity of the melt, probably through the above equilibrium. Similar studies of WO_3 in KCl–NaCl at 700° [116] had also found that the addition of O^{2-} to such a melt also resulted in the formation of $W_2O_7^{2-}$ and $WO_3 \cdot 2O^{2-}$, in addition to WO_4^{2-}.

In addition to the studies cited above, a great deal of effort has been reported in both the Russian and Japanese literature, but unfortunately in very obscure publications for which not even abstracts are available. It is likely that the more successful works would find their way into more accessible journals.

In summary, it appears that the only fully successful tungsten plating is that developed by Senderoff and Mellors [8] for fluoride melts, and perhaps the Bureau of Mines process for borate melts. None of the work on pure chloride melts has yet resulted in a useful coating. In view of the excellent results obtained by Koyama *et al.* [94–97] for plating molybdenum from fluoride–borate melts, a systematic study of tungsten plating in this system might produce useful results. The work of Ito *et al.* [115] suggests the importance of acid–base chemistry in these melts, but these aspects also remain to be explored.

REFERENCES

1. Hocking, M.G., Vasantasree, V. and Sidky, P.S. (1989) *Metallurgical and Ceramic Coatings*, Longman, London.
2. Inman, D. and White, S.H. (1978) *J. Appl. Electrochem.*, **8**, 379–90.
3. Sethi, R.S. (1979) *J. Appl. Electrochem.*, **9**, 411–26.
4. Acherman, W.L., Carter, J.P., Kenahan, C.B. and Schlain, D. (1966) *U.S. Bur. Mines Rep. Invest. 6715.*

5. Bunshah, R.F. *et al.* (1982) *Deposition Technologies for Films and Coatings,* Noyes, Park Ridge, NJ.
6. Davis, G.L. and Gentry, C.H.R. (1956) *Metallurgia,* **53**, 3–17.
7. Stern, K.H., McCollum, M.E. (1985) *Thin Solid Films,* **124**, 129–34.
8. Mellors, G.W. and Senderoff, S. (1964) Canadian Patent 688, 546.
9. Kipouros, G.J. and Sadoway, D.R. (1985) *Sagamore Army Mat. Res. Conf. Proc.,* **30**, 493–503.
10. Stern, K.H. and Meador, S.E. (1965) *J. Res. Natl Bur. Stds,* **69A**, 553–6.
11. Stern, K.H. (1968) *J. Phys. Chem.,* **72**, 1963–75.
12. Stern, K.H. (1970) *J. Phys. Chem.,* **74**, 1327–32.
13. Labrie, R.J. and Lamb, V.A. (1959) *J. Electrochem. Soc.,* **106**, 895–99.
14. Jenkins, H.H., Mamantov, G. and Manning, D.L. (1968) *J. Electroanal. Chem.,* **19**, 385–9.
15. Stern, K.H. (1989) *J. Electrochem. Soc.,* **136**, 439–42.
16. For LiCl–KCl see Laitinen, H.A., Ferguson, W.S. and Osteryoung, R.A. (1957) *J. Electrochem. Soc.,* **104**, 516–20.
17. For this and other purification procedures for fluorides see Mamantov, G. (ed.) (1969) in *Molten Salts,* Marcel Dekker, New York.
18. Stern, K.H. and Hess, R.M. (1993) U.S. Patent 5, 242, 563.
19. Plambeck, J.A. (1976) *Fused Salt Systems,* Vol. X of *Encyclopedia of Electrochemistry of the Elements* (ed. J.A. Bard), Marcel Dekker, New York.
20. Sibert, M.E. and Steinberg, M.A. (1955) *J. Electrochem. Soc.,* **102**, 641–7.
21. Wurm, J.G., Gravel, L. and Potvin, R.J. (1957) *J. Electrochem. Soc.,* **104**, 301–8.
22. Bright, N.F. and Wurm, J.G. (1958) *Canad. J. Chem.,* **36**, 615–22.
23. Fortin, B.J., Wurm, J.G., Gravel, L. and Potvin, R.A. (1959) *J. Eletrochem. Soc.,* **106**, 428–33.
24. Clayton, F.R., Mamantov, G. and Manning, D.L. (1973) *J. Electrochem. Soc.,* **120**, 1193–9.
25. Clayton, F.R., Mamantov, G. and Manning, D.L. (1973) *J. Electrochem. Soc.,* **120**, 1199–1201.
26. Nardin, M. and Lorthior, G. (1977) *J. Less Common Metals,* **56**, 269–76.
27. Chassaing, E.M., Basile, F. and Lorthior, G. (1979) *J. Less Common Metals,* **68**, 153–8.
28. Chassaing, E., Basile, F. and Lorthior, G. (1981) *J. Appl. Electrochem.,* **11**, 187–91.
29. Ferry, D.M., Noyon, E. and Picard, G. (1984) *J. Less Common Metals,* **97**, 331–41.
30. Rolland, W.K., Haarberg, G.M. *et al.* (1987) Abstract 3.17 of the International Society of Chemistry, 38th Meeting.
31. Horstik, A.J., vanEyden, G.J.M. and Honders, A. (1987) Abstract 8.88 of the International Society of Electrochemistry 38th Meeting.
32. Popov, B., Wagner, J.B., Wendt, H. and Koneska, Z. (1989) *Advanced Metallic and Ceramic Materials.* Proceedings of the Joint EC-Yugoslavia Coloquium on Advanced Materials, pp. 76–93.
33. Rolland, W., Steden, A. and Thonstad, J. (1984) *Proceedings of the International Symposium on Molten Salts* (ed. M. Blander), The Electrochemical Society, Vol. 84-2, pp. 775–87.
34. deLepinay, J. and Paillere, P. (1984) *Electrochim. Acta,* **29**, 1243–50.
35. Robin, A., deLepinay, J. and Barbier, M.J. (1987) *J. Electroanal. Chem.,* **230**, 125–41.
36. deLepinay, J., Bouteillon, J. *et al.* (1987) *J. Appl. Electrochem.,* **17**, 294–302.
37. Tokumoto, S., Tanaka, E. *et al.* (1975) German Patent 2623740.
38. Winand, P. (1962) *Electrochim. Acta,* **7**, 475–508.
39. Manning, D.L. and Mamantov, G. (1963) *J. Electroanal. Chem.,* **6**, 328–9.
40. Baboian, R., Hill, D.L. and Bailey, R.A. (1965) *J. Electrochem. Soc.,* **112**, 1221–4.

41. Inman, D., Hills, G.J., Young, L. and Bockris, J.O.M. (1960) *Ann. N.Y. Acad. Sci.*, **79**, 803–29.
42. Basile, E., Chassaing, E. and Lorthior, G. (1981) *J. Appl. Electrochem.*, **11**, 645–51.
43. Mellors, G.W. and Senderoff, S. (1966) *J. Electrochem. Soc.*, **113**, 60–6.
44. Kuznetsov, S.A., Kuznetsova, S.V. and Stangrit, P.T. (1990) *Elektrokhim.*, **26**, 63–8.
45. Guang-Sen, C., Okida, M. and Oki, T. (1990) *J. Appl. Electrochem.*, **20**, 77–84.
46. Poinso, J.Y., Bouvet, S., Ozil, P. *et al.* (1993) *J. Electrochem. Soc.*, **140**, 1315–21.
47. Baboian, R., Hill, D.L. and Bailey, R.A. (1965) *J. Electrochem. Soc.*, **112**, 1221–4.
48. Martinez, G.M., Wong, M.M. and Couch, D.E. (1969) *Trans. Metall. Soc. AIME.*, **245**, 2237–42.
49. Mellors, G.W. and Senderoff, S. (1964) Canadian Patent 688, 546.
50. Schlain, D., Kenahan, C.B. and Acherman, W.L. (1960) *Corrosion*, **16**, 70t–72t.
51. Laitinen, H.A. and Rhodes, D.R. (1962) *J. Electrochem. Soc.*, **109**, 413–19.
52. Laitinen, H.A. and Chessmore, R.B. (1975) *J. Electrochem. Soc.*, **122**, 238–44.
53. Stern, K.H. (1992) *J. Appl. Electrochem.*, **22**, 717–21.
54. Lei, K.P.V., Campbell, R.E. and Sullivan, T.A. (1973) *J. Electrochem. Soc.*, **120**, 211–15.
55. Cattoir, F.R. and Baker, D.H. (1960) *U.S. Bur. Mines Rep. Invest. 5360.*
56. Sullivan, T.A. and Cattoir, F.R. (1965) *U.S. Bur. Mines. Rep. Invest. 6631.*
57. Baker, D.H. and Ramsdell, J.D. (1960) *J. Electrochem. Soc.*, **107**, 985–9.
58. Mellors, G.W. and Senderoff, S. (1965) *J. Electrochem. Soc.*, **112**, 266–72.
59. Mellors, G.W. and Senderoff, S. (1966) *J. Electrochem. Soc.*, (1966) **113**, 66–71.
60. Decroly, C., Mukhtar, A. and Winand, R. (1968) *J. Electrochem.Soc.*, **115**, 905–12.
61. Cohen, U. (1981) *J. Electrochem. Soc.*, **128**, 731–9.
62. Capsimalis, G.P., Chen, E.S., Peterson, R.E. and Ahmad, I. (1987) *J. Appl. Electrochem.*, **17**, 253–60.
63. Taxil, P. and Mahenc, J. (1987) *J. Appl. Electrochem.*, **117**, 261–9.
64. Saeki, Y. and Suzuki, T. (1965) *J. Less Common Metals*, **9**, 362–6.
65. Lantelme, F., Baboian, A. and Chevalet, J. Private communication.
66. Kuznetsov, S.A., Morachevskii, A.G. and Stangrit, P.T. (1982) *Sov. Electrochem.*, **18**, 1357–61.
67. Chemla, M. and Grinevitch, V. (1973) *Bull. Soc. Chim. France*, 853–9.
68. Barhoun, A., Berghoute, Y. and Lantelme, F. (1992) *J. Alloys and Compds*, **179**, 241–52.
69. Drossbach, P. and Petrick, P. (1957) *Z. Elektrochem.*, **61**, 410–15.
70. Efros, I.D. and Lantranov, M.F. (1963) *Russ. J. Pract. Chem.*, **36**, 2577–83.
71. Senderoff, S., Mellors, G.W. and Reinhart, W.J. (1965) *J. Electrochem. Soc.*, **112**, 840–5.
72. Ahmad, I., Spiak, W.A. and Janz, G.J. (1981) *J. Appl. Electrochem.*, **11**, 291–7.
73. Thompson, J.F. and Pan, S.K. (1990) Report of the U.S. Army ARDEC Laboratory, Vatervliet, New York.
74. Espinola, A., Dutra, J.B. and Silva, F.T. (1991) *Anal. Chim. Acta.*, **251**, 53–8.
75. Suzuki, T. (1970) *Electrochim. Acta.*, **15**, 303–13.
76. Konstantinov, V.I., Polyakov, E.G. and Stangrit, P.T. (1978) *Electrochim Acta*, **23**, 713–16.
77. Baimakov, A.N., Kuznetsov, S.N., Polakov, E.G. and Stangrit, P.T. (1985) *Sov. Electrochem.*, **21**, 541–6.
78. Lantelme, F., Barhoun, A., Li, G. and Besse, J.P. (1992) *J. Electrochem. Soc.*, **139**, 1249–55.
79. Chen, E.S., Capsimalis, C.P. and Weigle, G.R. (1987) *J. Appl. Electrochem.*, **17**, 315–21.
80. Laitinen, H.A. and Bankert, R.D. (1967) *Anal. Chem.*, **39**, 1790–5.
81. Stern, K.H. and Rolison, D.R. (1990) *J. Electrochem. Soc.*, **137**, 178–83.

82. Lei, K.P.V., Hiegel, J.M. and Sullivan, T.A. (1972) *J. Less Common Metals*, **27**, 353–65.
83. Levy, S.C. and Reinhardt, F.W. (1975) *J. Electrochem. Soc.*, **122**, 200–4.
84. Inman, D., Legey, J.C. and Spencer, R. (1975) *Electroanal. Chem.*, **61**, 289–301.
85. Vargas, T. and Inman, D. (1987) *J. Appl. Electrochem.*, **17**, 270–82.
86. Inman, D., Vargas, T., Duan, S. and Dudley (1984) *Proceedings of the International Symposium on Molten Salts*, The Electrochemical Society, pp. 545–58.
87. Yoko, T. and Bailey, R.A. (1984) *J. Electrochem. Soc.*, **131**, 2590–5.
88. Senderoff, S. and Brenner, A. (1954) *J. Electrochem. Soc.*, **101**, 16–32.
89. Senderoff, S. and Mellors, G.W. (1967) *J. Electrochem. Soc.*, **114**, 556–60.
90. White, S.H. and Twardoch, U.M. (1987) *J. Appl. Electrochem.*, **17**, 225–42.
91. Selis, S.M. (1968) *J. Phys. Chem.*, **72**, 1442–6.
92. Kipouros, G.J. and Sadoway, D. (1988) *J. Appl. Electrochem.*, **18**, 823–39.
93. Senderoff, S. and Mellors, G.W. (1967) *J. Electrochem. Soc.*, **114**, 586–9.
94. Koyama, K., Hashimoto, Y., Omori, S. and Terawaki, K. (1984) *Trans. Jpn. Inst. Metals*, **25**, 265–75.
95. Koyama, K., Hashimoto, Y., Omori, S. and Terawaki, K. (1984) *Trans. Jpn. Inst. Metals*, **25**, 804–9.
96. Koyama, K., Hashimoto, Y. Omori, S. and Terawaki, K. (1986) *J. Less Common Metals*, **123**, 223– 31; **132**, 57–67.
97. Baraboshkin, A.N., Martemyanova, Z.S., Plaksin, S.V. and Esina, N.O. (1977) *Sov. Electrochem.*, **13**, 1558–62.
98. Baraboshkin, A.N., Martemyanova, Z.S., Plaksin, S.V. and Esina, N.O. (1978) *Sov. Electrochem.*, **14**, 6–11.
99. Esina, N.O., Tarasova, K.P. and Baraboshkin (1987) *Sov. Electrochem.*, **23**, 103–7.
100. Hallopeau, L.A. (1898) *Compt. Rend.*, **127**, 755–6.
101. Brenner, A., Burkhead, A.P. and Seegmiller, E. (1947) *J. Res. Natl Bur. Stds.*, **39**, 351–83.
102. vanLiempt, J.A.M. (1925) *Z. Elektrochem*, **31**, 249–55.
103. Baraboshkin, A.N., Kaliev, K.A. and Aksentev, A.G. (1978) *Sov. Electrochem.*, **14**, 1592–5.
104. Baraboshkin, A.N. and Plaksin, S.V. (1983) *Dokl. Akad. Nauk SSR*, **270**, 348–50.
105. Banks, E., Fleischman, C.W. and Meites, L. (1970) *J. Solid State Chem.*, **1**, 372–5.
106. Fredlein, R.A. and Damjanovic, A. (1972) *J. Solid State Chem.*, **4**, 94–102.
107. Randin, J.P. (1973) *J. Electrochem. Soc.*, **120**, 1325–30.
108. Fink, C.G. and Ma, C.C. (1943) *Trans. Electrochem. Soc.*, **84**, 33–63.
109. McCawley, F.X., Kenahan, C.B. and Schlain, D. (1964) *U.S. Bur. Mines Rep. Invest. 6454*.
110. Senderoff, S. (1966) *Metall. Rev.*, **11**, 97.
111. Senderoff, S. and Mellors, G.W. (1966) *J. Electrochem. Soc.*, **114**, 586–90.
112. Hayashi, H., Hayashi, N., Uno, K. and Takehara, Z., (1987) *Proceedings of the International Symposium on Molten Salts* (ed. M. Blander), The Electrochemical Society, Vol. 87-7, pp. 814–24.
113. Ito, Y., Shimada, T., Yabe, H. and Oishi, J. (1987) *Proceedings of the International Symposium on Molten Salts*, (ed. M. Blander), The Electrochemical Society, Vol. 87-7, pp. 825–33.
114. Yabe, H., Ito, Y., Ema, K. and Oishi, J. (1987) *Proceedings of the International Symposium on Molten Salts*, (ed. M. Blander), The Electrochemical Society, Vol. 87-7, pp. 804–13.
115. Yabe, H., Ema, K. and Ito, Y. (1990) *J. Electrochem. Soc.*, **137**, 1739–43.
116. Shepoval, V.I., Grischenko, V.F. and Zarubitskaya, L. (1973) *Ukr. Khim. Zhuro.*, **39**, 867–875.

3

Metalliding

Kurt H. Stern

3.1 INTRODUCTION

While Mellors and Senderoff at Union Carbide were developing their molten salt process for electroplating the refractory metals (Chapter 2), Newell C. Cook and his associates at the General Electric Research and Development Center serendipitously invented metalliding. As described by Cook [1], they were engaged in trying to synthesize fluorocarbons by electrolytically fluorinating graphite in molten alkali metal fluorides. Their process involved passing a current between a graphite anode and a platinum cathode through the salt at $525\,°C$, while passing gaseous SiF_4 through the salt. The intent was to provide Si(IV) cations which would deposit on the platinum and prevent the vaporization of alkali metals. As expected, fluorocarbons were deposited on the anode but, contrary to expectation, the silicon did not form a film or dendrites on the cathode, but diffused smoothly into the platinum to form a hard surface layer. Further work showed that similar processes occurred with other substrates and dissolved metal-containing compounds, and that in many cases the result was the hardening of the substrate's surface region.

Between 1962 and 1972 approximately 20 patents were obtained by Cook on producing diffusion coatings by a process generally called 'metalliding'. A description was published by Cook in 1969 [1]. For at least a decade virtually no publications appeared in the literature. Since then, only a few studies relating to the process, both fundamental and applied, have been published.

Metallurgical and Ceramic Protective Coatings. Edited by Kurt H. Stern.
Published in 1996 by Chapman & Hall, London. ISBN 0 412 54440 7

This is not necessarily from a lack of interest, but may only reflect proprietary activity. For example, the author is familiar with the Metalliding Institute at Gannon University in Erie, Pennsylvania, which provides consulting and research services to industrial concerns, but has not published any results.

In what follows, the basic features of the metalliding process will be described, both with respect to the similarities and the differences with molten salt electroplating. Details were given by Cook in his patents and summarized in his *Scientific American* article. His approach is largely Edisonian, i.e. it consists of a description and systemization of the process and the properties of the metallided surfaces produced. Only recently has there been any interest in the kinetic aspects of metalliding, and in the control of the surface region structure by adjustment of the chemical and electrochemical parameters.

3.2 THE METALLIDING PROCESS

Since many of the procedural details are contained in the patent literature, and this is not always readily available, the content of the patents with respect to common features has been summarized in this section. Table 3.1 lists the patents granted to N.C. Cook, all of which have been assigned to the General Electric Company. Table 3.2 lists the atomic numbers of the substrates into which various elements can be diffused. In general, as the metalliding element lies further to the right in the periodic table, the number of elements into which it can be diffused diminishes. For example, the number of elements into which cobalt, nickel and iron can be diffused is less than the number for beryllium, boron and silicon.

Table 3.1 Metalliding patents by N.C. Cook *et al.*

Patent number	Date	Subject
3,024,175	6 Mar 1962	Berrylliding
3,024,176	6 Mar 1962	Boriding
3,024,177	6 Mar 1962	Siliciding
3,232,856	1 Feb 1966	Chromiding
3,479,158	18 Nov 1969	Zircon-, hafniding
3,479,159	18 Nov 1969	Titaniding
3,489,536	13 Jan 1970	Scandizing
3,489,537	13 Jan 1970	Aluminiding
3,489,538	13 Jan 1970	Yttriding
3,489,539	13 Jan 1970	Manganiding
3,489,540	13 Jan 1970	Nickel-, cobalt, ironiding
3,514,272	26 May 1970	Vanadiding
3,567,598	2 Mar 1971	Tinniding, tungsteniding
3,701,639	31 Oct 1972	Soldering applications
3,814,673	4 June 1974	Tantalliding, nickeliding

Table 3.2 Metalliding by element (from patents)

Metalliding element	Substrate (by atomic number)		
Be	21–29	39–47	57–79
B	23–29	41–47	73–79
Si	23–29	41–47	73–79
Cr	26–29	42, 44–47	74–79
Ti	23–29	41–47	73–79
Zr, Hf	22–29	40–47	72–79
V	24–29	41–47	73–79
Sc	25–29	43–47	75–79
Co, Ni, Fe	27–29	42–47	74–79
Mn	23, 24, 26–29	41–47	73–79
Li	24–29	46, 47	78, 79
Al	23–29	41–47	73–79
Y, rare earths	4, 21, 22, 25–29	40, 43–47	72, 75–79
Ta, Nb	23–29	41–46	73–79

The purpose of metalliding is to 'improve' the surface characteristics of the substrate and thereby extend its range of applications. Major improvements frequently cited are increased hardness and increased corrosion resistance. Although, in a sense, each combination of substrate and metalliding element (metallider) is unique, some generalizations can be made. For example, boron primarily increases hardness, whereas silicon increases corrosion resistance. Details will be described under the specific metallider.

3.2.1 Metalliding Procedures – General

Metalliding shares many important procedures with electroplating:

(a) The medium is nearly always a mixture of molten alkali metal fluorides. Alkaline earth fluorides may also be used in the mixed solvents. Above 900 °C pure LiF is frequently recommended because of its low vapor pressure. Some special cases in which chlorides are used will be mentioned below.

   The solute is generally a fluoride of the metallider.

(b) All the requirements of melt purity and absence of dissolved oxides are the same for molten salt plating. Carbon, such as residues from anodes used in the melt purification, must be absent when elements that form carbides are to be plated.

(c) Oxygen must be absent; generally an oxygen-free cover gas is used.

(d) The requirements for containers are the same as for plating. Fluorides require metal containers not attacked by the melt, e.g. nickel or nickel alloys.

The most important difference between metalliding and plating is that in metalliding there is, in principle, no coating–substrate boundary and the speed of the process is determined by the diffusion rate of the metallider into the substrate. Thus, if no potential were applied and no diffusion were to occur, the process would stop when the cathode was covered, since this would make both electrodes identical. It can only proceed if the metallider concentration at the cathode surface is continually lowered by diffusion, i.e. the alloying rate is proportional to $(time)^{1/2}$ and will therefore slow down with time, whereas plating is only dependent on the charge passed. In this ideal case there is also no distinct coating–substrate boundary. In practice, published cross sections frequently exhibit distinct boundaries. A major reason is that specific compounds are often formed by the diffusion process and there is then a specific boundary which marks the composition limit of the compound. For example, the ceriding of aluminum [2, 3] produces layers of different Ce–Al alloys with distinct boundaries between them. Micro-photographs of borided steel [1] show 'fingers' of borides extending unevenly into the substrate, rather than a smooth diffusion front. In this case, the fingers serve to anchor the borided region to the substrate. It is thus clear that each case needs to be studied individually to optimize the metalliding parameters.

For active alloying elements, such as silicon and boron, no applied potential is required and the silicided or borided surface region grows by diffusion only. Consequently there is very little dimensional change, in contrast to plating, for which the size of the cathode increases. However, as has also been discussed in Chapter 2, there is some overlap between plating and metalliding, since an element plated at a lower temperature under an applied potential may be made to diffuse into the substrate by raising the temperature.

In general, the more active elements can be diffused into a wider variety of substrates than those less active, i.e. those toward the right-hand side of the periodic table. Different elements have different effects on the properties of a particular substrate. For example, boriding primarily increases hardness, whereas siliciding increases corrosion resistance. In the interfacial region, properties such as hardness vary with distance from the surface because the composition changes. These compositions may represent intermetallic compounds or solid solutions. This particular aspect, i.e. the detailed composition of the surface region, is not addressed in the Cook patents, although there is some mention of it in the *Scientific American* article [1] and more recently published papers. Since many features of the metalliding process are the same for different metalliders and substrates, it seems useful to summarize them here and to note exceptions.

Metallider

This is always the element, except when the use of an alloy is required. For example, the low melting point of aluminum, 660 °C, which lies below the optimum temperature for aluminiding, necessitates the use of a higher melting aluminum alloy.

Substrate

The list of elements which can be metallided by a particular element is given in Table 3.2. In addition, a variety of alloys has also been metallided. This includes various steels and nickel alloys, such as Inconel and Monel.

Solvents

As discussed previously, a wide variety of molten fluorides is mentioned by Cook as being suitable. For the more active metalliders, such as beryllium, boron and silicon, temperatures in the 600–800 °C range are used and solvents such as FLINAK are suitable. For less active elements, Cook recommends a temperature range of 900–1100 °C and pure LiF as solvent. In a few instances, such as aluminum, chloride solvents are preferred.

Solutes

In nearly all cases the solute is a fluoride of the metallider. The exception is aluminiding, which is carried out in a chloride melt and therefore requires $AlCl_3$. The solute concentration range can be very wide, but high concentrations are wasteful and the recommended range is a few (1–5) mol%. In most cases the pure solute is a solid, but for silicon it is gaseous SiF_4, which is bubbled through the melt. An alternative is K_2SiF_6, which is a solid and readily available.

Temperature

See solvents.

Electrical Parameters

Voltage

For the more active elements no voltage needs to be applied, since the electrochemical driving force is sufficiently large to produce anodic dissolution and cathodic reduction. For less active elements a small voltage may be applied to increase the metalliding rate. When the substrate is more anodic than the metallider, e.g. titaniding, zirconiding and hafniding of aluminum, the cathode must be maintained at a negative potential until it is removed from the melt; this is to avoid its dissolution. Unless it is desired to study the electrochemical aspects of metalliding, no reference electrode is required, and metalliding is carried out between anode and cathode. Basically, the metallider dissolves anodically and is reduced to the element on the (cathodic) substrate. In some recent work on ceriding aluminum [2, 3], a three-electrode system was used in order to insure a constant potential, and hence constant cerium concentration at the cathode.

Current density

During the metalliding process it is most important that the rate of anodic dissolution does not exceed the rate of metallider diffusion into the substrate. Since the voltage is often fixed by the electrochemical gradient, the chief means for adjusting the anodic dissolution rate is to control the current. In general, this means adjusting the cell resistance. Although this can be done by changing various aspects of cell geometry, it is most easily accomplished by introducing a variable resistance in the external circuit. For most elements, a suitable CD is $0.1-1.0$ A/dm^2 ($1-10$ mA/cm^2). This adjustment is mostly made by trial and error.

3.3 PROPERTIES OF METALLIDED LAYERS

3.3.1 Thickness

In principle there should not be any metallided layers, since an ideal diffusional structure would exhibit a metallider concentration decreasing with distance from the substrate surface according to Fick's laws of diffusion. However, cross-sectional micrographs in the literature [1, 2] frequently show distinctive 'coating'–substrate boundaries. These may sometimes represent fairly abrupt transitions from one intermetallic compound to another. Such is the case for boride coatings mentioned by Cook [1], which show boride 'fingers' extending into the substrate. It is thus clear that actual metallided surfaces may not result from ideal Fickian processes. There is apparently a practical thickness which is mentioned by some authors or which can be inferred from published cross sections. This seems to range from tenths of microns to about a mil (10^{-3} inch). In any particular case, the coating thickness matters less than the advantageous properties conferred on the substrate, chiefly hardness and corrosion resistance.

3.3.2 Adherence

Since the metallider diffuses into the substrate and simply produces a change in properties with distance from the surface, there is, in principle, no separate coating, as there is in electrodeposition, and the problem of adherence should not arise. However, in practice, the situation may be distinctly different. For example, Menezes *et al.* [2] have shown that, in ceriding aluminum, distinct layers are formed: an aluminum-rich overlayer, probably $CeAl_4$–$CeAl_3$, and an inner layer which is separated from the outer layer by a gap containing Cl, in addition to Ce and Al. Although the hardness of the coating could be measured and was much greater than that of either element, coating adherence was poor. This was attributed to the coating formation by reductive precipitation:

$$15e + 4CeAlCl_6 \text{ (melt)} = CeAl_4 + 24Cl + 3Ce^{3+}$$

Although very few metalliding processes have been studied in such detail, it is clear that each case must be considered unique and ideal diffusion coatings may not be common.

3.3.3 Hardness

A perusal of the Cook patents (Table 3.1) shows that he found virtually all metallider–substrate combinations to confer improved hardness on the substrate. This holds even when the main purpose is to improve the corrosion resistance, as it is for siliciding. The outstanding example of improved hardness is boriding. On steel Knoop hardness values as high as 3000 have been obtained, and above 4000 on molybdenum. These coatings are so hard that they can be polished only slowly, even with diamond. Hardness values are listed for other metallided surfaces in some of the Cook patents. These are consistent with the above conclusion. Cook *et al.* [1] have also published a microphotograph of borided molybdenum which clearly shows the hardness rising from the outer (boron-rich) suface to a maximum in the metallided region and then falling again toward pure molybdenum.

3.3.4 Corrosion Resistance

According to Cook, one of the major reasons for metalliding (besides improved hardness) is to provide improved corrosion resistance. Indeed, the following patented processes have 'corrosion-resistant' in the title: boriding, berylliding, siliciding and chromiding. Several others mention improved corrosion resistance without providing quantitative data.

   A detailed study [4] of tantalided and hafnided nickel and steel samples in strong electrolyte solutions, including hot H_2SO_4, HNO_3 and H_3PO_4, were reported by workers at the General Electric Company, who measured corrosion currents as a function of applied voltage. The best corrosion resistance was obtained for tantalided nickel, about the same as for pure tantalum. Tantalided steel was corrosion resistant only in a highly oxidizing acid in the absence of halide ions. Hafniding did not provide useful corrosion protection.

3.4 INDIVIDUAL METALLIDERS

The metalliding patents of N.C. Cook listed in Table 3.1 cover the period 1962–1974; they are organized by metalliding element, i.e. each patent covers a metalliding element and lists the substrates into which it was diffused, the conditions under which it was done, and the results. The sequence followed in this section is the chronological order in which the patents were issued.

A search of the literature shows that nearly all the work reported since then has been on only two elements: 232 publications and patents for boron and 112 for silicon. A major problem in trying to systematize this work is that terms like 'boriding', or 'boroniding' have been extended by various authors to cover processes other than Cook's molten salt diffusion. Another problem is that much of the literature is contained in Russian and Japanese patents and publications are not readily accessible, except for abstracts. Details will be discussed in section 4.2.

3.4.1 Berylliding

Cook patent 3,024,175 describes the berylliding of nickel, yttrium, copper, uranium and titanium from alkali fluoride melts containing BeF_2. Temperatures were in the 600–800 °C range to minimize the vaporization of this salt. Since beryllium is an active metal, no voltage needs to be applied, although a voltage may be applied to keep the CD in the 1–5 A/dm^2 range. In several cases the metallided layers contained intermetallic compounds of definite composition, e.g. $NiBe_{22}$, YBe_{13}. When the melt was clean, current efficiencies close to 100% were achieved.

In some cases, qualitative and quantitative hardness measurements as a function of coating thickness were reported. The increase was particularly notable for copper; from Knoop hardness 75 to values above 600.

No work on berylliding seems to have been reported in the literature since Cook's original patent.

3.4.2 Boriding

The diffusion of boron into metals, particularly steel, is highly effective in increasing the hardness of the substrate. Cook patent 3,024,176 describes the process which employs an alkali metal fluoride solvent, either binary system or ternary FLINAK, containing BF_4^- or BF_3 (g) as solute. In each case, the temperature should not exceed 800°C because BF_3 becomes increasingly volatile above this temperature:

$$BF_4^- \text{ (sol)} = BF_3 + F^-$$

Since boron is electrochemically very active, no potential need be applied, and the electrolysis rate is controlled by the diffusion of boron into the substrate.

Since the original Cook patent, there have been many publications and patents, but the literature is confused by terms like 'borizing' and 'boronizing', which may or may not mean the same thing as 'boriding'. Some of these processes [5] involve pack cementation. Borax has also been used as solute [6]; its main advantage is its low volatility. The main problem with a study of the voluminous literature is that most of it is in rather obscure Russian and Japanese journals (for an example see references in Danek and Matiasovsky [8]), or in equally inaccessible patents.

Brookes *et al.* [7] attempted an electrokinetic study of boriding, but not much additional information was obtained, except that even partial substitution of chloride for fluoride greatly interferes with boride deposition, probably because of the volatility of BCl_3. Danek and Matiasovsky [8] confirmed that the rate of boride formation depends almost entirely on the temperature, i.e. the diffusion rate, and on the kinetics of the reaction of boron and iron to form Fe_2B and FeB. Incidentally, all studies agree that the boriding process is of the 'reactive diffusion' type, i.e. the formation of compounds, with a distinct boundary between the reacted zone and the unreacted substrate. There is still some question as to whether FeB or Fe_2B is formed first, and there is at least one report [9] that Fe_2B is formed before FeB, and that growth is controlled by the diffusion of boron into Fe_2B.

The chief motivation for boriding is the increase in hardness and wear resistance [6], for which factors of 3–10 have been reported. In addition, the melting points of the borides are much higher than those of the substrates, and the same goes for oxidation resistance.

The process is so useful that it has become commercial, but this has not been documented in the open literature, partly because of proprietary considerations.

3.4.3 Siliciding

Cook patent 3,024,177 describes the siliciding of vanadium, chromium, cobalt, molybdenum, copper, tantalum, Inconel and other alloys as well as the noble metals. Since one of the major benefits of the process is to improve the corrosion resistance of the substrate, the application to easily corroded metals is probably the most significant; some hardening also seems to occur.

Because silicon diffuses so easily into many metals, only relatively low temperatures, 600–800 °C, are required. The melt can be FLINAK or a binary alkali metal fluoride. Since the conductivity of silicon is rather low, Cook recommends increasing the anode surface area by using chunks held in a nonreactive basket of porous carbon or silver mesh, rather than a solid rod.

No voltage needs to be applied and the anodic dissolution rate is limited by the diffusion of silicon into the substrate.

One of the major applications is the siliciding of molybdenum, which prevents the oxidation of the base metal and produces a material highly useful in the production of heating elements and other articles needing protection from high temperature oxidation. Surface silicon is oxidized to SiO_2, which limits further oxidation.

The detailed structure of silicided molybdenum has been studied by a variety of techniques [10]. In contrast to substrates in which the metallider concentration varies smoothly from the surface into the interior, silicided molybdenum is of the 'reactive diffusion' type, i.e. diffusion results in the formation of a stoichiometric compound, $MoSi_2$, which forms a definite boundary with the substrate.

3.4.4 Chromiding

Cook patent 3,232,835 describes the chromiding of iron and steel from a bath consisting of CrF_2 in FLINAK or NaF–CaF_2 at temperatures of 900–1100 °C. Substrates include carbon, steel, cast iron, molybdenum, platinum, nickel and tantalum. If a higher-valent chromium compound was used, no chromiding occurred until it had been reduced to Cr(II). Coating thickness ranged from a few tenths of a mil to somewhat over a mil at current efficiencies close to 100%. A small voltage (< 100 mV) was applied only when necessary to maintain the current at reasonable levels. Coatings are described as hard and corrosion resistant, but no quantitative measurements are reported.

A variant of the Cook method was reported by Mukerjee *et al.* [11], in which steel was chromided by dipping it into a slurry of chromium powder in an organic solvent and heating the steel to 800–900 °C. Alternatively, a paste of chrome ore in aqueous NH_4F was used. Chromium diffuses into steel with very little formation of intermetallic compounds. Multiple metalliding results in increased corrosion resistance. Vacuum annealing in hydrogen removes residual stress and improves the mechanical properties.

3.4.5 Zirconiding and Hafniding

Cook patent 3,479,158 reported the zirconiding of several metals from either NaF–LiF or pure LiF in the 800–1000 °C range, using ZrF_4 as solute. Coatings up to 1 mil thick were reported on the following metals, generally with current efficiencies above 50% (tungsten was only 22%): nickel, cobalt, vanadium, platinum, copper, molybdenum, tungsten, niobium and titanium. Coatings were generally smooth, hard and flexible. The temperature was usually 1100 °C.

No specific experiments are reported for hafniding, but it is claimed that this process can be carried out as with zirconiding by using a hafnium anode and HfF_4 as solute.

3.4.6 Titaniding

Cook patent 3,479,159 reported the titaniding of several metals from a molten LiF–TiF_3 bath at 1100 °C. The following metals were plated, generally at high current efficiencies: cobalt, steel, vanadium, chromium, copper, tantalum, molybdenum, niobium, palladium and platinum. The hardness of the coating seems not to be an intrinsic property of the substrate–metallider combination, since duplicate samples are described as soft and moderately hard, respectively. Thus hardness may depend partly on the details of the metalliding process. A titanided nickel strip was described as having a Knoop hardness of 700 and as being resistant to corrosion by concentrated HNO.

3.4.7 Scandizing

Cook patent 3,489,536 reported the scandizing of nickel, copper, Monel and palladium from an $LiF-ScF_3$ melt near 1000 °C with fairly high current efficiency. Very few experimental details are given.

3.4.8 Aluminiding

Aluminiding is an important industrial process, and besides Cook patent 3,489,537, several methods have been used, such as dipping in molten aluminum and pack cementation. Because the melting point of aluminum is much lower than the preferred metalliding temperature, the aluminum anode must either be in liquid form – a graphite basket shielded by a tightly woven Monel screen is suggested as a container – or a solid aluminum–nickel alloy. Most metalliding was carried out at 1000–1100 °C from an $LiF-AlF_3$ melt.

The following metals were aluminided with high current efficiency in the usual way: steel, nickel, cobalt, vanadium, chromium, molybdenum, niobium, platinum and palladium.

Titanium, zirconium and hafnium are more active electrochemically than aluminum and ordinarily displace it from its melts. For successful aluminiding it is necessary to impose on them a cathodic potential of 1–2 V while they are immersed in the melt.

3.4.9 Yttriding and rare earthiding

Although Cook patent 3,489,538 purports to include all the rare earths, only yttrium and gadolinium were actually tried, although the other rare earths would be expected to behave similarly. The melt was LiF with a trifluoride of the metal as solute, and the temperature range was 900–1100 °C. The yttriding of nickel was studied in some detail, and intermetallic compounds were found in the coating: YNi at low CD and YNi at higher values. Because lithium is anodic to lithium, an appropriate voltage must be applied to yttrided samples until they are removed from the melt. The yttrided metals were nickel, cobalt, titanium, zirconium, rhodium, platinum and palladium. Mechanical properties differed with the substrate, some hard and others soft. Current efficiencies varied from 5 to 95%. Results were similar for gadolinium.

The ceriding of aluminum has been studied for both KF–LiF and KCl–LiCl melts containing Ce(III) [2, 3]. A three-electrode cell was used in order to maintain a constant cerium activity at the cathode surface. Interestingly, the depositions seem to be different in the two melts: Al–Ce intermetallic compounds form in the fluoride melt, solid solutions in the chloride melt. The compounds formed were mainly $CeAl_3$ and $CeAl_4$. Although the metallided region is harder than either of the two pure metals, it is not very adherent. The authors postulate a reductive precipitation mechanism for compound formation on the aluminum surface:

(1) aluminum dissolves anodically in the chloride melt:

$$6Cl + Al = AlCl_6^{3-}(melt) + 3e$$

(2) formation of a complex in the melt:

$$AlCl_6^{3-} + Ce^{3+} = CeAlCl_6(melt)$$

(3) reduction of the complex in the reducing, electronically conducting melt and precipitation on the metal:

$$15e(melt) + 4CeAlCl_6 = CeAl_4 + 24Cl^- + 3Ce^{3+}$$

Mechanisms such as this show that metalliding does not always consist of simple Fickian diffusion and they account for the observation that metallided layers frequently have distinct boundaries and/or may not be very adherent.

3.4.10 Manganiding

Cook patent 3,489,539 reports the manganiding of several metals in an $LiF–MnF_2$ melt over the 1000–1100 °C range and at current efficiencies generally between 50 and 100%. The following metals were manganided: titanium, nickel, cobalt, vanadium, niobium, molybdenum, palladium, platinum, copper, silver, gold, Inconel and steel. The coatings, 0.5–2.0 mil thick, were described as bright and smooth, and usually flexible and soft. The main application is probably improved corrosion resistance, but no further work has been reported.

3.4.11 Nickel-, cobalt-, ironiding

Cook patent 3,489,540 has shown that nickel, cobalt and iron can be diffused into various substrates – copper, platinum, gold and molybdenum from $NaF–LiF$ melts containing the respective fluoride of the metallider (NiF_2, CoF_2, FeF_2). Temperatures were in the 800–900 °C range. Most of the coatings were produced with high current efficiency. The metallided materials appear to be soft and thus offer no distinct advantages over unmetallided articles. No data on corrosion resistance are provided.

3.4.12 Lithiding

The chief reason for lithiding metals is to improve their corrosion resistance. Cook patent 3,489,659 mentions this effect as being particularly applicable to metals which oxidize to *p*-type semiconducting oxides: Cr (Cr_2O_3), Co (CoO), Ni (NiO), Cu (CuO), Fe (FcO) and Mn (MnO).

Lithium is produced by the electrolysis of LiF, using a carbon anode. The lithium which is produced at the cathode, diffuses into the metal. The details of the diffusion process have been studied by Tedmon and Hagel [12], who find that substantial diffusion occurs only into platinum, accompanied by a pronounced increase in the hardness of the lithided layer. Current efficiency was near 100%. For the other metals various diffusion structures were observed, depending on the details of the plating process, but current efficiencies for copper, silver and chromium were less than 10%. In spite of this low CE, lithiding is still useful in protecting the underlying metal from corrosion. Oxidation studies of lithided chromium show that the kinetics are parabolic, and that the rates are much less for lithided samples than for unprotected samples. In addition to slowing oxidation rates, lithiding increases resistance to spalling and enhances resistance to nitrification during air oxidation.

The effect of lithiding is attributed to the Verwey–Hauffe mechanism, wherein Li^+ ions substitute in the cation sublattice of the Cr_2O_3, reducing the Ce^{3+} vacancy concentration. Since the oxidation rate is probably controlled by cation diffusion, a decrease in the cation vacancy concentration would lower this diffusion rate. This explanation should be applicable to the other metals cited in Cook's patent, but it seems that detailed corrosion studies are yet to be done.

3.4.13 Vanadiding

Cook patent 3,514,272 reported the vanadiding of several metals from $LiF-VF_3$ melts at 1000–1100 °C. Coatings up to 5 mil in thickness were reported on the following metals: cobalt, chromium, niobium, molybdenum, tungsten, tantalum, gold, copper, platinum, nickel and steel. Current efficiencies ranged from 30 to 100%. Coatings were described as smooth, hard and adherent. Vanadium appears to diffuse into the substrates without the formation of intermetallic compounds, but this was not extensively investigated.

No subsequent work on vanadiding seems to have been reported in the literature.

3.4.14 Tinniding

The chief motivation for tinniding, as given by Cook patents 3,567, 598 and 3,701,639, is to provide a method for soft-soldering molybdenum and tungsten. These metals cannot be soldered – largely because they are not wet by solder – and, evidently, they cannot be alloyed with tin in bulk, although they can be tinnided by the usual diffusion-controlled procedure.

LiCl–KCl eutectic or pure LiF containing SnF_2 can be used as the bath. Since the metalliding temperature, 900 °C or higher, lies far above the melting point of tin, the molten tin anode is contained in a graphite crucible tightly surrounded by woven metal cloth to avoid introducing carbon particles into the melt. The tinnided metals are readily wet by soft solder and can be easily joined to other metals, such as copper, nickel and iron.

3.4.15 Tantalliding and niobiding

Cook patent 3,814,673 primarily gives examples of tantalliding. The corrosion resistance of tantalum to acid solutions is well known and the main application is to the protection of metals to be used in these media (tantalum plating is used for the same purpose). Thus one of Cook's examples is the tantalliding of expanded nickel screen to be used as current collector in concentrated phosphoric acid solutions at 150 °C.

Tantalliding was carried out from a K_2TaF_7–LiF melt at 900–1100 °C with high current efficiency. Similar work was also done with various steels, but the presence of carbon in the steel led to the formation of tantalum carbide.

The patent also mentions that niobium, i.e. K_2NbF_7, can be substituted for K_2TaF_7, and a niobium anode for a tantalum anode, but no results for this are reported.

Taxil and Yu [13] have recently reported the niobiding of nickel for use as insoluble anodes. FLINAK was used as solvent at the lower end of the 850–1050 °C temperature range, and LiF–NaF or NaF–KF at the higher end. Nb(IV) was generated *in situ* from the reaction.

$$4Nb(V) + Nb = 5Nb(IV)$$

by adding an excess of metallic niobium to K_2NbF_7.

The alloy layer which forms when anode and cathode are connected has the composition $NbNi_3$. It is homogeneous and exhibits no porosity. A cross section of the niobided nickel shows a sharp coating–substrate boundary, although the electrochemical parameters are indicative of diffusion.

3.4.17 Metalliding aluminum

Aluminum is an extremely useful metal for many applications, but suffers from its low hardness (Knoop hardness 30), and low melting point. The latter problem can only be addressed by bulk alloying, but Cook patent 3,522,021 describes metalliding techniques to increase surface hardness. Because aluminum is electrochemically very active, the alkali metal fluorides are to be avoided, since aluminum can displace volatile alkali metals from their melts. Lithium salts should be avoided if lithiding the substrate is not desired. The preferred melt constituents are chlorides and bromides, particularly those of the alkaline earths and their mixtures with alkali metal chlorides.

Since the melting point of aluminum is low (660 °C), metalliding must be carried out below 650 °C, sometimes as low as 400–500 °C. Metals which were successfully diffused into aluminum include lithium, magnesium, scandium, yttrium, the rare earths, hafnium and zirconium. Chlorides of the metalliding elements served as solutes. Coatings of more than 1 mil were formed with high current efficiency. The only hardness value given in the patent is for cerided aluminum, which was 15 times greater than for aluminum alone.

These results have recently been confirmed and extended by workers at Rockwell International [2, 3]. They used electrochemical impedance, voltammetric and surface analysis to study the ceriding of aluminum from an LiCl–KCl melt at 560 °C. Potential control, rather than current control, was used to provide constant activities at the surface. The coating composition was $CeAl_4$–$CeAl_3$, and its hardness was at least 20 times that of pure aluminum. However, one major weakness of the procedure is the poor adherence of the coating.

The proposed mechanism of the process is discussed in section 3.4.9.

3.5 CONCLUSIONS

Metalliding is an important technique for improving the surface hardness and corrosion resistance of metals without producing significant dimensional changes. Since the process is driven by diffusion of the metallider into the substrate, it slows with time, but reported thicknesses, microns through mils, are adequate for protection.

Contrary to the expected Fickian diffusion gradient, those metallided surfaces studied in detail exhibit 'reacted' diffusion zones; compounds of definite stoichiometry are formed and there is a sharp boundary between these zones and the substrate.

Although Cook characterized nearly all his metallided surfaces as providing increased hardness and corrosion resistance, only a few processes have been studied in detail, principally borided steel and silicided molybdenum, and to some extent, aluminiding. Further work on this method certainly seems warranted.

REFERENCES

1. Cook, N.C. (1969) *Sci. Amer.*, **221**, 38–46.
2. Menezes, S., Jeanjaquet, S., Raleigh, D. and Kendig, M. (1987) *J. Electrochem. Soc.*, **134**, 2997–3003.
3. Kendig, M., Menezes, S., Jeanjaquet, S. and Raleigh, D. (1989) *Mater. Manufact. Process.*, **4**, 385–409.
4. Danzig, I.F., Dempsey, R.M. and LaConti, A.B. (1971) *Corrosion*, **2**, 55–62.

5. Kunst, H. (1975) *J. Less Common Metals*, **43**, 307–16.
6. Teneva, D. (1979) *J. Less Common Metals*, **67**, 493–7.
7. Brookes, H.C. (1976) *Trans. Inst. Metal Finishing*, **54**, 191–5.
8. Danek, V. and Matiasovsky, K. (1977) *Surf. Technol.*, **5**, 65–72.
9. Bonomi, A., Habersaat, R. and Bienvenue, G. (1978) *Surf. Technol.*, **6**, 313–19.
10. Petrescu, N., Petrescu, M., Britchi, M. and Pavel, L. (1973) *Rev. Roumaine*, **18**, 1853–8.
11. Mukerjee, D., Venkatakrishna Iyer, S. and Guruvia, S. (1987) *J. Electrochem. Soc. India*, **83**, 259–61.
12. Tedmon, C.S. and Hagel, W.C. (1978) *J. Electrochem. Soc.*, **115**, 147–51.
13. Taxil, P. and Yu, Q.Z. (1985) *J. Appl. Electrochem.*, **15**, 947–52.

4

Electrodeposition of refractory compounds from molten salts – borides, carbides and silicides

Kurt H. Stern

4.1 INTRODUCTION

Chapters 2 and 3 describe the protection of structural metals from wear and corrosion by refractory metal plating and by metalliding. This chapter discusses a third method: the plating of a substrate by a refractory compound. This method has something in common with the preceding methods; it requires plating a refractory metal, and it also resembles metalliding in that a refractory compound is formed, though on the surface of the substrate, rather than by diffusing a second element into the substrate to form a compound within the surface region. Coating a metal with a refractory compound has proved to be very useful and there are many methods for accomplishing this. These generally fall into two groups: gas phase and condensed phase. Gas phase methods have been described in considerable detail in the book edited by Bunshaw [1]. In this chapter we discuss the electroplating of some refractory compounds. Compared to gas phase methods, electroplating is still in its infancy, but it has the potential to be superior for covering complex shapes, and it might allow lower temperatures to be used in some cases. So far only the compounds in the title have been plated; there are no methods for plating oxides and nitrides. There are several

Metallurgical and Ceramic Protective Coatings. Edited by Kurt H. Stern.
Published in 1996 by Chapman & Hall, London. ISBN 0 412 54440 7

sources that provide information on the properties of refractory compounds. The well-known book by Storms [2] gives phase diagrams of the respective metal–carbon phase diagrams, crystallographic data and information on hardness and chemical properties of the refractory carbides. Toth [3] relates structure (crystal, defect and electronic) to properties (thermodynamic, mechanical, electrical, magnetic and superconducting). The collections of papers on the physics and chemistry of carbides, borides and nitrides presented at a NATO workshop and edited by Freer [4] contains useful information on various aspects of structure and stoichiometry. The book edited by Matkovich [5] contains several useful articles on various aspects of boron and boride chemistry and the monograph by Aronson, Lunstrom and Rundquist [6] contains a critical review of the properties and crystal chemistry of borides, carbides and phosphides. In view of so much readily available information, this chapter will emphasize those aspects not yet systematized elsewhere, i.e. the electroplating of the refractory compounds from molten salts. Just as the refractory metals can only be plated from molten salts, it is obvious that the refractory compounds can only be plated from molten salts.

4.2 METHODOLOGY

Refractory compounds are plated from molten fluorides in the same temperature range as the refractory metals, i.e. near 750°C. Thus the methodology is the same as that discussed in Chapter 2. The new feature is that the desired nonmetal – boron, carbon or silicon – must be plated simultaneously with the metal and at the same potential, and must react to form the compound on the surface. When this occurs it lowers the temperature dramatically from that required to react the elements in bulk, since reaction occurs between individual atoms. However, the requirements imposed by the above restrictions mean that not all compounds can be plated. One of the objectives of current research is to find out which ones can be plated and what are the best methods.

Probably the earliest systematic work was carried out by Andrieux and Weiss during the 1940s [7]. Although it resulted in electrosynthesis rather than electroplating, it showed that many binary compounds could be synthesized electrochemically from molten salts. Specifically, the following compounds of tungsten and molybdenum were prepared by electrolysis from molten salts: borides, carbides, arsenides, sulfides and antimonides (only the first two groups are refractory). Baths generally were alkali metal borate–fluoride mixtures, excellent solvents for the oxides of molybdenum and tungsten. In a paper more than 70 pages long, Weiss [8] describes in detail the relation between melt composition, voltage, temperature and the stoichiometry of the product, which was usually obtained as a mass of crystals on the wall of the carbon crucible, serving as the cathode. Relevant parts of this work will be described in the appropriate sections.

4.3 BORON AND THE BORIDES

Although the contribution of surface and near-surface boron to increasing the hardness of metals is well established, surprisingly little work has been done on the electrodeposition of either elemental boron or refractory borides. The source of boron in nearly all cases has been KBF_4 and/or B_2O_3 or alkali metal borates. In many cases this work has been quite empirical. For example, the use of both KBF_4 and B_2O_3 in the same melt quite obscures the actual source of boron in the plated product. This question has been addressed only recently (see below).

4.3.1 Boron

Up to 1960 attempts to plate boron were reported by Andrieux and Weiss [7], Ellis [9], Miller [10], Murphy *et al.* [11], Stern and McKenna [12], Nies [13] and Cooper [14]. All of them used various combinations of alkali metal chlorides, fluorides, KBF_4 and B_2O_3. Of these, the highest purities, in excess of 90%, were obtained by Cooper [14] from $KF–KBF_4–B_2O_3$, $KCl–KBF_4$, and $KCl–KBF_4–B_2O_3$ melts. All of these efforts produced dendritic, spongy or compacted powder deposits.

In order to improve the process, Kellner [15] thoroughly investigated the effect of several factors on both the quality and the purity of plated boron. He found that oxygen and oxygenated compounds, including B_2O_3, must be rigorously excluded from the melt in order to produce a smooth, hard and adherent deposit. His solvent was 1:1 LiF–KF mixture, and boron was introduced as $BF_3(g)$. This gas is very soluble and the electroactive species is obviously BF_4^-. The temperature was 700 °C. Electrochemical and mass balance studies were consistent with a three-electron reduction:

$$BF_4^- + 3e = B + 4F^-$$

Because the reduction potential is close to that of potassium, small impurity levels of the alkali metal, up to a few percent, were codeposited. Best results were obtained at current densities below 50 mA/cm^2, and BF_3 concentrations above 20 g/cm^2. X-ray diffraction patterns of the boron coating were diffuse, consistent with 'amorphous' boron. The hardness is purity dependent; the highest hardness, 2055 kg/mm^2 is obtained for the purest boron.

Some support for Kellner's interpretation comes from mechanistic studies of boron deposition reported by Makyta, Matiasovsky and Fellner [16], who addressed the question of the relation between KBF_4 and B_2O_3 as precursors for boron deposition. In LiF–KF at 700 °C, when KBF_4 was the only source of boron, only one voltammetric wave was observed, consistent with a single three-electron process. Two reduction waves are obtained with B_2O_3 as the only boron source. The first of them is accounted for by the generation of electroactive BF_4^- by the chemical reaction

$$2B_2O_3 + 3LiF + KF = 3LiBO_2 + KBF_4$$

No definite conclusions were reached about the borate species and its reduction, but its diffusion constant is much lower than that of BF_4^-.

4.3.2 Borides

Very little work has been done on the electrodeposition of borides, and the field remains largely unexplored in spite of the known beneficial properties, e.g. hardness, which protect the substrate metal from wear. In contrast, boriding, the process invented by Cook [17] and described in the previous chapter, as well as methods which accomplish boriding by other means (see Chapter 6) are better established.

Weiss was probably the first to investigate the electrochemical preparation of refractory borides [8]. He found that MoB and Mo_2B could be deposited from a melt containing B_2O_3, Na_2O, NaF, and MoO_3, i.e. a borate–molybdate melt, at 1000 °C. The composition of the product, which was only obtained as crystals imbedded in the frozen salt matrix, depended on the percentage of both boron and molybdenum. In similar studies of tungsten boride, with WO_3, only WB was obtained as the product.

As part of their work on plating zirconium from molten fluorides, Mellors and Senderoff [18] found that the addition of KBF_4 to an $LiF–KCl–K_2ZrF_6$ melt at 800 °C produced excellent coherent coatings of ZrB_2. Similar results were obtained from $FLINAK–K_2ZrF_6$ melts, but not from $NaF–LiF$ melts. Evidently potassium is a necessary melt constituent.

Of all the refractory borides, TiB_2 has attracted the most attention. This is due to its great oxidation resistance (slight even at 1000 °C), Vickers hardness (4000), erosion resistance, corrosion resistance to hot concentrated brines and coating–substrate adhesion. Kellner *et al.* [19] extended Kellner's work on boron plating [15] from FLINAK by adding BF_3 to this melt. The titanium was added as the metal, but was probably transformed to its fluoride. At the cathode the most likely reactions are

$$2BF_4^- + 6e = 2B + 8F^-$$

$$\frac{TiF_6^{3-} + 3e = Ti + 6F^-}{2BF_4^- + TiF_6^{3-} + 9e = TiB_2 + 14F^-}$$

No electrochemical studies were carried out to elucidate the mechanism of this nine-electron process. However, from a practical point of view, excellent coatings were obtained.

A detailed study of tantalum boride plating was carried out by Rameau [20], who used voltammetric measurements to elucidate the mechanism of tantalum and boron deposition separately in KCl at 930 °C. The mechanism for tantalum agrees approximately with that found by Senderoff, Mellors and Reinhart [21] in FLINAK at 750 °C, except that the first step is attributed to

$$3Ta^{5+} + 2Ta = 5Ta^{3+}$$

instead of a three-electron step. This second step is an irreversible two-electron reduction. Boron is formed independently by the reduction of BF_4:

$$BF_4^- + 3e = B + 4F^-$$

The compound TaB is then formed by the direct reaction of the elements. TaB_2 can only be formed after the melt is exhausted of tantalum, and boron is reduced on the already formed TaB. The products were identified by X-ray diffraction, but the physical state of the coating was not described.

In contrast to the work in FLINAK, investigators at the U.S. Bureau of Mines found borate melts to have some advantages over molten fluorides: they are easier to dry and permit higher plating rates Most of this work is described in reports [22–24] and patents, but not in scientific journals. For TiB_2 the melt consisted of the following mixture:

	wt%
$LiBO_2$	58.62
$NaBO_2$	39.08
Na_2TiO_3	0.99
Li_2TiO_3	0.76
TiO_2	0.55

All of this work is concerned with the practical aspects of plating which was designed to protect valves and pipes used in transporting hot, geothermal brines from corrosion. Thus, although the plating procedures were carefully worked out and the properties of the coating studied, only the overall reactions were given. It is known that the electrolyte must be 'conditioned' by the addition of titanium, probably as Ti(III). Plating can be carried out with either Ti or TiB_2 anodes, which probably dissolve electrochemically according to

$$Ti = Ti^{3+} + 3e$$

$$TiB_2 = Ti^{3+} + 2B^{3+} + 9e$$

respectively. Since TiB_2 can be plated even when the anode is titanium metal, it is likely that the melt borate is the source of boron:

$$Ti^{3+} + 2BO_2^- + 9e = TiB_2 + 4O^{2-}$$

However, these proposals are merely reasonable interpretations of plating observations. Much further work would be required to elucidate the mechanism of what is likely to be a complex nine-electron reaction.

The wear characteristics of the coatings produced in borate melts were examined on hardened steel by pin-on-disk wear tests of TiB coated pins of molybdenum, nickel and Inconel [25]. A comparison of these coated surfaces with single-crystal Al_2O_3 and with steel indicates that the best application of TiB_2 coatings is to a high speed, unlubricated sliding environment, such as a high speed lathe.

Quite recently Matiasovsky and coworkers [26, 27] have investigated the detailed mechanism of TiB_2 deposition from halide melts. In a pure fluoride melt [26] (LiF–KF and FLINAK) Ti(IV) is reduced in two steps via Ti(III), and boron (from KBF_4) is reduced to the element in a single step. TiB_2 is formed by the direct combination of the elements. In a pure chloride melt (NaCl–KCl) [27] the reduction of titanium was studied with Ti(III) as the starting material. It proceeded in two steps via Ti(II). However, this process was complicated by the disproportionation of this intermediate. This could be eliminated by the addition of NaF, which promotes the direct reduction of Ti(III) to Ti(0). The reduction of BF_4^- proceeds to the element, similar to the process in a fluoride melt. Synthesis of TiB_2 is also similar. From a practical viewpoint a mixed chloride–fluoride melt is probably preferable since it avoids the corrosion problems associated with pure fluoride melts. However, the consequences of changing melt composition on the character of the TiB coating remain to be studied.

It thus appears that the only borides that have been prepared electrochemically are those of molybdenum, tungsten, zirconium and tantalum, of which only the last two were produced as coatings. The field is certainly ripe for exploitation, since an adaptation of successful plating methods should be effective for electroplating other borides. The choice of borides can be made on the basis of properties listed in the cited references.

4.4 SILICON AND THE SILICIDES

4.4.1 Silicon

Just as the plating of borides requires a process for plating boron, so the plating of silicides requires a process for plating silicon. Aside from silicide plating, a major motivation for plating silicon comes from the utility of this element in the manufacture of photovoltaic devices. For this application it is necessary that silicon does not bond with the substrate. Since this element diffuses rapidly into many metals [28], making 'siliciding' a simple procedure, substrates are generally limited to carbon and silver, elements which do not alloy with silicon.

Much of the early history of silicon electroplating is not entirely germane to the present discussion on silicide coatings because it tends to concern electrowinning, particularly of liquid silicon. And often the silicon was impure. An extensive description of this work was given by Elwell and Feigelson [29].

In an extensive series of studies Rao, Elwell and Feigelson investigated silicon plating from K_2SiF_6 dissolved in KF–LiF and FLINAK at 750 °C [30], and from SiO_2–BaO–BaF_2 melts at 1450 °C [31]. In the first of these studies it was shown that coherent, inclusion-free layers of silicon could be grown on silver and carbon as long as the current density and deposition voltage remained below specific limits. It was also shown that the K_2SiF_6 concentration and deposition voltage has a critical effect on the current efficiency. The work on silicate melts is of less interest in the current context because liquid silicon is produced, and thus cannot be used as a step in the formation of silicide coatings.

In contrast to the work just described, Cohen and Huggins [32, 33] at the same laboratory (Center for Materials Research, Stanford University) concentrated on the structure of the deposited silicon. Using a K_2SiF_6–LiF–KF melt at 750 °C with a dissolving silicon anode, epitaxial silicon layers were deposited on single-crystal silicon cathodes. Polycrystalline coatings were obtained on silver, tungsten and niobium. For tungsten and niobium, thin diffuse layers of WSi_2 and $NbSi_2$ were observed at the interface, but the deposition rate far exceeded the diffusion velocity.

The deposition mechanism is not entirely clear. Elwell and Rao [34], using cyclic voltammetry in a K_2SiF_6 melt, describe the deposition process as 'quasi-reversible'; the rate-determining step is the charge-transfer process coupled to a volume diffusion stage. Si(II) may also be an important species in the melt. Rao *et al.* pointed out that a chemical step

$$Si + Si(IV) = 2Si(II)$$

may be responsible when powdery or spongy deposits are obtained. This conclusion was also reached by Boen and Boutellion [35], who interpreted their voltammetric data as the above chemical reduction followed by the electrochemical reduction of Si(II):

$$Si(II) + 2e = Si$$

In further work [36] it was found that Si(IV) is also reduced electrochemically to Si(II), so that the entire mechanism can be described as

$$Si(IV) + e = Si(II) \quad \text{reversible}$$

$$Si(II) + 2e = Si \quad \text{quasi-reversible}$$

$$Si(IV) + Si = 2Si(II)$$

Excellent, smooth and very pure silicon coatings were obtained on carbon when the solvent was purified and the potential step was kept small. A pulsed current was found beneficial to keep the Si(II) concentraton near the surface constant and to aid in the dissolution of impurities.

4.4.2 Silicides

As pointed out previously, the plating of refractory silicides is severely inhibited by the rapid diffusion of silicon into many metals. As an example of such diffusion, the results of attempts to plate silicon on Inconel, a widely used alloy of nickel, chromium and iron, may be cited [37]. Plating was carried out from a FLINAK melt at 750 °C, with K_2SiF_6 as the solute. An elemental analysis of the cross section of a plated specimen showed that, although the surface region was higher in silicon than in the alloy constituents, the relative concentrations of the alloy constituents and the silicon varied from the surface approximately 20 μm into the interior, before silicon was completely absent. A careful analysis of the near-surface region indicated that the metals had counterdiffused into the silicon during the plating process.

Because the diffusion of silicon into many potential substrates is quite rapid, successful silicide plating requires either a metal substrate with which silicon alloys only slowly, or a barrier coating of such a metal which prevents silicon penetration into the base metal. Examples of the first case are tungsten and niobium, into which silicon diffused so slowly that pure silicon could be plated on them (see above). Silicides of tungsten and niobium could be plated to protect the substrate from corrosion, e.g. WSi_2 on tungsten. Silver and carbon do not alloy with silicon at all, but they are not practical barriers for useful applications. Tantalum seems to be useful, particularly if its silicide is to be plated (see below).

Dodero [38] was probably the first to study the formation of a 'silicon–titanium alloy' by molten salt electrolysis of a TiO_2–alkali metal silicide–fluoride solution. Oxygen is evolved at the anode, and the crystals formed on the cathode may result from the reduction of Si(IV) and Ti(IV) by an alkali metal. Similar results were obtained by Beaudoin [39], who used a K_2SiF_6 eutectic containing TiO_2 at 725 °C. $TiSi_2$ was formed as crystals with a current efficiency of 22%.

The problem of $TiSi_2$ deposition was examined in much greater detail by Delimarskii and coworkers [40]. They pointed out that the addition of TiO_2 to a melt containing SiF_6^{2-} produced M_2TiF_6 through the equilibrium

$$M_2SiF_6 + TiO_2 = M_2TiF_6 + SiO_2$$

Using voltammetry in an equimolar KCl–NaCl melt containing K_2SiF_6 and Na_2TiF_6 at 685 °C, they concluded that silicon and titanium are formed independently on the cathode – silicon probably in a two-step process – where they combine to form $TiSi_2$. No data on current efficiency or the character of the deposit were given.

The first study which examined the effects of melt composition and plating parameters on the character of the coating was carried out by Stern and Williams [41], who plated tantalum silicide from a FLINAK melt containing K_2TaF_7 and K_2SiF_6. The crystalline characteristics of the coating seemed to vary with both the plating voltage and the molar Ta/Si ratio in the melt, but all were dense and adherent. A thin tantalum layer preplated on the nickel cathode prevented silicon from diffusing into the substrate. The coating hardness was 1200 kg/mm^2, in general agreement with that reported for Ta$_5$Si$_3$ [42], a coating composition also consistent with Auger analysis. The coating is thermally stable in air up to 400 °C, and shows only small changes in crystal shape up to 600 °C. Above this temperature, the coating structure degrades and spalls.

The procedure described for tantalum silicide seems to be generally applicable to the plating of refractory silicides, only the appropriate metal fluoride being required [43]. For example, MoSi$_2$ has been found to be highly resistant to oxidation up to 1000 °C [44] and would be an excellent coating for protecting metals from corrosion, but no plating studies on this or other refractory fluorides have yet been reported.

4.5 CARBON AND THE CARBIDES

4.5.1 Carbon

The plating of refractory carbides requires a process for plating carbon. However, in contrast to the plating of boron and silicon, which have potential practical applications, the plating of carbon is not of any great interest in itself. The requirements for a suitable precursor are that (a) it is soluble in the chosen melt, e.g. FLINAK, and (b) that it is reducible at a potential no greater than the reduction potentials of the solvent cations, preferably close to that of the metal precursor being reduced simultaneously. An obvious additional requirement is that the deposited carbon should react with the simultaneously plated metal to form the carbide.

The only precursors that have been investigated for carbon deposition are cabonate ion and carbon dioxide. They are connected by the equilibrium

$$CO_3^{2-} = CO_2 + O^{2-}$$

and detailed mechanistic studies are required to ascertain which species is actually reduced. Since the relative concentrations of CO_3^{2-} and CO_2 obviously depend on the O^{2-} concentration, the basicity of the melt probably affects some aspect of carbon deposition.

Since the reduction and oxidation of CO_3^{2-} are important for the operation of molten carbonate fuel cells, the most thorough studies have been carried out for pure molten alkali metal carbonates and carbonate mixtures. Bartlett and Johnson studied these processes both theoretically – based on thermodynamic data [45] – and experimentally [46]. There are potentially four possible reduction processes:

(a) the formation of carbon and oxide

$$CO_3^{2-} + 4e = C + 3O^{2-}$$

(b) the deposition of the alkali metal

$$M^+ + e = M(0)$$

(c) the formation of carbon monoxide and oxide

$$CO_3^{2-} + 2e = CO + 2O^{2-}$$

(d) the formation of carbide and oxide

$$2CO_3^{2-} + 10e = C_2^{2-} + 6O^{2-}$$

Which of these processes actually occurs depends on both the alkali metal cation and the temperature. However, the reduction of CO_2 is always favored; if it does not occur, this is due to the low activity of CO_2 in a pure carbonate melt. The situation is obviously more complicated when carbonates are solutes in other melts, such as halides, and some assumptions about solution thermodynamics would be required to determine the reduction products theoretically. It is not surprising, therefore, that so far only experimental studies have been carried out. Delimarskii and coworkers [47] showed that carbon was not liberated by the electrolysis of a 1:1 solution of K_2CO_3 and Na_2CO_3 in KCl, but was produced when a sufficiently high CO_2 pressure was introduced above the halide melt. These results were confirmed in a more detailed study [48] of CO_2 electrolysis in a 1:1 KCl–NaCl melt in the 700–800 °C range. It should be noted, however, that one of the reduction products is O^{2-}, i.e.

$$CO_2 + 4e = C + 2O^{2-}$$

and that this may form CO_3^{2-} near the electrode

$$O^{2-} + CO_2 = CO_3^{2-}$$

The authors could not observe the effect of this reaction on the reduction of CO_2 and concluded that it must be slow.

Not much work has been done on the reduction of carbonate ion in fluoride melts. Stern *et al.* [49] could not observe a reduction peak for this ion before the cathodic reduction of the solvent cation and concluded that CO_3^{2-} is reduced by the alkali metal produced at the cathode:

$$4Me + Me_2CO_3 = C + 3Me_2O$$

However, Selman and Topor did observe such a reduction peak close to the cathodic limit of the melt [50]. What is most interesting is that the direct reduction is highly temperature dependent: at 600 °C no reduction occurs before the reduction of the melt; a reduction peak becomes noticeable at 700 °C and pronounced at 800 °C. Reduction does occur even at 600 °C, but only indirectly – via the alkali metal formed at the cathodic limit of the melt. Another result reported by the authors is that although carbon and molybdenum can be formed simultaneously on the cathode they apparently do not react to form the carbide below 750 °C (see below). The situation may be somewhat unsettled, but there is no question that carbon is the reduction product and that therefore one of the requirements for carbide plating is met.

Another requirement for carbide plating is the thermal stability of carbonates in the fluoride solvent, since at temperatures near 800 °C these salts might be expected to decompose [51]. In a study of thermal decomposition and electrochemical reduction of carbonate in this melt [49], it was found that the melt slowly lost carbonate, both by thermal decomposition, i.e. $CO_3^{2-} = CO_2 + O^{2-}$, and by vaporization of an alkali metal carbonate molecule, i.e. $M_2CO_3(sol) = M_2CO_3(g)$. However, both these processes were too slow to substantially deplete the melt of carbonate during the plating process.

4.5.2 Carbides

As pointed out previously, the elecrodeposition of a refractory carbide requires that the metal-containing precursor and the carbon-containing species be reducible at approximately the same potential, since there is only one potential at which deposition of metal and deposition of carbon can occur simultaneously. However, many successful depositions have been carried out empirically without any electrochemical studies of the deposition process. Weiss [8] was probably the first to deposit refractory carbides (tungsten and molybdenum), from a B_2O_3–LiF–Na_2O solvent containing Na_2CO_3 as a carbon source and WO_3 or MoO_3 as the source of the metal. The mono- and dicarbides were obtained in the form of small crystals in a boule of frozen melt. Gomes and Wong [52] obtained powdered WC from a melt containing NaCl, Na_2WO_4, $Na_2B_2O_4$ and NaOH.

Most of the reported work on carbide plating has been semiempirical, i.e. reasonable combinations of precursors have been reduced electrochemically to see if a carbide would result.

A more systematic approach was taken by Shapoval *et al.* [53], who studied the decomposition potentials of tungstates, titanates, carbonates and borates of alkali and alkaline earth metals. They found that the potentials of WO_4^{2-} and MoO_4^{2-} were close to that of CO_3^{2-} and predicted that the carbides of tungsten and molybdenum should be platable, as had already been observed empirically, but that titanium carbide should be unplatable if a titanate is used as the precursor. Moreover, because oxyanions are involved in equilibria such as

$$WO_4^{2-} = WO_3 + O^{2-}$$

any substance which affects this equilibrium might be used to shift the deposition potential closer to that for carbon plating. Using these principles, they plated tungsten carbide from an NaCl–KCl melt containing Na_2WO_4, $MgCl_2$ serving to adjust the acidity, with CO_2 under pressure serving as the carbon source [54]. Molybdenum carbide was similarly plated [55]. This process was studied in more detail by examining the plated surface at different stages of the plating process. A crystallization overpotential was found to exist for molybdenum carbide deposition, and nucleation occurred at many sites to form a compact layer.

In addition to metallic oxyanions, complex fluorides, e.g. K_2TaF_7, have also been used as precursors. Which one is preferable for a particular metal depends on the chemistry and electrochemistry of the two species. In the next section the current state of affairs with respect to carbide plating of the nine refractory metals will be described. As will become obvious, progress has reached very different stages for the different metals.

4.5.3 Individual carbides

The sequence of carbides described in this section is the same as in Chapter 2 for the refractory metals. Since the properties of carbides – phase diagrams, thermodynamic properties, crystallographic properties, hardness, and chemical reactivity – are described in detail by Storms [2], the main emphasis in this chapter is on the electroplating of coatings. In general, all the refractory carbides are quite hard, but they differ considerably in their oxidation resistance. The choice of which carbide to plate for a specific application depends on its exposure conditions; generally these are wear, erosion and high temperatures. In general, the oxidation resistance of carbides does not extend to as high a temperature as that of silicides and borides; this is because the carbon – usually held interstitially – oxidizes to a gas.

Titanium

The Ti–C system shows only one compound, TiC, which exists over a very wide composition range. Vickers hardness ranges from 1600 to 2800. Oxidation resistance is low, formation of TiO_2 beginning at 450 °C.

No attempts to plate TiC or to synthesize it electrochemically have been reported. Stern [56] considered both K_2TiO_3 and K_2TiF_6 as potential starting materials. K_2TiO_3 is only slightly soluble in FLINAK at 750 °C (<0.2 mol%) and is not reducible below the cathodic limit of the solvent. K_2TiF_6, which is used to plate the metal, reacts as an acid with carbonate as a base in a reaction evolving CO_2:

$$K_2TiF_6 + 3K_2CO_3 = K_2TiO_3 + 3CO_2 + 6KF$$

This is a general reaction that interferes in the electrodeposition of all refractory carbides which use the complex fluoride of the metal as reactant. However, as has been studied in detail for tantalum carbide (see below), the reaction generally does not go to completion and the carbide can still be plated by a judicious choice of conditions. For titanium this has not yet been done.

Zirconium

As for titanium, only a single compound, ZrC, exists in this system. It oxidizes rapidly above 500 °C. No attempts to plate this compound have been carried out, but Stern [56] has reported some preliminary studies of combining ZrF_4, a precursor for zirconium plating [57] with K_2CO_3. For a Zr/C ratio near 1 the solution turns cloudy because an insoluble material is formed, most likely ZrO_2:

$$ZrF_4 + 2K_2CO_3 = ZrO_2 + 2CO_2 + 4KF$$

Further work is required to see if this reaction goes to completion. Zirconates are also insoluble and therefore cannot serve as reactants.

Hafnium

HfC appears to be the only carbide in the Hf–C system. The U.S. Bureau of Mines has studied the preparation and properties of melted HfC and found that both its melting point and hardness vary with the carbon content [58]. So far, no studies related to the electrodeposition of this material have been reported.

Vanadium

Two carbides, VC and V_2C are known in the V–C system. Although the hardness varies somewhat with composition, it generally lies between 2000 and 3000, making it one of the hardest materials known. Powdered V_2C reacts slowly with air at ambient temperature, but the monocarbide is probably stable up to several hundred degrees.

Vanadium carbide has been prepared metallurgically [59], but not electrochemically. Stern [56] has reported some preliminary studies of vanadate reductions in FLINAK. The solubilities of both $NaVO_3$ and Na_3VO_4 were high enough to permit reduction studies. The $NaVO_3$ was clearly reducible, but its reduction resulted in the formation of a dark material; it was not the metal, but it could not be identified.

Niobium

This metal, called columbium in the older literature, forms two carbides, NbC and Nb_2C. The hardness depends on composition, but is probably near $2000\,kg/mm^2$. NbC is quite unreactive, and air corrosion only becomes severe above 1000 °C.

Hockman and Feigelson [60] synthesized NbC electrolytically from a melt containing $Na_2B_4O_7$, NaF and KF as solvent, and Nb_2O_5 and Na_2CO_3 as solutes. Composition and electrochemical parameters had to be carefully adjusted to prevent the formation of other products, e.g. borides, but in any case the product was obtained only as discrete crystals in a matrix of frozen melt. Stern [56] studied the behavior of both $LiNbO_3$ and K_2NbF_7 in FLINAK at 750 °C. The oxyanion is fairly soluble and could be reduced below the cathodic limit of the melt; nevertheless, the product was not the metal but probably some lower-valent compound. The fluoride reacted with the carbonate in the acid–base reaction common to all the refractory metal fluorides:

$$K_2NbF_7 + 3K_2CO_3 = KNbO_3 + 3CO_2 + 7KF$$

However, CO_2 evolution stops when the Nb/C ratio reaches 1, indicating that the reaction does not go to completion.

In view of the desirable properties of vanadium carbides, and the fact that the acid–base reaction does not prevent the electrodeposition of tantalum carbide (see below), further work on the electrodeposition of niobium carbide coatings seems warranted.

Tantalum

The tantalum–carbon phase diagram shows two carbides, TaC and Ta_2C. The hardness of TaC is in doubt, values between 1500 and 2000 having been reported. The hardness of Ta_2C seems to be near $1000\,kg/mm^2$. Burning occurs in pure oxygen above 800 °C, suggesting that the compound is stable to nearly that temperature in air.

Hockman and Feigelson [60] prepared small crystals of TaC under the same conditions as NbC (see above). Stern and Gadomski [61] seem to have been the first to electroplate adherent coatings of tantalum carbide, using FLINAK as solvent and K_2TaF_7 and Na_2CO_3 as solutes. The composition of the coating –Ta, TaC, Ta_2C and their mixtures – depended both on the solute concentration and the voltage applied between the tantalum anode and the nickel cathode. Thermogravimetric tests showed that the coating was stable into the 500–600°C range. Hardness was measured as 1600 ± 100 kg/mm^2, and mechanical tests showed that the coating is highly adherent, probably because it is anchored to the substrate by a thin tantalum–nickel interdiffusion zone. The process is covered by a patent [62].

After the acid–base reaction between refractory metal fluorides and carbonates was discovered, the effect of this reaction on tantalum carbide plating was investigated by various techniques [63]. In common with other fluorides (see above), CO_2 is evolved from $TaF_5^{2-} - CO_3^{2-}$ melts at temperatures as low as 500 °C. Also in common with other fluorides the reaction

$$K_2TaF_7 + 3K_2CO_3 = KTaO_3 + 3CO_2 + 7KF$$

does not go to completion and sufficient reactant concentrations remain in the melt if the initial Ta/C ratio is carefully chosen. Detailed voltammetric studies showed that when the C/Ta ratio is 3, the peaks characteristic of tantalum reduction completely disappear, indicating that the $KTaO_3$ which is formed completely passivates the electrode. Even at ratios as low as 1, carbonate affects the voltammetric tantalum reduction peaks. However, near this ratio Ta_2C can be plated successfully because sufficient carbonate for carbide formation remains in the melt, and the partial passivation of the cathode for Ta(V) reduction impedes the direct plating of tantalum metal. Similar detailed studies would probably be useful to permit the plating of other refractory carbides.

Chromium

The Cr–C phase diagram shows three carbides: $Cr_{23}C_6$, Cr_7C_3 and Cr_3C_2. The hardness is in the 1000–1400 range for Cr_3C_2. The carbides resist oxidation into the 800–1000 °C range.

Although chromium plating from molten salts has been studied by several workers [64–68], only one study of chromium carbide plating has been reported [69]. In this work two possible precursors were studied: CrO_4^{2-} was found to be quite soluble in FLINAK at 750 °C, but could not be reduced below the cathodic limit of the melt; based on previous work on chromium metal plating, CrF_2 was then chosen as precursor. Addition of carbonate to this melt leads to the usual CO_2 evolution reaction

$$CrF_2 + K_2CO_3 = CrO + CO_2 + 2KF$$

which does not go to completion. Since plating experiments for the metal showed that a temperature of a least 850 °C was required to avoid dendrite formation, a temperature range of 850–900° was adopted for carbide plating. Chromium carbide coatings were obtained for Cr/C ratios near 1, but the current efficiency did not exceed 10%. No metal and no carbide were obtained if Cr(III) was used. Since all the carbides exhibit closely spaced X-ray diffraction lines, the stoichiometry of the coating could not be determined. Although the low current efficiency is disappointing, the fact that chromium carbide has been successfully plated suggests that improvements would result from the optimizing of concentration and electrochemical parameters.

Molybdenum

There are two carbides of molybdenum, MoC and Mo_2C. The monocarbide is thermodynamically unstable at ambient temperatures, but it can be prepared at high temperatures then cooled without decomposing. The carbides are oxidized in air at 700–800 °C, and the hardness of the dicarbide has been reported to be near 1500. Their main application is probably the protection of metals from wear up to temperatures of a few hundred degrees.

The first attempt to electrodeposit these carbides was made by Weiss [8], who formed small crystals from a borate–halide melt. Which carbide was formed depended on both the Mo/C ratio and the electrochemical parameters. Since then there have been reports of several other studies whose objective was the electrolytic preparation of Mo_2C. Heinen, Barber and Baker reported [70] obtaining a firmly adherent plate of the dicarbide, consisting of dendritic crystals, from the electrolysis of a melt consisting of NaF, KF, $Na_2B_4O_7$, Na_2CO_3 and molybdenite. Similar results were also reported by Indian workers [71] using nearly the same melt. In both cases the objective was to use the carbide as a starting material for producing the metal.

Successful attempts to plate adherent and continuous coatings have been made only recently. It is noteworthy that all of them have used molybdates, which are easily reducible, rather than complex halides. This is despite the fact that the metal has been successfully plated from chloride melts [72]. However, the molybdenum chemistry in these melts is quite complex [73], and molybdates offer a more straightforward path to reduction. (For a more extended discussion of this topic see Chapter 2).

Shapoval *et al.* [53] predicted the feasibility of plating molybdenum carbide by the simultaneous reduction of MoO_4^{2-} and CO_2 in an NaCl–KCl melt acidified with Mg(II). Their evidence that the reduction occurs was based on voltammetry, rather than plating. However, in later work [55] the formation of molybdenum carbide from tungstate–molybdate melts was observed, but mainly from the point of view of observing the initial stages of crystallization.

Successful attempts to plate Mo_2C have recently been reported by Topor, Selman and and Aladjov [74, 75]. They used FLINAK as solvent, and K_2CO_3 and Na_2MoO_4 as solutes at a temperature of 850 °C. Good adhesion resulted from a nickel–molybdenum interdiffusion layer where nickel was the substrate. Additions of $Na_2B_4O_7$ and the use of current reversal techniques further improved the quality of the coatings. In a subsequent mechanistic study [50] the authors find molybdate reduction to be quite complex, involving several intermediate oxidation states and chemical reactions. As already discussed in secion 4.5.1, carbonate reduction is highly temperature-dependent, occurring only above 700 °C, the product being carbon. The elements then react on the surface to form the compound.

Tungsten

The tungsten–carbon phase diagram shows two carbides, WC and W_2C. Early hardness measurements are affected by questions of which carbide was being measured, but recent results indicate values near 1500 for W_2C and near 2000 for WC; WC is stable in air up to 700 °C.

Despite the useful properties of tungsten carbide, very little work has been done to electroplate it. Early work by Weiss [8] showed that both WC and W_2C could be plated from borate–fluoride melts, but only as dispersed crystals. Which carbide was plated depended primarily on the melt composition. Gomes and Wong [76] studied many melt compositions in NaCl containing various proportions of Na_2WO_4, $Na_2B_2O_4$, NaOH and Na_2CO_3. The composition of the deposit depended on the melt composition, but in all cases the deposit was powdery. Shapoval *et al.* [77] electrodeposited tungsten carbide powder from a $KCl–NaCl–Na_2WO_4$ melt containing Mg(II), under a pressure of CO_2. The gas served both as a source of carbon and an adjuster of acidity. Dense and adherent coatings of W_2C were obtained by Stern and Deanhardt [78] from a FLINAK melt containing Na_2WO_4 and Na_2CO_3 on nickel substrates at -1.0 to -1.5 V relative to a platinum quasi-reference electrode. The detailed microstructure of the coating depended on both the melt composition and the voltage, but it was always dense and columnar. It followed the lines of an engraving 25 μm wide and deep, so perhaps it could be used to harden engraving plates.

Silicon carbide

Silicon is not a metal but its carbide, SiC, has long been known as an extremely hard and erosion-resistant material. It would also be useful for semiconducting devices which can tolerate high temperatures and radiation levels. The electrodeposition of the compound obviously requires the simultaneous deposition of the elements. Silicon was discussed in section 4.4 and carbon in section 4.5. The only reported attempt to plate SiC was by Elwell, Feigelson and Simkins [79], who reduced K_2SiF_6 and Li_2CO_3. Although SiC was plated from both fluoride and borate–fluoride solvents, the best results were obtained from a pure melt of the two reactants at $1000–1050\,^{\circ}C$. At this temperature the vaporization of the carbonate and reactions of the melt with the container begin to be troublesome. Otherwise, even higher temperatures seem to be beneficial.

Some of the problems cited by the authors are associated with the desired objective to produce crystals with a particular structure (SiC exists in several crystalline modifications). For refractory coatings these problems might not exist, and further work on the process would seem to be warranted.

Boron carbide

Boron carbide, B_4C, is another nonmetallic carbide which is extremely hard and would seem to be desirable as a coating material. Methods for electrodepositing boron and carbon are well established and have been described in this chapter. However, a literature search shows no previous work in this field. Studies to determine under what conditions the elements react on the surface would seem to be of considerable interest.

4.6 CONCLUSIONS

The electrodeposition of refractory compounds is clearly an underdeveloped field, both with respect to fundamental research and the development of practical applications, although many of the 27 refractory borides, silicides and carbides (not counting multiple stoichiometries) would surely make excellent coatings, based on their known properties.

Only for carbides is enough known to at least attempt a systematization of preparative procedures [56]: the ease of carbide plating increases from group IVA to VIA of the periodic table, and with increasing atomic weight of the metal. All complex fluorides can be reduced to the metal (see Chapter 2), but they react with carbonates to produce CO_2. However, since this reaction does not go to completion, the carbides can still be plated by a judicious choice of conditions. This has been demonstrated so far only for tantalum [62, 63]. The utility of oxyanions for carbide plating depends on the solubility of their alkali metal salts, which appears to be inversely related to their melting points. So far, only the carbides of chromium [69], molybdenum [74, 75] and tungsten [78] have been successfully plated; oxyanions of the other six, with the possible exception of vanadium and niobium, are either too insoluble or not reducible.

Only five borides – tungsten [8], molybdenum [8], tantalum [20], zirconium [18] and titanium [22–24] – have been studied. Of these, only the last two have been produced as coatings. Further work on these hard, corrosion-resistant materials seems warranted.

Refractory silicides have received even less attention. Tantalum silicide has been successfully plated [41], but even preliminary studies have only been carried out on titanium [38–40]. Considering the known corrosion resistance of these materials at high temperatures, this is certainly surprising.

REFERENCES

1. Bunshaw, R.F. *et al.* (1982) *Deposition Technologies for Films and Coatings*, Noyes, Park Ridge, NJ.
2. Storms, E.K. (1967) *The Refractory Carbides*, Academic Press, New York.
3. Toth, L.E. (1971) *Transition Metal Carbides and Nitrides*, Academic Press, New York.
4. Freer, R. *et al.* (1989) *The Physics and Chemistry of Carbides, Nitrides and Borides*, Kluwer, Boston.
5. Matkovich, V.I., *et al.* (1977) *Boron and Refractory Borides*, Springer-Verlag, New York.
6. Aronson, B., Lundstrom, T. and Rundqvist, S. (1965) *Borides, Silicides, and Phosphides*, Methuen, London. Reprinted by Wiley, New York.
7. Andrieux, J. and Weiss, G. (1948) *Mem. Soc. Chim.*, 598–9.
8. Weiss, G. (1946) *Ann. Chim.*, **12**, 446–523.
9. Ellis, R.B. (1957) U.S. Patent 2,810,683.
10. Miller, G.T. (1959) *J. Electrochem. Soc.*, **106**, 815–19.
11. Murphy, H.F., Tinsley, R.S. and Meenaghan, G.F. (1957) *Bull. Virginia Polytech. Inst. Eng. Exp. Station Ser. No. 115.*

12. Stern, D.R. and McKenna, Q.H. (1959) U.S. Patent 2,892,762.
13. Nies, N.P. (1960) *J. Electrochem. Soc.*, **107**, 817–20.
14. Cooper, H.S. (1951) U.S. Patents 2,572,248 and 2,572,249.
15. Kellner, J.D. (1973) *J. Electrochem. Soc.*, **120**, 713–16.
16. Makyta, M., Matiasovsky, K. and Fellner, P. (1984) *Electrochim. Acta*, **29**, 1653–7.
17. Cook, N.C. (1969) *Sci. Amer.*, **221**, 38–46.
18. Mellors, G.W. and Senderoff, S. (1966) *J. Electrochem Soc.*, **113**, 60–5.
19. Kellner, J.D., Croft, W.J. and Shephard, L.A. (1977) *Army Materials and Mechanics Center Report TR-17, ADA 047956.*
20. Rameau, J.J. (1971) *Rev, Int. Hautes Temp. Refract.*, **8**, 59–70.
21. Senderoff, S., Mellors, G.W. and Reinhart, W.J. (1965) *J. Electrochem. Soc.*, **112**, 840–5.
22. Schlain, D., McCawley, F.X. and Smith, G.R. (1976) *U.S. Bur. Mines Rep. Invest. 8146.*
23. Flinn, D.R., McCawley, F.X., Smith, G.R. and Needham, P.B. (1979) *U.S. Bur. Mines Rep. Invest. 8332.*
24. McCawley, F.X., Wyche, C. and Schlain, D. U.S. Patents 3,697,390 (1972); 3,827,954 (1974).
25. Kirk, J.A., Flinn, D.R. and Lynch, M.J. (1981) *Wear*, **72**, 315–23.
26. Makyta, M., Matiasovsky, K. and Taranenko, V.I. (1989) *Electrochim. Acta*, **3**, 861–6.
27. Taranenko, V.I., Zarutskii, I.V., Shapoval, V.I. *et al.* (1992) *Electrochim. Acta*, **37**, 263–8.
28. *Constitution of Binary Alloys* (a) (1958) Hansen, M., ed. (b) (1965) Elliott, R.P., ed., (c) (1969) Shunk, F.A., ed.
29. Elwell, D. and Feigelson, R.S. (1982) *Solar Energy Mater.*, **6**, 123–45.
30. Rao, G.M., Elwell, D. and Feigelson, R.S. (a) (1980) *J. Electrochem. Soc.*, **127**, 1940–4; (b) (1981) *J. Electrochem. Soc.*, **128**, 1708–10; (c) (1981) *Surf. Technol.*, **13**, 608–11.
31. Mattei, R.C., Elwell, D. and Feigelson, R.S. (a) (1981) *J. Electrochem. Soc.*, **128**, 1712–14; (b) (1981) *Proceedings of the Symposium on Electrocrystallization* (eds R. Weil and R.G. Baradas), The Electrochemical Society, Pennington, NJ, pp. 323–9.
32. Cohen, U. and Huggins, R.A. (1976) *J. Electrochem. Soc.*, **123**, 381–3.
33. Cohen, U. (1977) *J. Electronic Mater.*, **6**, 607–43.
34. Elwell, D. and Rao, G.M. (1982) *Electrochim. Acta.*, **27**, 673–6.
35. Boen, R. and Bouteillon, J. (1983) *J. Appl. Electrochem.*, **13**, 277–88.
36. DeLepinay, J., Bouteillon, J., Traore, S. *et al.* (1987) *J. Appl. Electrochem.*, **17**, 294–302.
37. Stern, K.H. and McCollum, M.E. (1985) *Thin Solid Films*, **124**, 129–34.
38. Dodero, M. (1952) *J. Chim. Phys.*, **49**, 12–14.
39. Beaudouin, J. (1966) *C.R.Acad. Sci. Paris*, 263C, 993–6.
40. Chernov, R.V., Nizov, A.P. and Delimarskii, Y.K. (1971) *Ukr. Khim. Zhur.*, **37**, 422–6.
41. Stern, K.H. and Williams, C.E. (1986) *J. Electrochem. Soc.*, **133**, 2157–60.
42. Kieffer, R., Nowotny, H., Bemesovsky, F. and Schachner, H. (1933) *Z. Metallk.*, **44**, 242–6.
43. Stern, K.H. (1987) U.S. Patent 4,662,998.
44. Beidler, E.A., Powell, C.F., Campbell, I.E. and Yntema, L.F. (1951) *J. Eletcrochem. Soc.*, **98**, 21–5.
45. Bartlett, H.E. and Johnson, K.E. (1966) *Canad. J. Chem.*, **44**, 2119–29.
46. Bartlett, H.E. and Johnson, K.E. (1967) *J. Electrochem. Soc.*, **114**, 457–61.
47. Delimarskii, Y.K. (1968) *Dokl. Akad. Nauk SSSR*, **183**, 1332–4.
48. Kushkov, K.B., Shapoval, V.I. and Novoselova, I.A. (1987) *Elektrokhim.*, **23**, 952–6.
49. Deanhardt, M.L., Stern, K.H. and Kende, A. (1986) *J. Electrochem. Soc.*, **133**, 1148–52.
50. Topor, D.C. and Selman, J.R. (1993) *J. Electrochem. Soc.*, **140**, 352–61.
51. Stern, K.H. and Weise, E.L. (1969) National Bureau of Standards, *Natl Stand. Data Ser. No. 30.*

52. Gomes, J.M. and Wong, M.M. (1969) *U.S. Bur. Mines Rep. Invest. 7247.*
53. Shapoval, V.I., Kushkov, K.B. and Novoselova, I.A. (1988) *Ukr. Khim. Zhur.*, **48**, 738–42.
54. Kushkov, K.B., Novoselova, I.A., Supatatsvhvilii, D.G. and Shapoval, V.I. (1990) *Elektrokhim.*, **24**, 48–51.
55. Polishchuk, V.A., Kushkov, K.B. and Shapoval, V.I. (1990) *Elektrokhim.*, **26**, 305–10.
56. Stern, K.H. (1992) *J. Appl. Electrochem.*, **22**, 717–21.
57. Mellors, G.W. and Senderoff, S. (1966) *J. Electrochem. Soc.*, **113**, 60–5.
58. Adams, R.P. and Beall, R.A. (1963) *U.S. Bur. Mines Rep. Invest. 6304.*
59. Downing, J.H. and Merkert, R.F. (1965) French Patent 1,422,683.
60. Hockman, A.J. and Feigelson, R.S. (1983) *J. Electrochem. Soc.*, **130**, 221–4.
61. Stern, K.H. and Gadomski, S.T. (1983) *J. Electrochem. Soc.*, **130**, 300–5.
62. Stern, K.H. (1984) U.S. Patent 4,430,170.
63. Stern, K.H. and Rolison, D.R. (1989) *J. Electrochem. Soc.*, **136**, 3760–7.
64. Inman, D., Legey, J.C.L. and Spencer, R. (1975) *J. Electroanal. Chem.*, **61**, 289–301.
65. Vagas, T. and Inman, D. (1987) *J. Appl. Electrochem.*, **17**, 270–82.
66. Smith, J.F. (1982) *Thin Solid Films*, **95**, 151–60.
67. Yoko, T. and Bailey, R.A. (1984) *J. Electrochem. Soc.*, **131**, 2590–5.
68. Yoko, T. and Bailey, R.A. (1986) *J. Appl. Electrochem.*, **16**, 737–42.
69. Stern, K.H. and Rolison, D.R. (1990) *J. Electrochem. Soc.*, **137**, 178–83.
70. Heinen, H.J., Barber, C.L. and Baker, D.H. (1965) *U.S. Bur. Mines Rep. Invest. 6590.*
71. Suri, A.K., Mukherjee, T.K. and Gupta, C.K. (1973) *J. Electrochem. Soc.*, **120**, 622–4.
72. Senderoff, S. and Brenner, A. (1954) *J. Electrochem. Soc.*, **101**, 16–32.
73. White, S.H. and Twardoch, U.M. (1987) *J. Appl. Electrochem.*, **17**, 225–42.
74. Topor, D.C. and Selman, J.R. (1988) *J. Electrochem. Soc.*, **135**, 384–7.
75. Aladjov, B., Topor, D.C. and Selman, J.R. (1992) *Proceedings of the 8th International Symposium on Molten Salts*, The Electrochemical Society, Pennington NJ, Vol. 92-16, pp. 488–99.
76. Gomes, J.M., Wong, M.M. (1969) *U.S. Bur. Mines Rep. Invest. 2747.*
77. Shapoval, V.I., Kushkov, K.B. and Novoselova, I.A. (1985) *Zhur. Prikl. Khim.*, **58**, 1027–30.
78. Stern, K.H. and Deanhardt, M.L. (1985) *J. Electrochem. Soc.*, **132**, 1891–5.
79. Elwell, D., Feigelson, R.S. and Simkins, M.M. (1982) *Mater. Res. Bull.*, **17**, 697–706.

5

Laser assisted surface coatings

J. Mazumder

Lasers provide the unique tool for high quality surface modification. With the development of continuous wave (CW) CO_2 lasers of high output power, surface modification technology is rapidly growing in the identification of new and improved processing methods. Many surface-related failures by mechanisms involving wear, corrosion, erosion or high temperature oxidation can be minimized by laser surface modification techniques such as laser surface alloying, laser cladding, laser chemical vapor deposition (LCVD) and laser ablation. Laser ablation is a relatively new application for laser assisted coating and is used extensively for high T_c superconducting films; it has almost become a field in itself. This chapter does not cover laser ablation.

The main topics of this chapter are alloying, cladding and chemical vapor deposition. These processes allow the surface properties of a structure to be tailored to the surface requirements of the application without sacrificing the bulk characteristics of the structure. The additional advantage of this is that strategic materials, which are expensive or scarce, can be conserved. The inherently rapid heating and cooling rates associated with the laser surface modification technique produce metastable and nonequilibrium phases and offer the possibility of new materials with novel properties.

Metallurgical and Ceramic Protective Coatings. Edited by Kurt H. Stern.
Published in 1996 by Chapman & Hall, London. ISBN 0 412 54440 7

5.1 LASER SURFACE ALLOYING

5.1.1 Principles of laser surface alloying (LSA)

Laser surface alloying (LSA) is a process whereby a thin layer at the surface of a metal is melted by a laser beam with the simultaneous addition of the desired alloying element, thus changing the surface chemical composition of the metal. The concept of laser surface alloying is shown in Fig. 5.1. This shows the cross-sectional sequence of LSA [1].

Figure 5.1a shows that the substrate (B) coated with alloying element (A) is irradiated with the laser beam. For most metals and laser wavelengths used for materials processing, a large fraction of the laser energy will be lost by specular and diffuse reflections. Part of the energy is absorbed by the substrate via inverse bremsstrahlung interaction of substrate free electrons and laser beam photons. The absorbed energy raises the surface temperature above the melting point almost instantaneously ($\approx 1\,\mathrm{ps}$) and the liquid–solid interface moves (solid arrow) through the alloy element layer towards the substrate, as indicated in Fig. 5.1b. Within a short period, the liquid–solid interface reaches the substrate, and interdiffusion begins between the alloying element and the substrate (Fig. 5.1c). By the time the maximum melt depth is achieved (Fig. 5.1d) the laser beam has swept across the spot and solidification starts (Fig. 5.1e). Solidification completes very quickly and forms a surface alloy of $A_x B_{1-x}$ (Fig. 5.1f). The solidification is so rapid that diffusion in solid state can be neglected. This rapid solidification often leads to novel metastable phases and sometimes even metallic glasses. The modified surface thus processed can have superior chemical, physical or mechanical properties. The depth of the alloyed zone, which coincides with the diffusion depth of the alloying elements, can be controlled by the laser power and the dwell time of the laser beam. Depending on the choice of alloy design at the surface, a less expensive base material, such as AISI 1020 steel, can be locally modified to increase resistance to corrosion, erosion, wear and high temperature

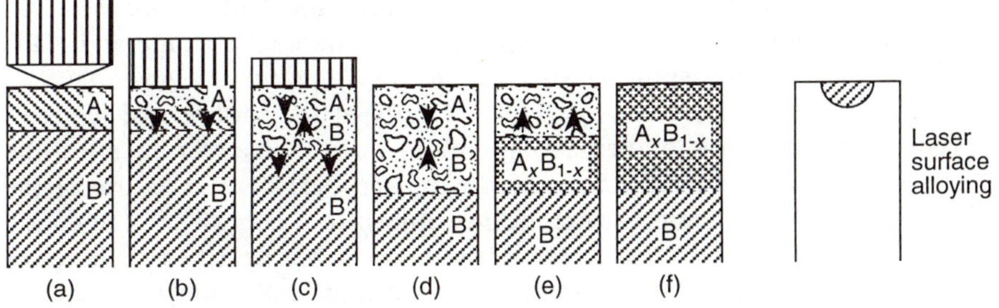

Figure 5.1 Stages of laser surface alloying [1].

oxidation; only those surfaces locally modified will possess properties characteristic of high performance alloys.

5.1.2 Process variables

The major independent process variables for laser surface alloying are as follows: (1) laser power, (2) beam diameter, (3) traverse speed, (4) absorptivity, (5) amount of alloying element added, (6) thermophysical properties of the substrate and alloying element and (7) degree of overlapping for surface alloying of laser areas. The major dependent variables are (1) depth, (2) width, (3) solute content and (4) microstructure of the laser alloyed zone (LAZ).

There is a great deal of data available determining the feasibility of laser surface alloying of various binary, ternary and quarternary systems [2, 3, 4] but no significant study has been carried out to determine the effect of independent process variables on dependent process variables except one report by Chande and Mazumder [5]. With the help of factorial experimental design, they determined the effect of process parameters on dependent variables and also addressed the issue of reproducibility. The important observations are summarized below.

Depth of laser alloyed zone

The LAZ depth is influenced by the power, diameter, speed and the diameter times speed (DXS) interaction. The power does not seem to interact with diameter or speed over the tested levels of the two factors reported by Chande and Mazumder [5]. Hence, its effect may be interpreted independently, whereas the interpretation of the effect of speed or diameter must also include the level of the other.

An increase in the power increases the depth irrespective of the level of diameter and speed. An increase in total power will supply more energy at the surface that will enhance heat flow into the substrate, increasing the LAZ depth.

An increase in both speed and diameter caused the depth to increase. An increase in the speed restricts the laser material interaction time and leads to a decrease in the depth. An increase in the diameter, at a given level of power and speed, decreases the power density at the surface and tends to make the temperature gradients more gradual. This will adversely affect the rate of heat flow under the irradiated spot and perhaps the nature of the laser–material interacion as well. Both will lead to a decrease in the depth. However, decrease in beam diameter, leading to power density exceeding the keyhole threshold, will increase the depth dramatically due to enhanced coupling by total internal reflection. For LSA it is advisable to operate in the power density range below the keyhole threshold.

Width of laser alloyed zone

Speed and power are two important variables affecting the LAZ width. The two variables do not seem to interact [6] and, therefore, their effects may be interpreted individually. An increase in total beam power increases the LAZ width, whereas increasing the speed decreases the width.

With an increase in total power, for a given range of interaction times (estimated as beam diameter/traverse speed), more energy will be dumped on the surface, which will steepen the temperature gradients. This will strengthen fluid flow driven by surface tension [7], increasing convective heat transfer and thus increasing the width. On the other hand, for a given incident beam power, an increase in traverse speed will restrict the duration of laser material interaction and the substrate will absorb less of the incident power, leading to a decrease in the width. In the upper limit, as speed is monotonically increased, the laser–material interaction will have too little time to produce any appreciable effect and the substrate and laser are then said to decouple. This is to be seen in the data of Weinman, DeVault and Moore [8].

Solute concentration in LAZ

Various methods can be used to add the alloying elements during laser surface alloying [3]. Some of the methods are given below.

1. Electroplating [5, 8–12]
2. Vacuum evaporation [13–18]
3. Preplaced powder coating [19–23]
4. Thin foil application
5. Ion implantation [16, 17]
6. Diffusion, e.g. boronizing [24, 25]
7. Powder feed [26–29]
8. Wire feed [30]
9. Reactive gas shroud, e.g. C_2H_2 in Ar or N_2 [31, 32].

The first six methods can be classified as predeposition, where an alloying element is placed as a separate step before laser treatment. The last three methods are classified as codeposition, where the alloying element is added during the laser treatment [2]. The deposition technique will directly affect the solute concentration, but hardly any comparative study is available. To date, there has been only one publication directly comparing the different predeposition techniques [33].

For laser surface alloying of electroplated nickel on low carbon steel (AISI 1020) [55] an increase in the power level decreases the solute content, whereas an increase in diameter and speed increases the solute content. This is also related to the melt volume. An increase in the power increases both the width and the depth of the LAZ. As the amount of alloying element (determined by the thickness of the coating and the irradiated area) must now be dispersed through a greater melt volume, a lower average solute content results. An increase in the diameter significantly increases the width, whereas an increase in the speed decreases the depth and the width. Both result in lower melt volumes and greater average solute contents.

An increase in the diameter causes a greater increase in nickel content at lower power levels. This is related to the melt volume; a reduced melt volume results from a decrease in power density at low power levels. One sees that melt volumes and average solute contents are inversely related.

At high power levels, energy is still available at a sufficient rate to melt relatively large volumes with lower interaction times at higher speeds than at low power levels. Although an increase in the speed does decrease the melt volume and does increase the solute content at both low and high powers, the effect on solute content is greater at low powers than at high powers, which is consistent with the picture of reduced melt volumes and sensitivity transitions mentioned above. Chande and Mazumder [5, 6] also studied the fluctuations about the average composition by analyzing the composition printed out every 10 s during an EPMA trace of the LAZ. On analysis of this data, the fluctuations about the mean composition were found to increase as the diameter and speed increased. Thus, the uniformity in the composition increased as the diameter and speed decreased. A decrease in speed implies that the interaction time was greater; the pool was thus molten for a longer time and diffusion on a local scale could be expected to help level out composition gradients to a greater degree, leading to lower fluctuations about the average composition. A decrease in beam diameter steepens the temperature gradients above the melt pool, which enhances the vigor of fluid flow in the pool. This produces a finer dispersion of solute-rich pockets, and local diffusion works to produce uniformly alloyed laser alloyed zones [7]. The effects of independent process variables on dependent variables are summarized in Table 5.1.

Reproducibility of LSA determined by replication in a statistically designed experiment was reported to be $\pm 9\%$ [5, 6]. Repeatability of the process depends upon the stability of the laser and its support systems and the mode of supply of the alloy element, but stability is more dominant.

5.1.3 Applications

The primary reason for laser surface alloying is common to all coating processes. Coating allows the surface properties of a structure to be tailored to the surface requirements of the application without sacrificing the bulk characteristics of the structure. An additional advantage is that materials which are expensive or scarce can be conserved. Surface alloying by laser produces coating thicknesses of 1–2000 μm at a coverage on the order of

Table 5.1 Summary of effect of independent process variables on dependent variables

Dependent variables	Independent variables[a]	Comments
1. Width of LAZ	↑ when P ↑ ↓ when S ↑	No interaction between variables
2. Depth of LAZ	↑ when P ↑ ↓ when S ↑ ↓ when D ↑	D and S interact Change in absorption mechanism
3 N; content of LAZ	↑ when D ↑ ↑ when S ↑ ↓ when P ↑	P and D interact Related to melt volume
4. Fluctuation of average composition	↑ when D ↑ ↑ when S ↑	

[a] P = power; S = speed; D = beam diameter; ↑ = increase; ↓ = decrease.

$1\,cm^2/s$. In light of the concern in the United States and European Union (EU) over strategic materials, the usefulness of LSA cannot be overemphasized (Fig. 5.2) [34]. LSA can play a key role in conserving strategic materials such as Co and Cr.

Since the alloys are produced by surface melting, a high degree of adhesion is achieved in LSA. The high cooling rate [2] (10^5 to 10^{11} K/s) characteristic of the laser surface alloying also produces a metastable phase [35, 36] and thereby offers the possibility of formation of new materials.

Use of a high power laser beam as a heat source for the alloying process produces an additional cost benefit. Because the heat input from the laser beam is localized to the areas to be processed, distortion of the workpiece is substantially reduced. Belforte [37] reported that a cost analysis of laser alloying versus conventional hardfacing techniques suggests that cost savings of up to 80% can be achieved.

Ferrous alloys

The first attempt to study the feasibility of laser surface alloying was made by using a pulsed ruby laser [38]. But the attempt was only partially successful due to limitations in the availablity of laser power. Mirkin [39] reported that a low carbon, iron workpiece surface coated with graphite powder was melted and subsequently hardened to 1400 kg/mm^2 up to a depth of 0.02 mm by using a 1.05 μm wavelength laser beam with an energy of 35 J and pulse duration of 10^{-3} s. Solid solutions containing 15–18 at% tungsten in iron with high crystal lattice distortion were observed after laser melting whereas

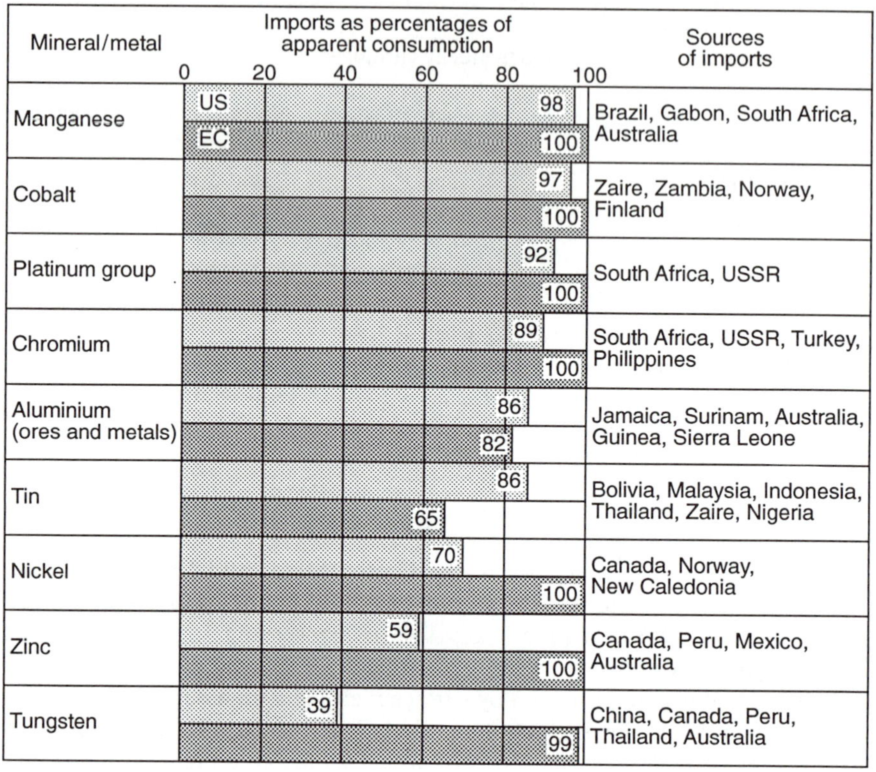

Mineral/metal	Imports as percentages of apparent consumption		Sources of imports
Manganese	US 98	EC 100	Brazil, Gabon, South Africa, Australia
Cobalt	97	100	Zaire, Zambia, Norway, Finland
Platinum group	92	100	South Africa, USSR
Chromium	89	100	South Africa, USSR, Turkey, Philippines
Aluminium (ores and metals)	86	82	Jamaica, Surinam, Australia, Guinea, Sierra Leone
Tin	86	65	Bolivia, Malaysia, Indonesia, Thailand, Zaire, Nigeria
Nickel	70	100	Canada, Norway, New Caledonia
Zinc	59	100	Canada, Peru, Mexico, Australia
Tungsten	39	99	China, Canada, Peru, Thailand, Australia

Figure 5.2 Dependence of United States and European Community on foreign imports for commercially important elements; numbers represent percentage imported relative to amount consumed [34].

under equilibrium conditions a maximum of 13 at% tungsten is soluble in α-Fe [35, 36], Tungsten carbide was dispersed into selected regions in a high speed tool steel to a depth of 0.5–1.25 mm by using a pulsed ruby laser with a nominal capacity of 5 J. Service life of the cutting tools after tungsten carbide dispersion increased by 100–200% because of wear reduction [40].

The scope for technically and commercially feasible laser surface alloying applications has increased greatly since the development of multikilowatt CW CO_2 lasers around 1970. Seaman [41] and Gnanamuthu [23] reported the effect of temperature on hardness and microstructure of the laser surface alloyed casing on steel using a 10 kW CW CO_2 laser. Similar work was also reported by Hella [42]. Moore and Weinman [9] reported laser surface alloying of steel (AISI 1018) and discussed the processing condition and the diffusion during the laser surface alloying.

Many reported works to date on laser surface alloying primarily discussed the processing parameters for chromium addition to steel, AISI 1018 [9, 23, 41, 42]. A chromium concentration of up to 50% at the alloyed casing is reported to be achieved by Gnanamuthu [23]. With an 18 mm × 18 mm beam size, a 12.5 kW laser power and a 1.69 mm/s traverse speed, Seaman and Gnanamuthu [41] generated an alloy casing 21 mm wide and 1.95 mm deep with 16% Cr on AISI 1018, which shows a martensitic structure of Rockwell C hardness 53. An alloy casing of composition 43% Cr and 4.4% C on AISI 1018 generated by a 6.4 mm × 19 mm oscillating (690 Hz) beam with 5.8 kW power and 21.17 mm/s traverse speed is reported to have a Rockwell C hardness of 64 [23]. In both cases the mild steel surface was coated with a slurry of chromium powder then melted using a laser beam.

Moore and Weinman [8] carried out their laser surface alloying experiments with sputter-deposited chromium layers (1–20 μm in thickness) on AISI 1018 steel. They used a traverse speed of 10–400 cm/s for a laser power range of 2–9 kW. The chromium content of the alloy casing was reported to be 11% for 7.5 kW laser power, 38 cm/s traverse speed and 8 μm chromium coating thickness; using the same processing parameters but with a 1.4 μm chromium coating thickness produced a chromium content of only 2% in the alloy casing.

The electron microprobe study carried out by Moore and Weinman [9] revealed that the mixing of the alloying element is substantial throughout the melt pool for a sample with 0.5 ms interaction time (7.5 kW, 50 cm/s). Moore and Weinman [9] also reported a cooling rate of 10^5 to 10^7 K/s (measured from dendrite arm spacing), which is comparable to splat quenching [43].

Molian and Wood [44–46] reported microstructural observations on Cr surface alloys produced with a CW CO_2 laser on commercial low carbon (0.2%) steel and high purity iron.

Electrochemical studies of laser surface alloyed Fe–Cr and Fe–Cr–Ni alloys are also available [9, 26, 47, 48] and the general consensus is that the corrosion properties are at least as good as those of 304 stainless steel.

Also enjoying attention is the making of surface layers for ferrous metals. Lin and Spaepen [49] have produced laser-alloys of modulated Fe–Fe_3B films with 30 ps laser pulses from a mode-locked Nd: YAG laser. Fe–B alloys of 5–24 at% B were found for LSA, compared to 12–28 at% B amorphous alloys produced by splat quenching. Gorogina *et al.* [50] studied the crystalline amorphous transition in seven binary, ternary and quarternary Fe–metalloid (B, Si, P, C) glass-forming alloys. They used continuous wave (CW) CO_2 and transverse excitation atmospheric (TEA) CO_2 lasers. Their calculated cooling rate was below the cooling rate normally required for amorphization of these compositions [51, 52]. Inal *et al.* and Inal and Yost have published detailed transmission and field ion microscopy results for the effect of Q-switched ruby lasers on $Fe_{80}B_{20}$ amorphous alloy. The Fe–Cr–C–B system was studied by Bergmann *et al.* [31] and Bergmann and Mordike [20–22].

Nonferrous alloys

Most of the LSA studies of nonferrous alloys have been carried out on Al, Cu, Ni and Ti. Al and Cu are more difficult to process due to high thermal diffusivity and reflectivity. A recent bibliography on this subject is available [1]. Dates of various binary systems for an aluminum substrate were compiled by Picraux *et al.* [53a] and are presented in Table 5.2.

Draper and Ewing [1] published a chronological list of laser surface alloying literature which summarizes various alloying systems and analytical studies carried out in the field; it is expected to serve as a quick survey. There are also several review articles: Draper and Poate [2], Steen [3], Draper [4], Draper and Ewing [1], Draper [57], Draper [4a], Follstaedt [58], Sood [59], Draper and Poate [60], Snow, Breinan and Kear [61], and Breinan, Snow and Brown [62]. All are good sources for literature on the subject.

5.2 LASER SURFACE CLADDING

Laser surface cladding is a process similar to laser surface alloying except that dilution by the substrate material is kept to a minimum. Similar to LSA, it can also provide tailored surface properties.

As a cladding process, it exhibits many advantages over alternative methods such as plasma spraying and arc welding. These advantages include a reduction in dilution, a reduction in waste due to thermal distortion (very little energy is absorbed by the substrate in comparison with plasma spraying and arc welding); a reduction in deposit porosity and a reduction in postcladding machining costs because the material can be more accurately placed; thicknesses between 0.3 and 3 mm can be deposited in one pass; rapidly solidified microstructure; noncontact method of application; relatively

Table 5.2 Comparison of maximum equilibrium solid solubilities C_0 in Al to those achieved by ion implantation followed by Nanosecond pulsed melting C_P as determined by ion-channelling measurements

Element	C_0 (at%)[a]	C_P (at%)[b]	C_P/C_0	Ref.
Cr	0.4	7.0 (l)	18	55
Cu	2.5	2.1 (e)	1	56
Mo	0.07	1.5 (l)	20	57
Ni	0.023	0.4 (e)	17	58
Sb	0.03	0.9 (l)	30	59
Sn	0.02	0.7 (e)	35	58

[a] Maximum equilibrium solid solubility reported in the literature [54–56].
[b] Pulsed heating method of ruby laser (l) or electron beam (e).

smooth surface finish ($\approx 25\,\mu m$ RMS) and the process is omnidirectional and can be easily automated.

5.2.1 Principles of laser cladding

The principles of laser cladding are similar to LSA, except that dilution by the substrate material is kept to a minimum.

5.2.2 Process variables

Similar to LSA, however, one of the important observations is that pneumatic powder feed at the beam–substrate interaction point provides the highest cooling rate as opposed to preplaced powder or wire feed [63]. A good discussion of the process parameters is available in Weerasinghe and Steen [64], Steen *et al.* [65] and Steen [66]. The window of operation for laser cladding is relatively small compared to surface alloying. In order to minimize dilution, the combination power density and interaction time is critical. Below the optimum range, the metallurgical bond will not be possible or there will be an interface full of discontinuities and porosity. Above the optimum range, dilution will increase.

5.2.3 Applications

The applications for laser cladding are very similar to those of laser surface alloying. Unlike laser surface alloying, there is very little or no dilution of the clad material. This makes laser cladding more potent for tribological applications.

Its potential for tribological applications was demonstrated by Gnanamuthu [23] by laser cladding of Tribaloy T-800 on ASTM A387 steel and Haynes Stellite alloys on AISI 4815 steel. It was observed that composition and microstructure were uniform across the cladding thickness. Subsequent evaluation of the clad material was restricted to hardness measurement and microstructure characterization.

Belmondo and Castagna [67] studied laser cladding of a powder mixture, which consisted of chromium carbides, chromium, nickel and molybdenum. Wear tests were carried out by Belmondo *et al.* on reciprocating motion testing equipment. Test data revealed that laser clad surfaces have superior wear properties compared to plasma sprayed surfaces.

5.2.4 Synthesis of nonequilibrium metallic phases

Inherently rapid cooling rates in laser cladding and surface alloying are being used to engineer nonequilibrium microstructures which cannot be rivaled by other processes. One of the major applications is the design of materials with tailored properties [68].

What is nonequilibrium synthesis?

Nonequilibrium synthesis is a method to produce binary or higher order materials where the kinetics of the process affect the transport of the constituent elements during phase transformation, resulting in a composition or crystallographic configuration different than that observed in the equilibrium condition, when the elements arrange themselves with the lowest possible Gibbs free energy. Figure 5.3 illustrates the phenomenon. A phase diagram under equilibrium conditions is shown as the solid line whereas the nonequilibrium phase diagram is shown as the dotted line. One can observe the shrinkage of the phase field under nonequilibrium conditions. Any alloy composition between the solidus lines of the equilibrium and nonequilibrium phase diagrams will be a nonequilibrium alloy with extended solid solution.

Besides deviation of composition from the normal, crystallographic configuration may sometimes be a result of nonequilibrium processing. Martensite in hardened steel is an example.

Why nonequilibrium synthesis?

A wide variety of microstructures can be generated, including extended solid solution and amorphous phase. Many surface-related failures, such as corrosion, oxidation, wear and fatigue, may be minimized or eliminated using these techniques. In this process, compositionally graded structure can be produced to minimize thermal stress due to the difference in coefficient of expansion. Another advantage of this method is that enormous savings of

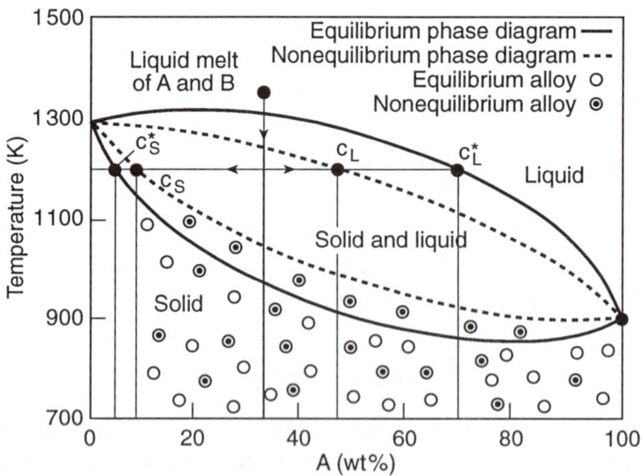

Figure 5.3 Nonequilibrium phase diagram.

alloying elements, which are often expensive or strategically important, can be achieved because only the surface is being modified.

Inherently rapid solidification involved in laser alloying and cladding provides a unique opportunity for nonequilibrium synthesis of novel alloys with extended solid solution. This has been experimentally observed for various binary, ternary and quaternary alloy systems. Compositions of alloys produced by laser aided synthesis often exceed the solid solubility limit, far beyond that expected from any equilibrium phase diagram. They offer the possibility of new materials with novel properties. The surface properties of a structure can now be tailored to the surface requirement of the application without sacrificing the bulk characteristics of the structure.

Moreover, the great control provided by the laser gives laser processing the potential of synthesizing nonequilibrium materials in near-net shape. In fact, such a goal has recently been realized and the proof of the concept is already available in processes such as laser glazing [69]. There are other methods [70] of producing nonequilibrium metallic materials using rapid solidification technology (RST) such as atomization, twin-roll quenching, and melt spinning. However, in order to make an engineering component, these materials produced by conventional RST need to be consolidated, but in the consolidation process, temperature and pressure cycles induce phase transition in a metastable material. Thus, laser processing has far-reaching implications for synthesis of nonequilibrium materials.

How to make tailored properties by nonequilibrium synthesis?

The potential of laser beams to produce metastable phases derives from their ability to induce rapid localized heating and melting and almost equally rapid quenching of the melt. Cooling rates up to 10^{14} K/s have been reported [71–73]. Laser processing with cooling rates of around 10^5 K/s is quite common [74]. In general, under such high cooling rates, two phases can be produced: glassy phases or metastable cystalline phases with composition and crystallographic configuration different than those predicted by the equilibrium phase diagram. Formation of glassy phases at cooling rates of around 10^6 K/s is only possible for typically eutectic alloys of three or more elements, often including metalloids. Higher cooling rates (10^{10} to 10^{14} K/s) enable glass formation in many binary and ternary alloys. But such versatility comes at the expense of the thickness of the quenched region.

Strategy for nonequilibrium synthesis for tailored properties

Selection of phases

A face-centered cubic (FCC) structure, for example, with a large number of available slip planes will be beneficial for ductility, whereas brittle noncubic phases with limited numbers of available slip planes will promote hardness and wear resistance. A combination of them with duplex phases is often needed to provide adequate toughness during service with reasonable wear resistance. Selection of phases is extremely important for desired mechanical properties.

Selection of elements

Selection of elements is important for promoting certain phases as well as for protection against chemical degradation such as corrosion. For example, chromium promotes a body-centered cubic (BCC) phase for ferrous alloys and Cr_2O_3 forms passive layers to inhibit corrosion at temperatures up to 800 °C. Reactive elements such as yttrium and hafnium are known to stabilize Al_2O_3 at temperatures above 800 °C, leading to high temperature oxidation resistance.

Selection of process parameters

Although it is well known that Y and Hf – reactive elements with a high atomic number – can improve high temperature properties, their dissolution is rather difficult. Processing techniques promoting atom trapping and extending solid solution are crucial to the incorporation of such alloys. As explained above, this is helped by the high cooling rates and strong convection inherent to laser melting and solidification. The important consideration is the synthesis of information from phase transformation kinetics and processing transport phenomena so that the chosen process parameters promote the desired phases and chemistry.

What was done before?

During the 1990s, considerable progress has been made in developing tailor-made materials by laser surface modification (LSM). Laser surface alloying (LSA) and laser cladding are two major LSM techniques for synthesis of unique materials. The review by Mazumder *et al.* [68] provides an excellent survey of the field. Mazumder *et al.* [75] and Singh and Mazumder [76–79] have reported various systems of metastable laser clad alloys, which include Fe–Cr–Mn–C (Mazumder *et al.* [75]; Singh and Mazumder [76, 77]; Choi and Mazumder [80]), Ni–Cr–Al–Fe–Hf [78], Ni–Cr–Al–Hf on Inconel 718 [81, 79], $Ni_{70}Al_{20}Cr_7Hf_3$ on nickel [82]. The ferrous alloys exhibited exceptional wear properties [80]. The wear scar for laser clad ferrous alloy was one-third that of the Stellite 6 after a lubricated line contact wear test (Fig. 5.4). The Ni alloys showed improved oxidation resistance up to 1200–1400 °C [83, 84] after a cyclic oxidation test. Commercial nickel superalloys under similar conditions lost significant weight (Fig. 5.5). These are good examples of how surfaces can be engineered for specific needs.

Various conventional metals and alloys are also laser clad, including Ni–Cr–Fe (stainless steel) on mild steel [85], 316 stainless on mild steel (En3)

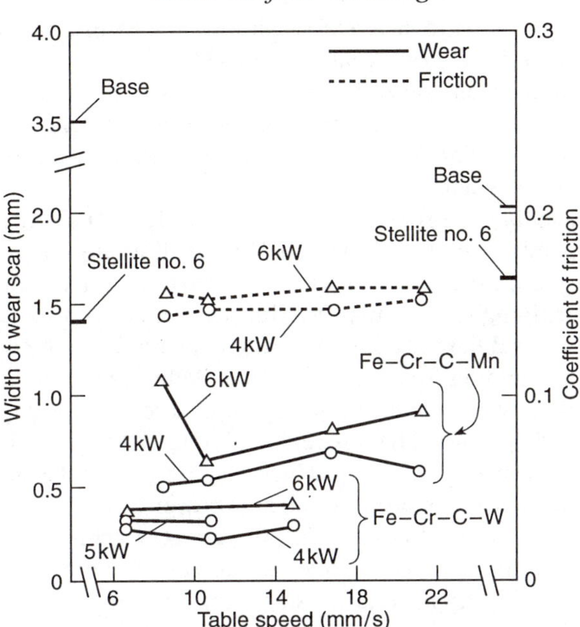

Figure 5.4 Friction and wear of laser clad surface as a function of table speed and power.

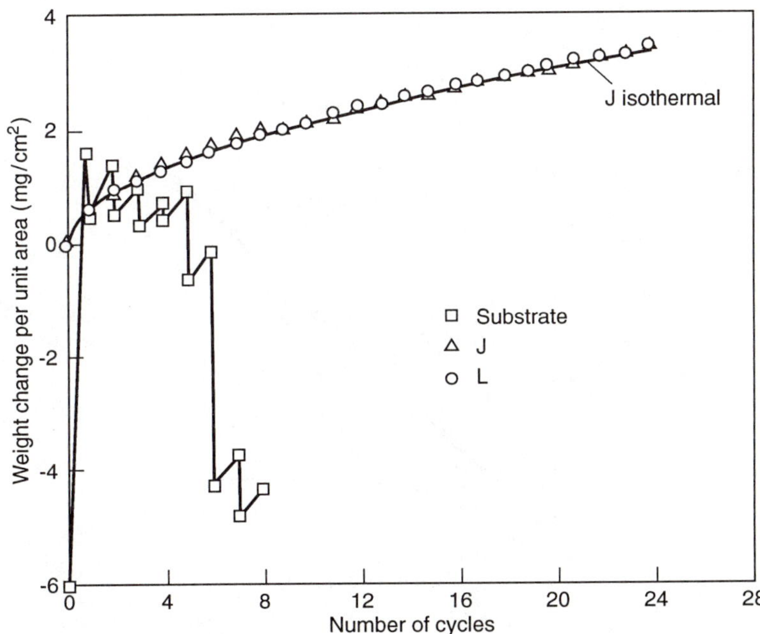

Figure 5.5 Weight change per unit area as a function of time during thermal cycling in slowly flowing air. J and L represents laser clad $Ni_{70}Al_{20}Cr_7Hf_3$ coating [83, 84].

[64], Wallex 6PC on mild steel [86], colmonoy 52M on mild steel [16, 17] and Stellite and Tribaloy on Nimonic 75 [87].

Steen [66] and Steen, Weerasinghe and Monson [65] have summarized their processing parameter data for laser cladding and developed an operating diagram as shown in Fig. 5.6. There are also some important modified laser cladding processes which are of interest.

Reverse machining, a process proprietary to General Electric, is used to repair aircraft components with Ni and Ti base alloy powder feed. Using the same powder composition as the substrate, damaged areas can be built up by reverse machining. The similar buildup technique with the same powder as the substrate or different alloy powders, depending on the application, is also applied for building up edges of turbine blades, as shown in the Fig. 5.7.

Laser cladding with *optical feedback system* can improve the process efficiency by 40% [3, 64]. This process is carried out under a hemispherical mirror, as shown in Fig. 5.8 [66], which reimages the reflected laser beam back to the clad pool, increasing the beam power utilization.

Substantial reduction of crack and porosity was achieved by *vibro laser cladding* [88]; cladding proceeds while the substrate is vibrated ultrasonically. However, this is only applicable for small components.

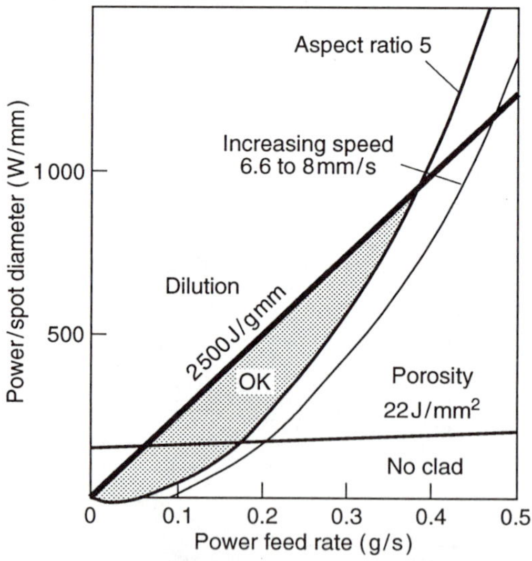

Figure 5.6 Approximate operating curves suggested: 'OK' signifies low dilution, good aspect ratio and effective fusion bond [66].

Figure 5.7 Turbine blade build up. (Courtesy Todd Rockstroh)

5.3 LASER CHEMICAL VAPOR DEPOSITION (LCVD)

The need for localized deposition of hardcoating and semiconducting materials gives the most important incentive for research activities in LCVD.

The advantages of LCVD include the following:

1. Control of area and depth heated
2. Rapid heating and cooling
3. Cleanliness
4. Spatial resolution and control of the deposited film
5. Limited distortion of the substrate
6. The ability to interface directly with laser annealing and diffusion of semiconductor devices.

                    Laser assisted surface coatings

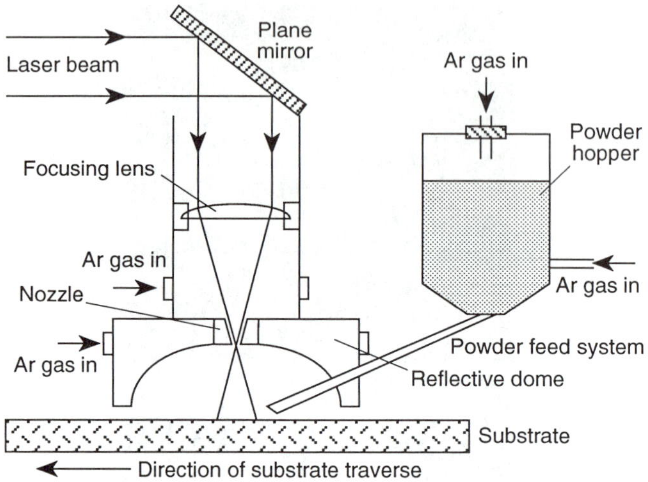

Figure 5.8 Arrangement for laser cladding by the blown powder technique [66].

5.3.1 Principles of LCVD

Laser chemical vapor deposition can be achieved either by photodissociation or by thermal dissociation. In photolysis, laser photons are absorbed directly in continuum (i.e. bound–free) electron transition [89, 90]. In pyrolysis, laser photons are absorbed by the substrate, raising its temperature; this includes endothermic reaction for a suitable gaseous atmosphere, leading to the desired deposit. Some typical LCVD reactions are given in Tables 5.3 and 5.4 [163].

LCVD by pyrolysis is best described by referring to Fig. 5.9. The laser is focused through a transparent window and the transparent reactants onto an absorbing substrate, creating a localized hot spot at which the reaction takes place. The absorptivities of the reactants and the substrate determine the choice of laser wavelength. As the deposition reaction is thermally driven, LCVD is exactly analogous and superficially similar to conventional CVD but actually differs in the following ways.

Localized deposition

The most obvious difference is that the deposition is localized because the heat source is localized; it is impossible to achieve using conventional CVD.

High deposition rate

LCVD deposition is fast; coatings on the order of 1 µm thick can be deposited in approximately 0.1 s. This translates into coating rates of mm/min. For

Table 5.3 Photochemical deposition

	Reactants or external medium	Laser wavelength	Process
		Gas phase	
Electronics Photochemistry (localized reactions)	$Fe(CO)_5$ $W(CO)_5$ $Cr(CO)_6$	257 nm	Photodeposition of Fe, W, Cr [119]
	$Cd(CH_3)_2$ $Cd(CH)_3$ $Zn(CH_3)_2$	257 nm	Photodeposition of Cd, Zn, Al [120, 121]
	$Al(CH_3)_6$		[122]
	$Cd(CH_3)_2$	157 nm	Photodeposition of Cd [123]
	$Ga(CH_3)_3$	257 nm	Photodeposition of Ga [124]
	$Mn_2(CO)_{10}$	337.4–356.4 nm	Deposition of Mn [125]
Electronics Photochemistry (large-area reactions)	$Pb(CH_3)_4$	254 nm (1 A)	Deposition of Pb [126]
	SiH_4/NH_3 SiH_4/N_{20}	193–257 nm	Deposition of SiO, SiN [127–129]
	$Zn(CH_3)_2/NO_2$	193 nm	Deposition of ZnO [129]
	$Al_2(CH_3)_6$	150 nm (1 A)	Deposition of Al [130]
	Si_4, GeH_4	193 nm, 248 nm	Deposition of Si, Ge [131]
	$Ga(CH_3)_3$ $In(CH_3)_3$	UV lamp	Deposition of Ga, In [132]
	$W(CO)_6$	193 nm, 248 nm	Deposition of W, Mo, Cr [133]
	$Mo(CO)_6$ $Cr(CO)_6$	260–270 nm	Deposition of W, Mo, Cr [133]
	Zn, Mg atoms	248 nm, 308 nm	Deposition of Zn, Mg [134]
Vibrational Photochemistry	SiH_4	9–11 μm	Deposition of Si [135, 136]
(large-area reactions)	SiH_4/NH_3 SiH_4/C_2H_4	9–11 μm	Deposition of Si_3N_4, SiC powders [135–137]
		Photochemical deposition	
Electronics	$Cd(CH_3)_2,$ $Zn(CH_3)_2$	257 nm	Prenucleation
Photochemistry (localized reactions)	$Al_2(CH_3)_6$	193 nm	Photodeposition [89, 90, 119, 122, 138]
	$TiCl_4$	257 nm	Catalyzed photodeposition [139]
	Methyl methacrylate	257 nm	Polymerization
	$Cd(CH_3)$	337–676 nm	Photodeposition [140]
Electronics Photochemistry (large-area reactions)	$Fe(CO_5)$	253.7 nm (1 A)	Photodeposition [141]
	$Zn(CH_3)_2,$ $Cd(CH_3)_2$	193 nm	Deposition [142]
	$Hg(CH_3)_2$	237 nm	
		Photoelectrochemical	
Photochemical reactions (localized)	Plating electrolytes	Band gap radiation	Plating onto InP, Si, GaAs [143, 144]
	Vacuum	Band gap radiation	Patterned evaporation on ZnS on CdS [145]

Table 5.4 Thermal deposition

	External medium	Laser wavelength	Process
Thermal reactions (UV/visible laser initiated, localized)	SiH_4	Visible	Deposition of Si [119, 146]
	Metal salt	Visible	Etching and plating [147, 150]
	Electrolytes		
	C_2H_2	488 nm	Deposition of C [151]
	$Al_2(CH_3)_6$	520–570 nm	Deposition of Al, An, Cd [140, 152–154]
	$Zn(CH_3)_2$		
	$Cd(CH_3)_2$		
	WF_6		Deposition of W on Si [155–157]
	$Ga(CH_3)$	530 nm	GaAs on GaAs [158]
	AsH_3		
	$Ni(CO)_4$	476–647 nm	Deposition of Ni [154, 159]
Thermal reactions (UV/visible laser initiated, large-area)	H_2O	193 nm	Formation of silanols [160]
Thermal reactions (IR laser initiated)	WF_6, $Fe(CO)_5$	9–11 µm	Deposition of Fe, W [116, 117, 156, 157]
	$Ni(CO)_4$	9–11 µm	Deposition of Ni, Ti, TiO_2, TiC [96]
	$TiCl_4$		
	$TiCl_4/CO_2$		
	$TiCl_4/CH_4$		
	$Cu(HCOO)_2$	1.06 µm	Deposition of Cu [161]
	Cobalt acetylacetonate	9–11 µm	Deposition of CoO [98]
	SiH_4/H_2	9–11 µm	Deposition of Si [118]
	$SiCl_4/H_2$	9–11 µm	Deposition of Si [162]

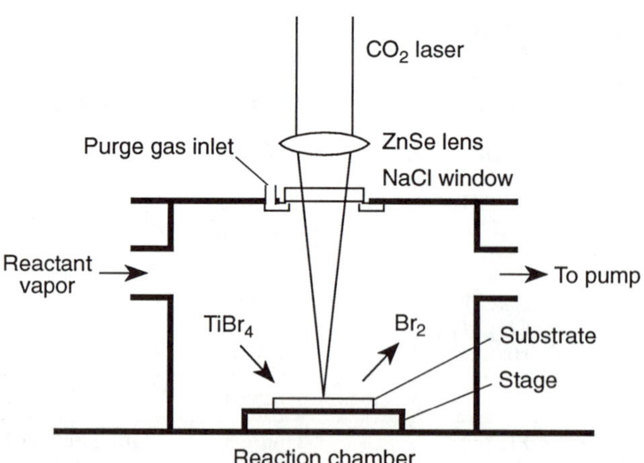

Figure 5.9 Experimental setup for laser chemical vapor deposition.

the reactions investigated to date, this rate is 2–4 orders of magnitude larger than standard CVD rates [91–93].

Heating and cooling rate

There is a wide range of heating and cooling rates available using LCVD. Of particular interest are the very rapid thermal rise and fall times achievable in LCVD, which make it possible to control the grain size of the deposited film by varying the irradiation waveform. For example, very fine-grained (possibly amorphous) material can be deposited using short laser pulses. In contrast, large-grained material is produced in equilibrium CVD because of the long dwell time at high temperatures.

Preheating of reactants

Generating very high surface temperatures in very short times in LCVD also avoids preheating of the reactants. In conventional CVD, the resulting premature gas phase reaction is prevented by the inclusion of a buffer gas. Unfortunately, the buffer gas also suppresses the surface deposition. Other possibilities in LCVD include the use of nonstandard reactants; in conventional CVD the substrate may experience unwanted etching by nonstandard reactants or their reaction products. The decrease in the time at the reaction temperature and the area heated using LCVD should eliminate this problem.

Gas flow rates

In conventional CVD, deposition rates and film properties are a strong function of gas flow rates and therefore the overall geometry of the substrate and deposition chamber. Because of the localized heating and the fast deposition rates, LCVD should be essentially independent of flow characteristics.

Surface reaction

Direct heating applied to the surface of the substrate, and the reactants themselves, will favor heterogeneous reaction paths. Reaction kinetics will differ from those observed in standard CVD.

Optical heat source

Since the energy source is optical, the results will depend on whether an absorbing film is deposited on an absorbing substrate, a reflective film on an absorbing substrate, an absorbing film on a reflective substrate or a reflective film on a reflective substrate. In other words, the deposition rate is a function not only of the thermal properties of the film as in standard

CVD, but also of the optical properties. We have evidence from preliminary experiments, for example, of self-limiting in the deposition of metal films on an absorbing substrate due to the reflection of most of the laser pulse by the initially deposited film. The opposite effect is observed when an absorber, TiC, is deposited on a reflective substrate, stainless steel.

Rough surfaces

The use of an optical heat source will also produce different results on surfaces which have scratches, digs, pinholes, etc., as compared with conventional CVD. Both the absorptivity and the thermal behavior of the surface and the nucleation mechanism of the deposition will depend on the details of the surface profile. Under properly chosen conditions it may be possible to produce a coating which is smoother than the original substrate. In particular, cracks and holes will serve as radiation traps and will be hotter than the surrounding surface. This phenomenon may be used to fill pinholes in previously deposited coatings.

LCVD by photolysis is generally slower than pyrolysis. For LCVD by photolysis the laser beam can be parallel to the substrate instead of normal to the substrate [94]. Excellent reviews of photolysis are available in [95, 89].

5.3.2 Process variables

The important independent process variables are (1) incident laser beam power, (2) beam diameter, (3) traverse speed or dwell time, (4) surface reflectivity, (5) gas pressure of the reactants in the reaction chamber, (6) chamber temperature and (7) substrate thermal properties. The corresponding dependent variables are (1) deposition rate, (2) composition and microstrucure and (3) deposit geometry, i.e. thickness and diameter.

Effects of the film and substrate

Allen [96] investigated the spot size diameter and the film thickness attainable as a function of the irradiation time for three cases of LCVD:

1. A reflecting film on an absorbing substrate, $Ni-SiO_2$
2. An absorbing film on an absorbing substrate, TiO_2 and $TiC-SiO_2$
3. An absorbing film on a reflecting substrate $TiC-$stainless steel.

These reactions were conducted using a 20 W laser run at 10 W nominal power.

For the deposition of Ni on SiO_2, it appears that the film thickness is proportional to the square root of the irradiation time (Fig. 5.10). Additionally, for the range of spot sizes and irradiation times used, the surface temperature was found to be proportional to $I_0 t^{1/2}$ where I_0 is the intensity of the beam and t is the irradiation time.

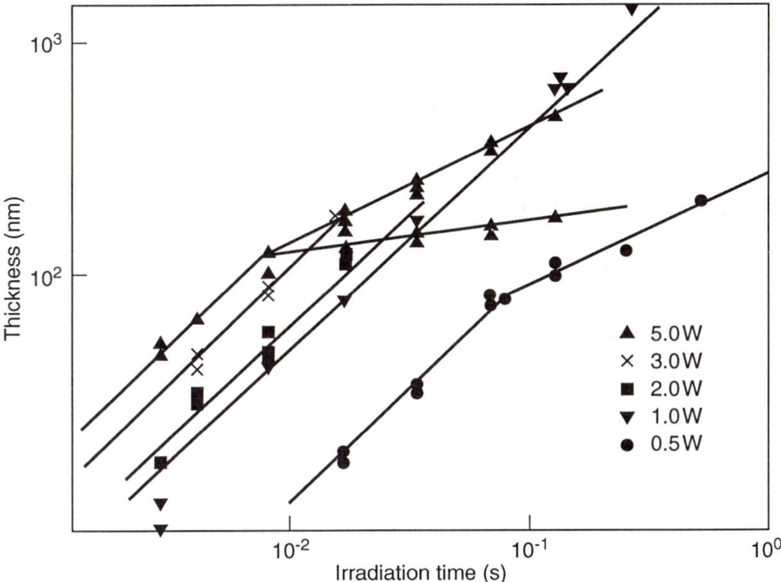

Figure 5.10 Thickness of LCVD Ni film as a function of irradiation time. For the 5.0 W data, the upper curve is the maximum thickness and the lower curve is the thickness at the center of the spot. Similar branching is observed in the 3.0 W and 2.0 W data but is not shown for reasons of clarity [96].

Figure 5.11 gives a plot of spot size diameter as a function of surface temperature. The diameter of the spot size increases rapidly above the threshold temperature, the temperature at which the reaction begins, but the rate levels off for longer irradiation times and higher laser powers. Also, the widths of the deposited films are less than the width of the depositing laser beam. It has been reported that the film diameters can be as small as 1/10 the diameter of the depositing beam diameter [97].

For increasing irradiation times, the deposition profile changes from a Gaussian distribution to a double-humped volcano. This point corresponds to the levelling off of the rate of spot diameter growth. In the future, this will be referred to as the onset of linear dependence. Several explanations have been given for this phenomenon: (1) convection is beginning to play a role in the deposition rate, favoring the edge of the spot, and (2) the center of the spot becomes too hot, and the sticking coefficient decreases for the reactants, favoring deposition at regions outside the center of the spot. Additionally, it has been theorized that depletion of the reactants is beginning to occur [97].

The thickness of the films as a function of irradiation time shows a linear relation in time t; however, the deposition rate decreases at all incident powers

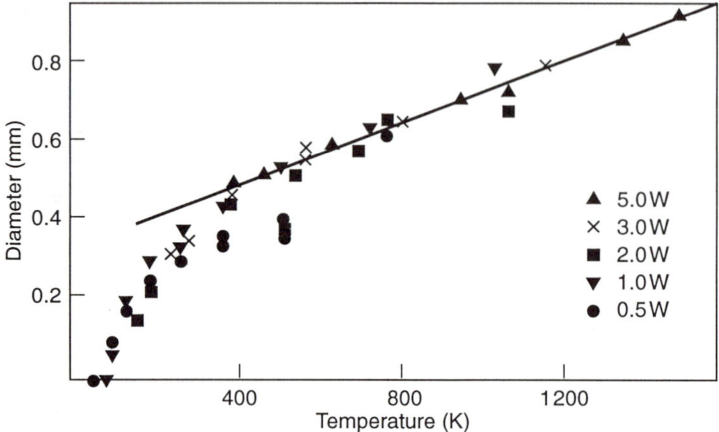

Figure 5.11 Spot diameter (full width) of Ni films deposited on quartz ($T \propto It^{1/2}$). Laser beam diameter, $D_{1/e^2} = 0.6\,\text{mm}$ [96].

with the onset of linear dependence. For multiple irradiations, it has been observed that the film thickness generated by n irradiations is less than n times the thickness of a single irradiation. This phenomenon, called optical self-limiting, occurs in many metal films and is produced because the reflecting film has an absorptivity less than the absorptivity of the substrate. As more film is deposited. the amount of energy absorbed by the surface decreases. This also lowers the surface temperature of the substrate and eventually leads to a decrease in the deposition rate. In certain cases, the deposition rate approaches zero [97].

For the deposition of TiO_2 on SiO_2, the range of conditions over which deposition can occur is much lower than for the deposition of Ni on SiO_2 because the higher reaction temperature (1200 °C) is close to the TiO_2 melting temperature. Film thicknesses of TiO_2 on SiO_2 are in general linear functions of the irradiation time. TiO_2 has a greater absorptivity than SiO_2, so it turns out that n irradiations produce a film thickness greater than n times the thickness of a single irradiation. Deposition rates ranged from 2 to 20 μm/min. The deposits obtained during this process showed no double humps and could be tailored by the radiation conditions. For irradiation conditions above the threshold temperature a sharply peaked profile was obtained. Multiple irradiations slightly above the threshold temperature produced a flat-topped profile, and for irradiation conditions between the two, broadly peaked films were obtained (Fig. 5.12).

Finally, for the deposit of TiC on both stainless steel and quartz, it was observed that the incident intensity necessary to produce deposition on steel was two orders of magnitude greater than for deposition on quartz. This occurs because the thermal conductivity of stainless steel is much higher than

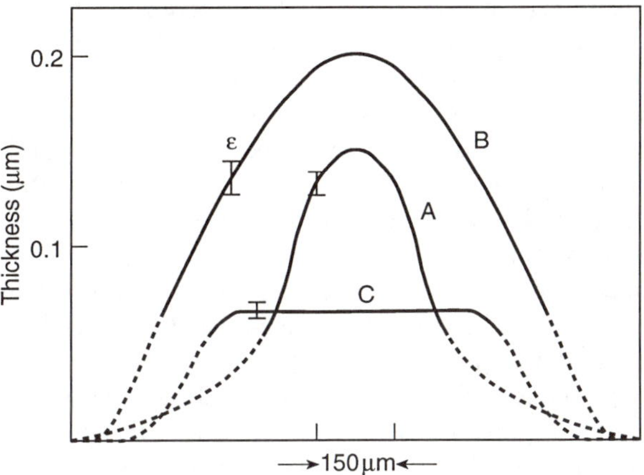

Figure 5.12 Thickness profiles of LCVD TiO_2 films using $P_0 = 10$ W and $D_{1/e^2} = 3.0$ mm: A is for a single irradiation of 0.5 s; B is for 10 irradiations of 0.25 s each; and C is for 20 irradiations of 0.12 s each [96].

the thermal conductivity of quartz. Multiple irradiations could be used to build up the coating thickness on both quartz and stainless steel; but for several cases, multiple irradiations at a single site produced melting, indicating that a greater coupling between the laser and the substrate is occurring once the TiC has been deposited.

Kinetics of LCVD

The deposition kinetics of LCVD were initially studied by Steen [98]. He concluded the following:

1. The deposition is directly dependent on the Reynolds number of the reactant gas flow.
2. The concentration of the reactants plays a large role.
3. The process is controlled by the reaction rate at low temperatures and by mass transfer at high temperatures. The reaction rate at low temperatures is found to be first-order Arrhenius.

Chen and Mazumder applied on-line laser induced fluorescence (LIF) spectroscopy and multiwavelength pyrometry to study the kinetics of LCVD for TiN on Ti–6 Al–4 V substrate with a CW CO_2 laser and a precursor gas mixture of $TiCl_4$, H_2 and N_2 [99–102].

It is found that the film growth rate, R_{TiN}, and the rate of Ti concentration increase, dC/dt, increase linearly with P_{TiCl_4}, when $P_{TiCl_4} < 27$ torr (3.6 kPa). Further increase in P_{TiCl_4} will result in a decrease in R_{TiN} and dC/dt. Similar observations for R_{TiN} have been reported for conventional CVD of TiN

[103–107]. The increase of deposition as P_{TiCl_4} increases is easy to understand, because $TiCl_4$ is the source of Ti. The decrease of deposition at higher P_{TiCl_4} is believed to be a result of competitive adsorption between reactant molecules $TiCl_4$, N_2 and H_2. After the initial growth, further film deposition will be onto the TiN-covered substrate. You, Nakanishi and Kato [108] showed that the N_2 and H_2 adsorption equilibrium constant is about 10^4 times smaller than that of $TiCl_4$. Hence, surface coverages using N_2 and H_2 are much smaller than using $TiCl_4$ at similar partial pressures. When P_{TiCl_4} is raised above a certain value, $TiCl_4$ takes up so many of the available surface sites that there may not be enough surface-adsorbed N and H to carry out the reactions, leading to a decrease in R_{TiN}.

To study the dependencies of R_{TiN} and dC/dt on the N_2 and H_2 partial pressures, P_{N_2} and P_{H_2}, experiments are performed in which P_{N_2} (or P_{H_2}) is fixed while changing P_{H_2} (or P_{N_2}) and adding inert Ar gas to maintain the total pressure P_{Total}. All other parameters are kept constant: the $TiCl_4$ partial pressure $P_{TiCl_4} = 27$ torr (3.6 kPa), the total pressure $P_{Total} = 600$ torr (80 kPa) and the CO_2 laser power of 400 W. The results are shown in Figs 5.13 and 5.14. TiN films of a few microns thickness can be deposited in less than 10 s. Evidently, the growth rate of LCVD is much greater than the typical rate of 1 nm/s for conventional CVD processes. All films are characterized by Auger electron spectroscopy (AES) as TiN_x ($x = 0.8 \pm 0.1$), with Cl contamination as low as 0.5 at% and O and Cl at%.

The behaviors of the Ti concentration and the film thickness are identical, indicating that the mechanism for producing the Ti species above the substrate may be the same as for film growth. Since LCVD of TiN is a surface reaction and TiN dissociates to its elements upon vaporization [109], the Ti atomic species in the gas phase above the substrate is believed to be a result of the

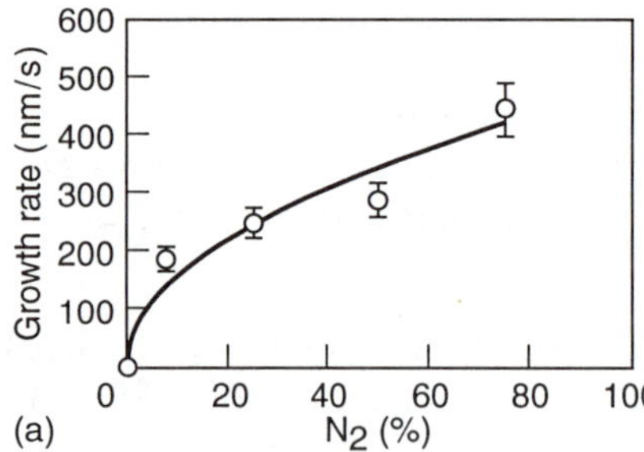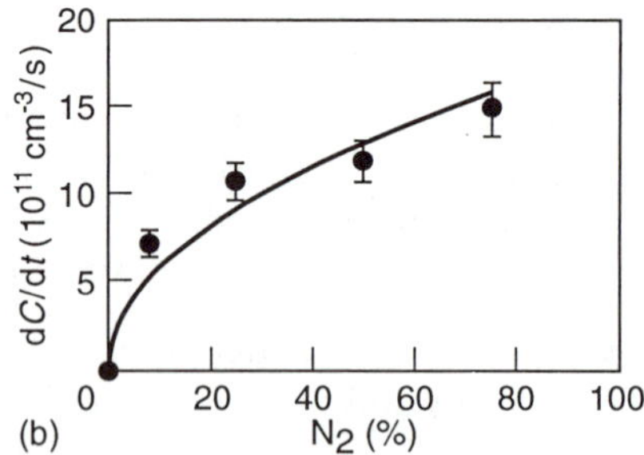

Figure 5.13 (a) Growth rates based on thickness measurements, and (b) rates of Ti concentration increase based on LIF measurements for different P_{N_2} when P_{H_2} is fixed at 25% of $(P_{N_2} + P_{H_2} + P_{Ar})$. P_{N_2} is varied as a percentage of $(P_{N_2} + P_{H_2} + P_{Ar})$. $P_{TiCl_4} = 27$ torr (3.6 kPa), $P_{Total} = 600$ torr (80 kPa) and laser power = 400 W.

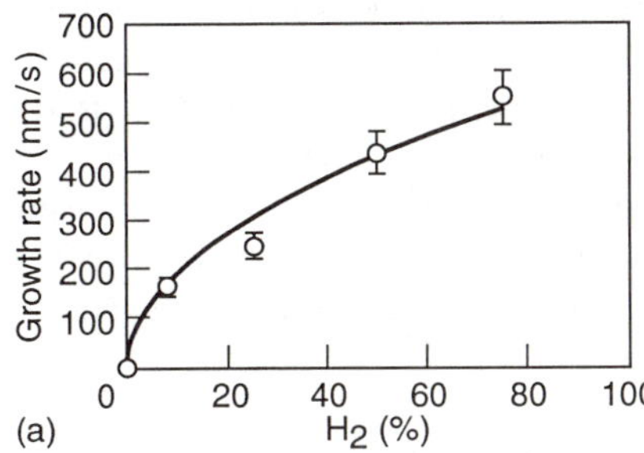
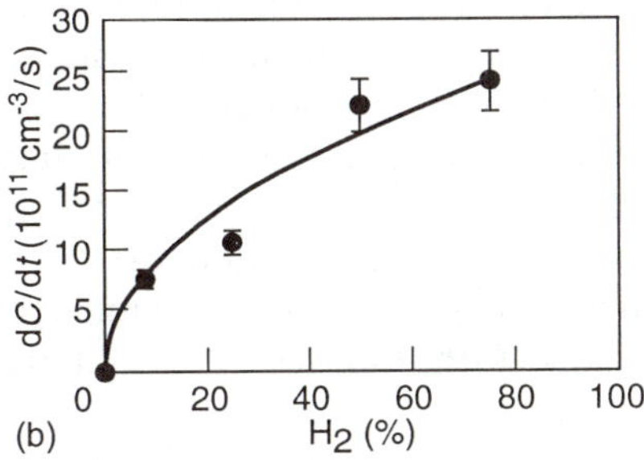

Figure 5.14 (a) Growth rates based on thickness measurements, and (b) rates of Ti concentration increase based on LIF measurements for different P_{N_2} when P_{H_2} is fixed at 25% of $(P_{N_2} + P_{H_2} + P_{Ar})$. P_{N_2} is varied as a percentage of $(P_{N_2} + P_{H_2} + P_{Ar})$. $P_{TiCl_4} = 27\,torr$ (3.6 kPa), $P_{Total} = 600\,torr$ (80 kPa) and laser power = 400 W.

TiN desorption from the deposited film. The concept of TiN desorption is also supported by the model [110] that incorporates TiN desorption to make excellent predictions on the film thickness profiles. The Ti concentrations detected by LIF are on the order of 10^{12} cm^{-3}. An equilibrium computation [111] of the Ti–H–N–Cl system at comparable experimental conditions shows that the Ti concentration produced by the gas phase reaction is six orders of magnitude smaller than that detected by LIF in our experiment. Hence, we ignore the gas phase production of Ti.

From Figs 5.13 and 5.14, R_{TiN} and dC/dt are found to be proportional to the square roots of P_{N_2} and P_{H_2}, respectively. The square root dependencies on P_{N_2} and P_{H_2} are also observed by other researchers in conventional CVD processes [111–114]. As to the detailed reaction pathways, there have been no studies before this research on the laser initiated CVD process. However, there have been different suggestions in the literature on conventional hot wall CVD [111–113, 103, 104], each explaining its own experimental observations. After careful examination, it is found that the possible surface reaction pathways listed by Nakanishi *et al.* [111] are the most suitable in interpreting our experimental results. Since the surface coverages of N and H atoms (when dissociatively adsorbed) are proportional to the square roots of P_{N_2} and P_{H_2} at low surface coverages, the results in Figs 5.13 and 5.14 indicate that the rate-limiting steps in the reaction process are the dissociative adsorptions of N_2 and H_2 molecules on the substrate surface.

The effect of changing CO_2 laser power on the LCVD process is also studied. The laser power is set at 250, 400 and 550 W. All other conditions are constant: $P_{TiCl_4} = 27\,torr$ (3.6 kPa), $P_{N_2}:P_{H_2} = 3:1$ and $P_{Total} = 600\,torr$ (80 kPa). Because the reactant gas composition remains unchanged for this set of experiments, and the only effect of the changing laser power is on the substrate surface temperature, the reaction kinetics can be studied using these results. R_{TiN} and dC/dt are plotted with respect to the substrate temperature in Fig. 5.15. The horizontal axis is chosen to be $1/T_{peak}$, where T_{peak} is the peak substrate temperature by simultaneous multiwavelength pyrometry for the different laser powers. Here T_{peak} is chosen because it is found to represent the substrate temperature at which the majority of the film growth occurs. With the previously determined reactant partial pressure dependencies, it follows from the Arrhenius equation that

$$R_{TiN} = K_1 P_{TiCl_4} P_{N_2}^{1/2} P_{H_2}^{1/2} \exp\left(-\frac{E}{RT}\right) \tag{5.1}$$

and

$$\frac{dC}{dt} = K_2 P_{TiCl_4} P_{N_2}^{1/2} P_{H_2}^{1/2} \exp\left(-\frac{E}{RT}\right) \tag{5.2}$$

where K_1 and K_2 are the Arrhenius constants, E is the apparent activation energy, and R the molar gas constant. From the slopes of the straight lines in Fig. 5.15, the apparent activation energy is calculated to be 104.3 kJ/mol from the film thickness data, and 125.7 kJ/mol from the LIF data, under the following conditions: $P_{TiCl_4} = 27$ torr (3.6 kPa), $P_{N_2} : P_{H_2} = 3{:}1$, $P_{Total} = 600$ torr (80 kPa) and $T = 1330{-}1520$ K. The similarity between the two values seems to confirm that the Ti species probed by the LIF technique are the direct result of the film growth on the surface. There are large discrepancies in the reported activation energy values for conventional hot wall CVD. Under different experimental conditions, they range from 39.3 kJ/mol to 308.9 kJ/mol [103–107, 111–114]. The values depend greatly on the ranges of substrate temperature at which the experiments are performed. At temperatures close to ours, Cao *et al.* [112] and Kato [103] reported a value of 120.4 ± 24.2 kJ/mol,

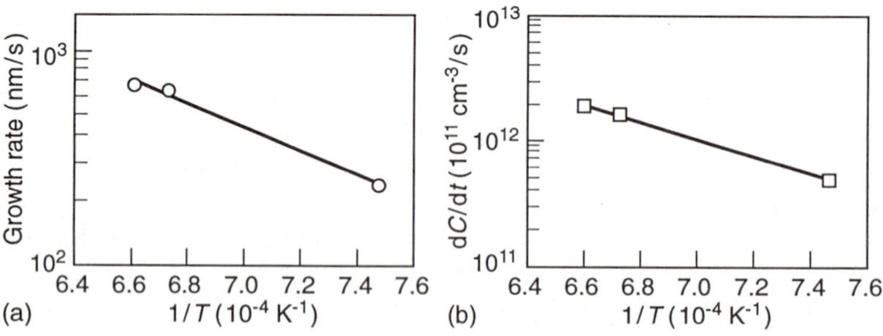

Figure 5.15 (a) Growth rates based on thickness measurements, and (b) rates of Ti concentration increase based on LIF measurements for different peak substrate temperatures. $P_{N_2} : P_{H_2} = 3{:}1$, $P_{TiCl_4} = 27$ torr (3.6 kPa) and $P_{Total} = 600$ torr (80 kPa).

which is in good agreement with our value. In another set of experiments, the partial pressure of $TiCl_4$ is kept constant at 27 torr (3.6 kPa). The gas ratio $N_2:H_2$ is also kept constant at 3:1. The only parameter that is changed is the amount of N_2 and H_2 mixture, leading to a total chamber pressure of 100, 350 and 600 torr, respectively (13, 47 and 80 kPa). CO_2 laser power is 400 W with a TEM_{00} beam diameter of 2 mm. The results are plotted in Fig. 5.16. Since the partial pressure of $TiCl_4$ is the same, the main reason for the higher deposition rate at a lower total pressure must be the reduced thermal convective loss to the ambient gas that results in a faster heating rate for the substrate. This is confirmed by the simultaneous multiwavelength surface temperature measurement (Fig. 5.17). The observation and analysis in this set of experiments provide additional evidence that the LCVD of TiN under

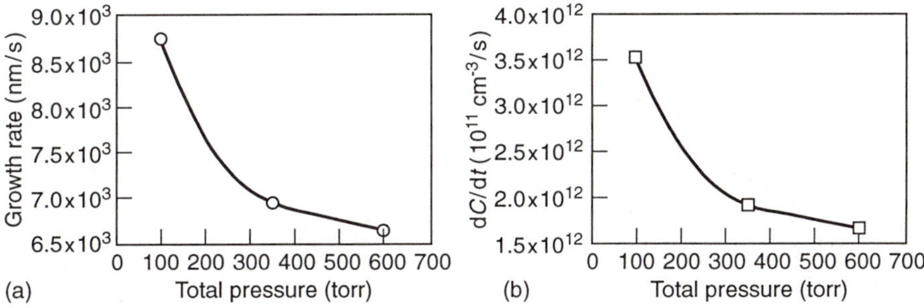

Figure 5.16 (a) Growth rates based on thickness measurements, and (b) rates of Ti concentration increase based on LIF measurements for different total pressures. $TiCl_4$ pressure = 27 torr (3.6 kPa), $N_2:H_2$ = 3:1, laser power = 400 W and TEM_{00} beam diameter = 2 mm [102].

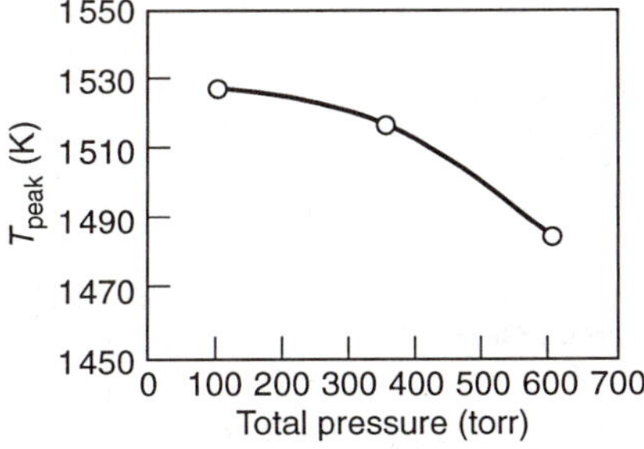

Figure 5.17 Peak temperatures on substrate surfaces measured by multiwavelength pyrometry for different total pressures. $TiCl_4$ pressure = 27 torr (3.6 kPa), $N_2:H_2$ = 3:1, laser power = 400 W and TEM_{00} beam diameter = 2 mm [102].

investigation is definitely driven by the laser heating of the substrate surface. In other words, it is indeed a pyrolytic LCVD process.

5.3.3 Applications

The deposition of material by thermal decomposition using a laser heat source was first reported for graphite from hydrocarbon vapor [115] at the Third International Conference on Chemical Vapor Deposition. At the same conference the following year, D.M. Mattox [116, 117] reported the LCVD of W by H_2 reduction on WF_6 using a multimode CW laser translated across an SiO_2 substrate. Film thickness and resistivity were measured as a function of the incident energy density. More recently, the LCVD of polycrystalline Si on SiO_2 was reported by Christensen and Larkin [118]. Using a CO_2 laser tuned to a minimum in the SiH_4 absorption spectrum corresponding to 10.33 μm, they deposited films up to 5 μm thick with a full width at half height (FWHH) of 400 μm. Si lines of unknown thickness and approximately 50 μm width were 'drawn' by translating the substrate during irradiation. Unlike conventional CVD of Si, where the addition of H_2 is necessary to prevent unwanted gas phase precipitation of Si particles, good quality LCVD films were producd without H_2. Titanium nitride was produced by using the reaction described in the earlier section on kinetics [99–102]. The typical microstructure of LCVD titanium is shown in Fig. 5.18; note how it is fine and crystalline.

REFERENCES

1. Draper, C.W. and Ewing, C.A. (1984) *J. Mater. Sci.*, **19**, 3815.
2. Draper, C.W. and Poate, J.M. (1985) *Int. Met. Rev.*, **30**(2), 85.
3. Steen, W.M. (1985) Surface engineering with a laser. *Metals and Materials*, **1**, 730.
4. Draper, C.W. (1981) *Appl. Opt.*, **20**, 3083.
4a. Draper, C.W. (1981) *Lasers in Metallurgy* (eds K. Mukherjee and J. Mazumder), The Metallurgical Society of AIME, Warrendale, PA, pp. 67–92.
5. Chande, T. and Mazumder, J. (1983) *Metall. Trans. B*, **14**, 181.
6. Chande, T. and Mazumder, J. (1983) *Opt. Eng.*, **22**, 362.
7. Chande, T. and Mazumder, J. (1982) *Appl. Phys. Lett.*, **41**, 42.
8. Weinman, L.S., DeVault, J.N. and Moore, P. (1979) *Application of Lasers in Materials Processing* (ed. E.A. Metzbower), ASM, Metals Park OH, p. 245.
9. Moore, P.G. and Weinman, L.S. (1979) Laser applications in materials processing, in *SPIE Proc.*, (ed. J.F. Ready), Vol. 198, p. 120.
10. Christodoulu, G. and Steen, W.M. (1983) *Lasers in Materials Processing* (ed. E.A. Metzbower) ASM, Metals Park OH.
11. Molian, P.A. (1982) *Scripta Metall.*, **16**, 65.
12. Molian, P.A. (1983) *Mater. Sci. Engng*, **50**, 241.
12a. Molian, P.A. (1983) *Mater. Sci. Engng*, **58**, 175.
13. Draper, C.W. *et al.* (1981) *Appl. Surf. Sci.*, **7**, 276–80.
14. Draper, C.W. *et al.* (1981) *Thin Solid Films*, **75**, 237.
14a Draper, C.W. *et al.* (1981) *Thin Solid Films*, **75**, 419–24.

12.0 µm

Figure 5.18 Scanning electron micrograph of LCVD TiN. CO_2 laser power $= 500\,$W, interaction time $= 0.5\,$s, beam diameter $= 1\,$mm, total chamber pressure $= 600\,$torr $(80\,$kPa$) - 25\,$torr $(3.3\,$kPa$)\ TiCl_4$, $144\,$torr $(19.2\,$kPa$)\ N_2$, $431\,$torr $(57.5\,$kPa$)\ H_2$.

15. Bykovskii, Y.A., Kulikauskas, V.S., Novolin, V.N. and Khabelashivili, T.D. (1982) *Ah. Tekh. Fiz.*, **52**(1): 61–3.
16. Jain, A.K. *et al.* (1982) *Rad. Effects*, **63**, 175.
17. Jain, A.K. *et al.* (1982) *Rad. Effects*, **53**, 175–81.
18. Draper, C.W. *et al.* (1982) *Laser and Electron Beam Interactions with Solids* (eds. B.R. Appleton and G.K. Celler), North-Holland, Amsterdam, pp. 413–18.
19. Walker, A.M. (1985) Ph.D. thesis, London University.
20. Bergmann, H.W. and Mordike, B.L. (1980) *Z. Metalldke*, **71**, 658–65.
21. Bergmann, H.W. and Mordike, B.L. (1981) *Z. Werkstofftech.*, **12**, 142.
22. Bergmann, H.W. and Mordike, B.L. (1981) *J. Mater. Sci.*, **16**, 863–9.
22a. Bergmann, H.W. and Mordike, B.L. (1981) *J. Mater. Sci.*, **16**, 497–503.
23. Gnanamuthu, D.S. (1980) *Opt. Eng.*, **19**, 783–92.

24. Man, C. (1984) Ph.D. thesis, London University.
25. Lamb, M., Man, H.C., Steen, W.M. and West, D.R.F. (1983) The properties of laser surface melted stainless steel and boronised mild steel, in *Proceedings of CISFFEL, Lyon*, CEA, Paris, Vol. 1, pp. 227–34.
26. Chande, T., Ghosh, A. and Mazumder, J. (1987) *Surf. Engng*, Jan. 1987.
27. Ayers, J.D., Schaefer, R.J. and Robey, W.P. (1981) *J. Met.*, **33**(8): 19–23.
28. Ayers, J.D. and Tucker, T.R. (1980) *Thin Solid Films*, **73**, 201–7.
29. Sepold, G. (1981) in *Proceedings of the International Conference on Surface Technology 1981* (ed. C. Rohrback), VDI-Verlag, Dusseldorf, pp. 323–30.
30. Megaw, J.H.P.C. (1980) *Surf. J.*, **22**, 6–11.
31. Bergmann, H.W., Bell, T. and Lee, S. (1985) Thermochemical treatment of titanium alloys with lasers (laser gas alloying), in *Proceedings of Laser '85, Munich*.
32. Walker, A.M., Folkes, J., Steen, W.M. and West, D.R.F. (1985) *Surf. Engng*, **1**(1).
33. Kovalenko, V.S., Volgin, V.I. and Mikhailov, V.V. (1977) *Tekh. Organ. Proizvod*, **3**, 50–2.
34. Swager, W.L. (1980) *Battelle Today*, **18**, 3–6.
35. Mirkin, L.I. (1971) Laser beam promoted saturation of iron by tungsten. *Br. Iron and Steel Inst. Trans.*, No. 9382 (translated from *Izvestiya Vysskilkh Uchebnykh Azvendnii Chernaya Metallurgiya*, **14**(2), 98).
36. Gasuko, I.V., Krapivin, L.L. and Mirkin, L.I. (1974) *Sov. Powd. Metall. Met. Ceram.*, **113**(1), 27.
37. Belforte, D.A. (1978) Laser surface treatment. Presented at the Colloquium on Lasers and Electro-Optical Equipment, Tokyo, Japan.
38. Cunningham, F.E. (1964) The use of lasers for the production of surface alloys. MSc thesis, MIT, Cambridge MA.
39. Mirkin, L.I. (1969) *Sov. Phys. Dok.*, **14**(5), 494.
40. Schmidt, A.O. (1969) *Trans. ASME B, J. Engng for Ind.*, **91**(3), 549.
41. Seaman, F.D. and Gnanamuthu, D.S. (1975) *Met. Prog.*, **108**(3).
42. Hella, R.A. (1978) *Opt. Eng.*, **17**(3), 198.
43. Brower, W.E. Jr., Strachan, R. and Flemings, M.C. (1970) *Cast Met. Res. J.*, **6**(4), 176.
44. Molian, P.A. and Wood, W.E. (1982) *Mater. Sci. Engng*, **56**, 271–7.
45. Molian, P.A. and Wood, W.E. (1982) *Scripta Metall.*, **15**, 1301
46. Molian, P.A. and Wood, W.E. (1983) *J. Mater. Sci.*, **18**, 2563–71.
47. Moore, P.G. and McCaffery, E. (1981) *J. Electrochem. Soc.*, **128**, 1391.
48. Lumsden, J.B., Gnanamuthu, D.S. and Moores, R.J. (1982) *Corrosion of Metals Processed by Directed Energy Beams* (eds C.R. Clayton and C.M. Preece), The Metallurgical Society of AIME, Warrendale PA, pp. 129–34.
49. Lin, C.J. and Spaepen, F. (1982) *Appl. Phys. Lett.*, **41**, 721–3.
50. Gorogina, G.G. *et al.* (1981) *Dokl. Akad. Nauk SSSR*, **259**, 826–9 (*Sov. Phys. Dokl.*, **26**, 761–3).
51. Inal, A.O. (1980) *J. Mater. Sci.*, **15**, 1947–52.
52. Samure, K.O., Shibur, K., Shingu, P.H. and Murakeme, Y. (1979) *J. Mater. Sci.*, **14**, 945–52.
53. Inal, O.T., Murr, L.E. and Yost, F.G. (1981) *Mater. Sci. Engng*, **51**, 101–10.
53a. Picraux, S.T. *et al.* (1981) *Laser and Electron-Beam Solid Interactions and Materials Processing* (eds J.F. Gibbons *et al.*), North-Holland, New York, pp. 575–82.
54. Elliott, R.P. (1965) *Constitution of Binary Alloys, First Supplement*, McGraw-Hill, New York.
55. Shunk, F.A. (1969) *Constitution of Binary Alloys, Second Supplement*, McGraw-Hill, New York.

56. Hansen, M. and Anderko, K. (1958) *Constitution of Binary Alloys*, McGraw-Hill, New York.

57. Draper, W.C. (1981) *Laser surface alloying: the state of the art*, in *Lasers in Metallurgy* (eds K. Mukherjee and J. Mazumder), The Metallurgical Society of AIME, New York, pp. 67–93; *J. Met.*, **34**(6), 24.

58. Follstaedt, D.M. (1982) Metallurgy and microstructures of pulse melted alloys, in *Laser and Electron-Beam Interactions with Solids*. Proceedings of the 1981 Materials Research Society, Vol. 4 (eds B.R. Appleton and G.K. Celler), Elsevier, New York, pp. 377–88.

59. Sood, O.K. (1982) *Rad. Effects*, **63**, 141.

60. Draper, C.W. and Poate, J.M. (1983) Laser surface alloying, in *Surface Modification and Alloying* (eds J.M. Poate, G. Foti and D.C. Jacobson), Plenum, New York, Ch. 13.

61. Snow, D.B., Breinan, E.M. and Kear, B.H. (1980) Rapid solidification processing of superalloys using high power lasers, in *Superalloys 1980*. Proceedings of the Fourth International Symposium on Superalloys (eds J.K. Tien *et al.*), Metals Park OH, pp. 183–203.

62. Breinan, E.M., Snow, D.B. and Brown, C.O. (1981) Program to investigate advanced laser processing of materials. *Final report to Office of Naval Research N00014-C-0387*.

63. Li, L.J. and Mazumder, J. (1985) *Laser Processing of Materials* (eds K. Mukherjee and J. Mazumder), The Metallurgical Society of AIME, Warrendale PA, p. 35.

64. Weerasinghe, V.M. and Steen, W.M. (1983) *Lasers in Materials Processing* (ed. E. Metzbower), ASM, Metals Park OH, p. 166.

65. Steen, W.M., Weerasinghe, V.M. and Monson, P. (1986) High power lasers and their industrial applications. *SPIE Proc. 650*, p. 226.

66. Steen, W.M. (1991) *Laser Materials Processing*, Springer-Verlag, London.

67. Belmondo, A. and Castagna, M. (1979) *Thin Solid Films*, **vol**, 249.

68. Mazumder, J., Liu, Y., Kar, A. and Shibata, K. (1994) Nonequilibrium synthesis with lasers: atom to automobile, in *Proceedings of the 8th Surface Modification Technologies Conference*, Emptek Publishing.

69. Breinan, E.M. and Kear, B.H. (1983) Rapid solidification laser processing at higher power density, in *Laser Materials Processing* (ed. M. Bass), North-Holland, Amsterdam.

70. Savage, S.J. and Froes, F.H. (1984) *J. Met.*, **36**(4), 20–33.

71. Bloembergen, N. (1979) *Laser Solid Interactions and Laser Processing* (eds S.D. Ferris, H.J. Leamy and J.M. Poate) AIP Conference Proceedings, vol. 50, pp. 1–20.

72. Von Allmen, M., Huber, E., Blatter, A. and Affostler, K. (1984) *Int. J. Rapid Solidification*, **7**, 15–28.

73. Affolter, K. and Von Allmen, M. (1984) *Appl. Phys. A*, **33**, 93.

74. Kar, A. and Mazumder, J. (1987) *J. Appl. Phys.*, **61**(7), 2645–55.

75. Mazumder, J., Cusano, C., Ghosh, A. and Eiholzer, C. (1985) *Laser Processing of Materials* (eds K. Mukherjee and J. Mazumder), The Metallurgical Society of AIME, Warrendale PA, p. 199.

76. Singh, J. and Mazumder, J. (July, 1986) *Mater. Sci. Technol.*, **2**, 709.

77. Singh, J. and Mazumder, J. (1987) *Metall. Trans. A*, **18**(2), 313–22.

78. Singh, J. and Mazumder, J. (1987) *Acta. Metall.*, **35**(8), 1995–2003.

79. Singh, J. and Mazumder, J. (1988) *Metall. Trans. A*, **19**(8), 1981–90.

80. Choi, J. and Mazumder, J. (1994) *J. Mater. Sci.*, **29**, 4460–76.

81. Singh, J., Nagarathnam, K. and Mazumder, J. (1987) in *Proceedings of the High*

Temperature Coatings Conference, Orlando FL, The Metallurgical Society of AIME, Warrendale PA.
82. Sircar, S., Ribaudo, C. and Mazumder, J. (1989) *Metall. Trans. A*, **20**(11), 2267–77.
83. Ribaudo, C. and Mazumder, J. (1989) *Mater. Sci. Engng*, **A121**, 531–8.
84. Ribaudo, C., Sircar, S. and Mazumder, J. (1989) *Metall. Trans. A*, **20**(11), 2489–97.
85. Takeda, T., Steen, W.M. and West, D.R.F. (1985) *ICALEO-84*, Laser Institute of America.
86. Jain, A.K., Kulkarni, V.N. and Sood, D.K. (1982) *Rad. Effects*, **53**, 183.
87. Steen, W.M. and Courtney, C.G.H. (1980) *Metals Technol.*, June 1980, pp. 232–7.
88. Powell, J. and Steen, W.M. (1981) *Lasers in Metallurgy*, The Metallurgical Society of AIME, Warrendale PA, pp. 94–104.
89. Ehrlich, D.J. and Tsao, J.Y. (1983) in *Proceedings of the European Physical Society Meeting*, Mauterndorf, Austria, Springer-Verlag, New York.
90. Ehrlich, D.J. and Tsao, J.Y. (1983) *J. Vac. Sci. Technol.*, **B1**(4), 969.
91. Mazumder, J. and Allen, S.D. (1979) *SPIE*, **198**, 73.
92. Powell, C.F., Cambell, I.E. and Gonser, B.W. (1955) *Vapor Plating*, Wiley, New York.
93. Carlton, H.E. and Oxley, J.H. (1966) *AIChE. J.*, **12**, 86.
94. West, G.A., Gupta, A. and Beeson, K.W. (1985) *Appl. Phys. Lett.*, **47**, 476–8.
95. Eden, J.G. (1992) *Photochemical Vapor Deposition*, Wiley, New York.
96. Allen, S.D. (1981) *J. Appl. Phys.*, **52**, 6501.
97. Allen, S.D. *et al.* (1982) *J. Vac. Sci. Technol.*, **20**, 469.
98. Steen, W.M. (1978) in *Proceedings of the International Conference on Advances in Surface Coating*, London, p. 175.
99. Chen, X. (1994) Ph.D. thesis, University of Illinois at Urbana-Champaign.
100. Chen, X. and Mazumder, J. (1994) *Appl. Phys. Lett.*, **65**(3), 298–300.
101. Chen, X. and Mazumder, J. (1994) *J. Appl. Phys.*, **76**(6), 3914–16.
102. Chen, X. and Mazumder, J. (1995) Laser chemical vapor deposition of titanium nitride. *Phys. Rev. B* **52**, 5947.
103. Kato, A. and Tamari, N. (1975) *J. Cryst. Growth.*, **29**, 55.
104. Tsao, C.J., Chen, E.B. and Miao, H.F. (1986) *Trans. Inst. Mining Met.*, **95**, 63.
105. Peterson, J.R. (1974) *J. Vac. Sci. Technol.*, **11**, 715.
106. Takahashi, T. and Suzuki, Y. (1974) *J. Jpn. Inst. Chem.*, **30**, 1043.
107. Kim, M.S. and Chun, J.S. (1983) *Thin Solid Films*, **107**, 129.
108. You. M.S., Nakanishi, N. and Kato, E. (1991) *J. Electrochem. Soc.*, **138**, 1394.
109. Davis, K.A., Brezinsky, K. and Glassman, I. (1991) *Combust. Sci. Technol.*, **77**, 171.
110. Conde, O., Kar, A. and Mazumder, J. (1992) *J. Appl. Phys.*, **72**, 754.
111. Nakanishi, N., Mori, S. and Kato, E. (1990) *J. Electrochem. Soc.*, **137**, 322.
112. Cao, Z.R., Du, Y.S. and Miao, H.F. (1989) *Surf. Engng*, **5**, 315.
113. Sadahiro, T., Cho, T. and Yamaya, S. (1977) *J. Jpn. Inst. Met.*, **41**, 542.
114. Yoshikawa, N., Aikawa, H. and Kikuchi, A. (1992) *J. Jpn. Inst. Met.*, **56**, 1132.
115. Lydtin, H. (1972) in *Proceedings of the Third International Conference on Chemical Vapor Deposition* (ed. F.A. Glaski), American Nuclear Society, Hinsdale, IL, p. 121.
116. Berg, R.S. and Mattox, D.M. (1972) in *Proceedings of the Third International Conference on Chemical Vapor Deposition* (ed. F.A. Glaski), American Nuclear Society, Hinsdale, IL, p.196.
117. Berg, R.S. and Mattox, D.M. (1972) in *Proceedings of the Fourth International Conference on Chemical Vapor Deposition* (ed. F.A. Glaski), American Nuclear Society, Hinsdale, IL, p. 121.
118. Christensen, C.P. and Larkin, K.M. (1978) *Appl. Phys. Lett.*, **32**, 254.
119. Ehrlich, D.J. *et al.* (1981) *Appl. Phys. Lett.*, **39**, 957.

120. Ehrlich, D.J. and Tsao, J.Y. (1982) *Appl. Phys. Lett.*, **41**, 297.
121. Ehrlich, D.J. *et al.* (1981) *Appl. Phys. Lett.*, **38**, 946.
122. Ehrlich, D.J. and Osgood, R.M. Jr. (1982) *Thin Solid Films*, **90**, 287.
123. Wood, T.H., White, J.C. and Thacker, B.A. (1983) *Appl. Phys. Lett.*, **42**, 408.
124. Rytz-Froideraux, Y., Salathe, R.P. and Gilgen, H.H. (1983) in *LaserDiagnostics and Photochemical Processing for Semiconductor Devices*, (eds R.M. Osgood *et al.*) North-Holland, New York, pp. 29–34.
125. Kitai, A. and Wolga, G.J. (1983) in *Surface Modification and Alloying* (eds J.M. Poate, G. Foti and D.C. Jacobson) Plenum, New York, pp. 141, 147.
126. Rigby, L.J. (1969) *Trans. Faraday Soc.*, **65**, 2421.
127. Boyer, P.K. *et al.* (1982) *Appl. Phys. Lett.*, **40**, 716.
128. Deutsch, T.F., Silversmith, D.J. and Mountain, R.W. (1983) in *Surface Modification and Alloying* (eds J.M. Poate, G. Foti and D.C. Jacobson), Plenum, New York, p. 129–34.
129. Solanki, R. and Collins, G.J. (1983) *Appl. Phys. Lett.*, **42**, 662.
130. Calloway, A.R., Galantowicz, T.A. and Fenner, W.R. *Aerospace Report AIR-82(8477)-1*; Calloway, A.R., Galantowicz, T.A. and Fenner, W.R. (1983) *J. Vac. Sci. Technol.*, **A1**, 534.
131. Andreatta, R.W. *et al.* (1982) *Appl. Phys. Lett.*, **40**, 183.
132. Aylett, M.R. and Haigh, J. (1983) in *Surface Modification and Alloying* (eds J.M. Poate, G. Foti and D.C. Jacobson), New York, pp. 177–82.
133. Solanki, R., Boyer, P.K. and Collins, G.J. (1982) *Appl. Phys. Lett.*, **41**, 1048; Solanki, R., Boyer, P.K., Mahan, J.E. and Collins, G.J. (1981) *Appl. Phys. Lett.*, **38**, 572.
134. Coombe, R.D. and Wodarczyk, F.J. (1980) *Appl. Phys. Lett.*, **37**, 846.
135. Hanabusa, M., Akira Namiki and Yoshihara, K. (1979) *Appl. Phys. Lett.*, **35**, 626.
136. Gattuso, T., Meunier, R.M., Adler, D. and Haggerty, J.S. (1983) in *Surface Modification and Alloying* (eds J.M. Poate, G. Foti and D.C. Jacobson), Plenum, New York, pp. 215–22.
136a. Haggerty, J.S. (1983) in *Proceedings of the International Conference on Ultrastructure Processing of Ceramics, Glasses and Composites* 13–17 Feb. 1983, Gainesville FL, Wiley, New York.
137. Cannon, W.R. *et al.* (1982) *J. Amer. Ceram. Soc.*, **65**, 324.
138. Ehrlich, D.J. and Osgood, R.M. Jr. (1981) *Chem. Phys. Lett.*, **79**, 381.
139. Tsao, J.Y. *et al.* (1983) *Appl. Phys. Lett.*, **42**, 559.
140. Rytz-Froidevaux, Y. *et al.* (1982) *Appl. Phys. Lett.*, **A27**, 133.
141. Bottka, N., Walsh, D.J. and Dalbey, R.Z. (1983) *J. Appl. Phys.*, **54**, 1104.
142. Chen, C.J. and Osgood, R.M. Jr. (1983) in *Surface Modification and Alloying* (eds J.M. Poate, G. Foti and D.C. Jacobson), Plenum, New York, pp. 169–75.
143. Karlicek, R.F.F., Donnelly, V.M. and Collins, G.J. (1982) *J. Appl. Phys.*, **53**, 1084.
144. Micheels, R.H. *et al.* (1981) *Appl. Phys. Lett.*, **37**, 418.
145. Arnone, C. *et al.* (1980) *Appl. Phys. Lett.*, **37**, 665.
146. Bauerle, D. *et al.* (1983) *Appl. Phys. Lett.*, **A30**, 147.
147. Von Gutfeld, R.J. *et al.* (1979) *Appl. Phys. Lett.*, **35**, 651.
148. Von Gurtfeld, R.J. *et al.* (1982) *IBM J. Res. Dev.*, **26**, 136.
149. Von Gutfeld, R.J. and Hodgson, R.T. (1982) *Appl. Phys. Lett.*, **40**, 352.
150. Von Gutfeld, R.J. and Puippe, J.C. (1981) *Oberflache-Surface*, **22**, 9.
151. Leyendecker, G. *et al.* (1981) *Appl. Phys. Lett.*, **39**, 921.
152. Rytz-Froideavaux, Y., Salathe, R.P. and Gilgen, H.H. (1981) *Appl. Phys. Lett.*, **84A**, 216.
153. Salathe, R.P., Gilgen, H.H. and Rytz-Froidevaux, Y. (1981) *IEEE J. Quantum Electron.*, **QE17**, 1989.

154. Herman, J.R. *et al.* (1983) in *Surface Modification and Alloying* (eds J.M. Poate, G. Foti and D.C. Jacobson), Plenum, New York, pp. 9–18.
155. McWilliams, B.M. *et al.* (1983) *Appl. Phys. Lett.*, **43**, 946.
156. Allen, S.D. and Trigubo, A.B. (1983) *J. Appl. Phys.*, **54**, 1641.
157. Allen, S.D., Trigubo, A.B. and Jan, R.Y. (1983) in *Surface Modification and Alloying* (eds J.M. Poate, G. Foti and D.C. Jacobson), Plenum, New York, pp. 207–14.
158. Roth, W. *et al.* (1983) in *Surface Modification and Alloying* (eds J.M. Poate, G. Foti and D.C. Jacobson), Plenum, New York, p. 193.
159. Krauter, W. *et al.* (1983) *Appl. Phys. Lett.*, **A31**, 13.
160. Muller, D.F., Rothschild, M. and Rhoades, C.K. (1983) in *Surface Modification and Alloying* (eds J.M. Poate, G. Foti and D.C. Jacobson), Plenum, New York, pp. 135–9.
161. Gerassimov, R.B. *et al.* (1982) *Appl. Phys. Lett.*, **B28**, 266.
162. Barananskas, V. *et al.* (1980) *Appl. Phys. Lett.*, **36**, 930.
163. Allen, S.D. (1990) Private communication.

FURTHER READING

Allen, S.D., Copley, S.M. and Edwards, R.H. (1982) DOE report.
Antona, P.L. and Bertani, A. (1981) Surface treatment of steel by alloying with a high-power laser, in *Proceedings of the 9th National Conference on Thermal Treatment*, Italian Association of Metallurgists, Milan, pp. 61–4.
Ayers, J.D. (1981) *Thin Solid Films*, **84**, 323.
Ayers, J.D. (1981) *Particles Composite Surfaces by Laser Processing* (eds K. Mukherjee and J. Mazumder), The Metallurgical Society of AIME, Warrendale PA.
Ayers, J.D., Tucker, T.R. and Schaefer, R.J. (1980) *Wear resistant surfaces by carbide particle injection, in Proceedings of the 2nd International Conference on Rapid Solidification Processing*, Claitors, Baton Rouge LA, pp. 212–20.
Battaglin, G. *et al.* (1982) *J. Appl. Phys.*, **53**, 3224–30.
Battaglin, G. *et al.* (1982) *Metastable Materials Formation by Ion Implantation* (eds S.T. Picraux and W.J. Choyke), North-Holland, New York, pp. 327–32.
Battaglin, G. *et al.* Mixing of Pb and Ni under ion bombardment and pulsed laser irradiation, in *Proceedings of the NATO-ASI on Surface Engineering*, Les Arcs, July 1983.
Battaglin, G. *et al.* Pulsed laser treatment of Eu and La implanted in Ni: surface alloying, trapping and damage, in *Proceedings of the 1983 European Materials Research Society*, Strasbourg, May 1983.
Becker, R., Rothe, R. and Sepold, G. Surface alloying by means of laser beams, in *Proceedings of the 3rd International Colloquium on Welding and Melting by Electrons and Laser Beam*, Sept. 1983.
Belotskii, A.V., Kovalenko, V.S., Volgin, V.I. and Pschenichnyi, V.I. (1977) *Fiz. I. Khim. Obrab. Mater*, **3**, 24.
Betaneli, A.I. *et al.* (1972) *Fiz. I. Khim. Obrab. Mater.*, **6**, 22.
Borovskii, I.B., Gorodoskii, D.D., Sharafeev, I.M. and Moryashchev, S.F. (1982) *Sov. Phys. Dokl.*, **27**, 259.
Brown, R. and Ayers, J.D. (1983) Solid particle erosion of Al 6061 with a laser melted and Tm particle injected surface layer, in *Wear of Materials 1983* (ed. K.C. Ludema), ASME, New York, pp. 325–32.
Buene, L. *et al.* (1980) *Appl. Phys. Lett.*, **37**, 385.

Buene, L. *et al. Appl. Phys. Lett.*, **37**, 583–90.

Buene, L. *et al.* (1981) Laser-pulse melting of hafnium-implanted nickel studied with TDPAC and RBS/channeling techniques, in *Nuclear and Electron Resonance Spectroscopy*. Proceedings of the 1980 Materials Research Society, Vol. 3 (eds E.N. Kaufmann and G. Schenoy), Elsevier, New York, pp. 391–6.

Cappeli, P.G. (1980) Laser surface alloying, in *NATO Lecture Series 106*, Advisory Goup for Aerospace Research and Development, Ch. 6, pp. 1–13.

Chan, C.L., Mazumder, J. and Chen, M.M. (1984) *Metall. Trans. A*, **15**, 2175.

Conti, C.A. *et al.* (1980) *Laser and Electron Beam Processing of Materials*, Academic Press, New York.

Copley, S.M. *et al.* (1979) Microstructures of surface alloyed Ag–Cu films produced by laser melt quenching, in *Rapidly Quenched Metals III*, Vol. 1 (ed. B. Cantor), The Metals Society, London, pp. 147–50.

Das, S., Dumler, I. and Mazumder, J. (1985) *LIM-2* (ed. M.F. Kimmit), p. 73.

Dekhtyar, I.Y., Nishchenko, M.M., Bukhalenko, V.V. and Kharitonskii, S.Y. (1979) *Fiz. Met. Metall.*, **47**, 887.

Della Mea, G. and Massoldi, P. (1979) Redistribution of Cu in polycrystalline and single crystal Al after laser irradiation, in *Laser–Solid Interactions and Laser Processing*. Proceedings of the 1978 Materials Research Society (eds S.D. Ferris, H.J. Leamy and J.M. Poate), AIP Conference Proceedings, New York, No. 50, pp. 212–13.

Della Mea, G., Donna Dalle Rose, L., Mazzoldi, P. and Miotello, A. (1980) *Rad. Effect*, **46**, 133.

Den Broeder, F.J.A., Vandenberg, J.M. and Draper, C.W. (1984) *Thin Solid Films*, **111**, 43.

Draper, C.W., Buene, L., Poate, J.M. and Jacobson, D.C. (1981) *Appl. Opt.*, **20**, 1730.

Draper, C.W. *et al.* (1980) Laser alloying of deposited metal films on nickel, in *Laser and Electron Beam Processing of Materials*, Proceedings of the 1979 Materials Research Society (eds C.W. White and P.S. Peercy), Academic Press, New York, pp. 721–7.

Ehrlich, D.J. *et al.* (1980) *Appl. Phys. Lett.*, **36**, 916.

Ehrlich, D.J. *et al.* (1980) *IEEE Electron Device Lett.*, **EDL**1.

Ehrlich, D.J., Osgood, R.M. Jr. and Deutsch, T.F. (1980) *Appl. Phys. Lett.*, **36**, 698.

Ehrlich, D.J., Osgood, R.M. Jr. and Deutsch, T.F. (1981) *Appl. Phys. Lett.*, **38**, 1018.

Ehrlich, D.J., Osgood, R.M. Jr. and Deutsch, T.F. (1981) *J. Electrochem Soc.*, **128**, 2041.

Ehrlich, D.J., Osgood, R.M. Jr. and Deutsch, T.F. (1982) *J. Vac. Sci. Technol.*, **21**, 23.

Greco, V.P. (1981) *Plat. Surf. Finish.*, **68**(3), 56.

Gurevich, M.E. *et al.* (1983) *Phys. Status Solidi*, **76**, 479.

Hussain, T. and Linker, G. (1982) *Solid State Commun.*, **44**, 745–9.

Ibbs, K.G. and Lloyd, M.L. (1983) *Pulsed laser treatment of virgin, self and europium implanted nickel: evidence of defect impurity interaction*, pp. 333–8.

Ibbs, K.G. and Lloyd, M.L. (1984) Thermal stability of metastable phases produced by laser treatment of aluminum implanted with chromium and molybdenum, in *Metastable Materials Formation by Ion Implantation*. Proceedings of the 1981 Materials Research Society, Vol. 7 (eds S.T. Picraux and W.J. Choyke), Elsevier, New York, pp. 327–32.

Jacobson, C.D. *et al.* (1981) *IEEE Trans. Nucl. Sci.*, **28**, 1828.

Jain, A.K., Kulkarni, V.N. and Sood, D.K. (1981) *Appl. Phys.*, **25**, 127.

Jain, A.K., Kulkarni, V.N. and Sood, D.K. (1981) *Nucl. Instrum. Meth.*, **191**, 151.

Jain, A.K., Kulkarni, V.N. and Sood, D.K. (1981) *Thin Solid Films*, **85**, 1.

Kovalenko, V.S. and Volgin, V.I. (1976) *Tekh. Organ. Proizvod*, **7**, 60.

Lin, C.-J. and Spaepen, F. (1983) Fe–B glasses formed by a picosecond pulsed laser quenching, in *Chemistry and Physics of Rapidly Solidified Materials* (eds B.J.

Berkowitz and R.O. Scattergood), The Metallurgical Society of AIME, New York, pp. 273–80.

Lin, C.-J. and Spaepen, F. (1983) *Scripta Metall.*, **17**, 1259.

Lin, C.-J., Spaepen, F. and Turnbull, D. (1984) *J. Non-Cryst. Solids*, **61–63**, 767.

Miotello, A. and Dona Dalle Rose, L.F. (1982) *Phys. Lett.*, **87A**, 317.

Miotello, A., Dona Dalle Rose, L.F. and Desalvo, A. (1982) *Appl. Phys. Lett.*, **40**, 135.

Molian, P.A., Wang, P.J., Khan, K.H. and Wood, W.E. (1983) *Mater. Sci. Engng*, **Vol**, 511–15.

Molian, P.A. and Wood, W.E. (1984) *Mater. Sci. Engng*, **62**, 271.

Mordike, B.L. and Bergmann, H.W. (1982) Surface alloying of iron alloys by laser beam melting, in *Rapidly Solidified Amorphous and Crystalline Alloys*. Proceedings of the 1981 Materials Research Society, Vol. 8, (eds B.H. Kear, B.C. Giessen and M. Cohen), Elsevier, New York, pp. 463–83.

Mordike, B.L., Bergmann, H.W. and Gros, N. (1983) *Z. Werkstofftech.*, **14**, 253.

Nilsson, Y. (1990) *J. Electrochem. Soc.*, **137**, 517–21.

Osgood, R.M. *et al.* (eds) *Laser Diagnostics and Photochemical Processing for Semiconductor Devices*, North-Holland, New York, 1983.

Peercy, P.S., Follstaedt, D.M., Picraux, S.T. and Wampler, W.R. (1982) *Laser and Electron Beam Interactions with Solids* (eds B.R. Appleton and G.K. Celler), North-Holland, New York, pp. 401–6.

Peters, J.W., Gebhart. F.L. and Hall, T.C. (1981) *Solid-State Technology*, Sept. 1980; Peters, J.W., in *Proceedings of the Electron Device Meeting*, pp. 240–3.

Pleiter, F. and Prasad, K.G. (1981) *Phys. Lett.*, **84A**, 345.

Rose, T.L. *et al.* (1983) *Appl. Phys. Lett.*, **42**, 193.

Sawchyn, I. and Draper, C.W. (1984) *Appl. Surf. Sci.*, **18**, 84.

Schaefer, R.J. *et al.* (1980) *Nucl. Instrum. Meth.*, **168**, 275.

Schaefer, R.J. and Mehrabian, R. (1983) Processing microstructure relationship in surface melting, in *Laser-solid Interactions and Transient Thermal Processing of Materials* (eds J. Narayan, W.L. Brown and R.A. Lemons), North-Holland, New York, pp. 733–44.

Sekula, S.T., Thompson, J.R., Beardsley, G.M. and Lowndes, D.H. (1983) *J. Appl. Phys.*, **34**, 6517.

Sepold, G. (1983) Hardening and alloying of steel surfaces by high-power laser beams, in *Physical Processes in Laser/Materials Interactions* (ed. M. Bertolotti), Plenum, New York, pp. 163–73.

Sepold, H.G. (1978) Hardening and alloying of steel surfaces by high-power laser beams, in *Gas-Flow and Chemical Lasers* (ed. J.F. Wendt), Hemisphere, Washington D.C., pp. 561–9.

Shibata, T. *et al.* (1980) *Appl. Phys. Lett.*, **36**, 566.

Silversmith, D.J. *et al.* (1983) Paper 14.3 presented at Materials Research Society Annual Meeting. November 1983.

Snow, D.B. (1980) Microstructure and mechanical properties of laser-processed Ni–Al–Mo based alloys, in *Superalloys 1980*. Proceedings of the Fourth International Symposium on Superalloys (eds J.K. Tien *et al.*), ASM, Metals Park OH, pp. 523–7.

Stritzker, B., Appleton, B.R., White, C.W. and Lau, S.S. (1982) *Solid State Commun.*, **41**, 321.

Takei, K., Fujimori, S. and Nagai, K. (1980) *J. Appl. Phys.*, **51**, 2903.

Ushakov, A.I., Gorovoy, A.M., Kazakov, V.G. and Il'Chuk, A.G. (1980) Alpha to gamma phase transition induced in Fe–Ni films by pulsed laser irradiation. *Fiz. Met. Metall.*, **50**, 440; English reference *Phys. Met. Metall.*, **50**, 189.

Wampler, W.R., Follstaedt, D.M. and Peercy, P.S. (1981) Pulsed laser annealing of aluminum, in *Laser and Electron-Beam Solid Interactions and Materials Processing*,

Proceedings of the 1980 Material Research Society, Vol. 1 (eds J.F. Gibbons, L.D. Hess and T.W. Signmon), Elsevier, New York, pp. 567–74.
Wang, Z.L., Westendorp, J.F.M. and Saris, F.W. (1983) *Nucl. Instrum. Meth.*, **209/210**, 115.
Wittmer, M. and Von Allmen, M. (1979) *J. Appl. Phys.*, **50**, 4786.

6

Sol–gel derived ceramic films – fundamentals and applications

C. Jeffrey Brinker, Carol S. Ashley,
Richard A. Cairncross, Ken S. Chen, Alan J. Hurd,
Scott T. Reed, Joshua Samuel, P. Randall Schunk,
Robert W. Schwartz and Cathy S. Scotto.

6.1 INTRODUCTION

Sol–gel processing begins with a colloidal dispersion, or sol, of particles or polymers in a liquid. Through subsequent chemical cross-linking, electrostatic destabilization, evaporation or some combination thereof, the fluid sol may be transformed into a rigid *gel*, which is a substance containing a continuous solid skeleton enclosing a continuous liquid phase. This *sol-to-gel* transition allows the solid phase to be shaped into films, fibers, microspheres or monoliths. Of these various forms, amorphous (or partially crystalline) thin films represent the earliest commercial application of sol–gel technology [1]. Thin films (normally less than 1 μm in thickness) use little in the way of raw materials and may be processed without cracking, overcoming the major disadvantage of sol–gel processing of bulk materials. Early applications of sol–gel coatings as optical films were reviewed by Schroeder [2]. Since then, many new uses of sol–gel films have appeared in electronic, protective, membrane and sensor applications [3–15]. Most often the as-deposited films

Metallurgical and Ceramic Protective Coatings. Edited by Kurt H. Stern.
Published in 1996 by Chapman & Hall, London. ISBN 0 412 54440 7

are amorphous, but depending on composition and thermal history, they may subsequently crystallize: ferroelectric PLZT (lead lanthanum zirconate titanate) and nonlinear optic $LiNbO_3$ are excellent examples of crystalline films derived from amorphous precursors [16–18].

Moisture and diffusion barriers, fracture resistance, scratch and wear resistance, thermal and electrical insulation and corrosion resistance are the most common functions of sol–gel ceramic coatings intended for protection. Optically clear silica-based sol–gels are routinely applied to transparent plastics to provide wear and scratch resistance [19]; for these and other optical applications the film microstrucure is tailored to provide a specific refractive index. Thermal insulation is best accomplished with highly porous films that possess a low thermal conductivity (e.g. aerogel films [20], closed-pore films [21]). One of the most popular applications in this area is in the protection of heat engine parts for which films have been traditionally deposited with plasma spray techniques [22]; however, there has been rising interest in using sol–gel processing as an alternative, less expensive method [23]. Metals are often coated with sol–gel films to protect surfaces from action by acids and oxidation. Corrosion resistance protection of steels with dense zirconium films [24] and of aluminium with silica films [25] are two recent examples that have been addressed in the literature. These coatings must be impermeable to corrosive substances and inert. Finally, fracture protection of ceramic substrates can be accomplished with sol–gel films. The function of the film is usually to fill surface flaws that can otherwise lead to fracture initiation [26–28]. Clearly, from these and other related protective functions, the properties of low thermal conductivity, chemical inertness, high-temperature stability, adjustable porosity and mechanical 'hardness' of sol–gel derived films make them amenable to many protective applications.

In sol–gel processing, film formation is accomplished through liquid film deposition methods, whereby a continuous liquid phase is forced to displace air at the substrate through a wetting process. The most commonly used techniques in this class of methods are dip-coating and spin-coating, which are covered in this chapter. Compared to more conventional routes of thin film formation, processes such as chemical vapor deposition, evaporation/condensation and sputtering, which are reviewed in other chapters of this book, sol–gel techniques require considerably less equipment, can be amenable to large and complex shapes, are applicable to substrates that cannot withstand high temperature and are potentially less expensive.

Most coatings are prepared by dip- or spin-coating. Dip-coating begins with immersing a substrate in a vessel filled with liquid. Withdrawal of the substrate from the liquid, if managed properly, results in a thin coherent film. Although it can be operated as a continuous process [29], dip-coating is essentially a batch process. Also, dip-coating is self-metered: the faster the substrate is withdrawn the thicker the entrained film. Thus, if dip-coating is the method of choice for a particular application then coating speed (e.g. for high throughput) may be a limiting factor to achieve thin films. Spin-coating can also be used to achieve remarkably thin and uniform films (20 nm with thickness variations less than 10%), but is only practical when the substrate is a 'small' disk, plate or bowl (less than about 100 cm^2). The main advantage gained by a sol–gel processing through spin, dip and related coating techniques is the ability to control both the composition and microstructure, including density, refractive index, pore size and surface area, *usually on molecular length scales* [30].

A protective coating must have desirable microstructure and acceptable uniformity to function (protect) as intended. This chapter reviews sol–gel thin film formation with the intent of establishing a link between the physics and chemistry of sol–gel film deposition and the resulting microstructure of the deposited films. We first review the aspects of sol–gel chemistry that govern the size and structure of the inorganic species which, along with any solvents, comprise the coating sol. In the following section we discuss the salient features of dip- and spin-coating processes, as revealed in many cases by *in situ* means of process characterization. In that section we provide further insight to the physics of the dip-coating and drying processes through modern computer-aided simulation. The next section addresses the drying portion of the coating process, which begins simultaneously with the fluid mechanics of the deposition, proceeds through a constant rate period in which liquid is removed from the gelled and shrinking film, and ends with a falling rate period during which the liquid recedes into the pore space. Finally, with a firm understanding of the physical rate processes involved during all stages of the sol–gel process, we take up the issue of microstructural control. We address the deposition of inorganic sols with regard to timescales and the effects of sol structure, capillary pressure, reaction kinetics and shear forces on such properties as refractive index, surface area and pore size of the deposited films. These properties determine a film's effectiveness as a protective layer for the intended application. An example of a sol–gel protective layer for space applications is then discussed in detail.

6.2 SOL–GEL CHEMISTRY

The sol–gel process utilizes inorganic or metal-organic precursors [31] to form metal salt, colloidal or polymerizable sols. In aqueous or organic solvents, the precursors are hydrolyzed and condensed to form inorganic polymers composed of M–O–M bonds. Although solution viscosity, solids content and substrate wetting are all considerations in sol preparation for thin film formation, the major requirement of the sol is that it remain stable with respect to precipitation or gelation, allowing uniform deposition by dipping, spinning, etc.

For inorganic precursors (salts) prepared in aqueous media, hydrolysis proceeds by the removal of a proton from an aquo ion $[MO_NH_{2N}]^{z+}$ to form a hydroxo (M–OH) or oxo (M=O) ligand:

$$[MO_NH_{2N}]^{z+} + pH_2O \longleftrightarrow [MO_NH_{2N-p}]^{(z-p)+} + pH_3O^+ \qquad (6.1)$$

Condensation reactions involving hydroxo ligands result in the formation of bridging hydroxyl (M–(μ-OH)–M) or bridging oxygen (M–O–M) bonds depending on the coordination number of M and the acidity of the bridging hydroxyl ligand. Normally, monomeric aquo cations are the only stable species at low pH, and various monomeric or oligomeric anions are the only species observed at high pH [32]. At intermediate pH, well-defined polynuclear ions are often the stable solution species, such as $[Zr_4(OH)_8(OH_2)_{16}]^{8+}$ or $[AlO_4Al_{12}(OH)_{24}(OH_2)_{12}]^{7+}$, but the metal solubility is normally limited there, usually leading to the precipitation of oxohydroxides, Al(O)OH or Fe(O)OH, or oxides, ZrO_2 or TiO_2 [32]. If dispersed or *peptized* by the addition of acid, or sterically stabilized by adsorption of organic molecules, these *particulate* sols are well suited for film deposition by dipping or spinning.

The most commonly used metal-organic precursors for sol–gel film formation are metal alkoxides $M(OR)_z$, where R is an alkyl group C_nH_{2n+1} [19]. Normally, the alkoxide is dissolved in alcohol and hydrolyzed by the addition of water under acidic, neutral, or basic conditions, although film formation is also possible by deposition of the neat alkoxides followed by exposure to moisture.

When alkoxide precursors are employed for sol preparation, hydrolysis and condensation reactions leading to M–O–M bond formation dictate the structure of the polymeric species formed in the solution. In the hydrolysis reaction, an alkoxide ligand is replaced with a hydroxyl ligand:

$$M(OR)_z + H_2O \longleftrightarrow M(OR)_{z-1}(OH) + ROH \qquad (6.2)$$

Condensation reactions involving the hydroxyl ligands produce polymers composed of M–O–M or M–(μ-OH)–M bonds, and in most cases, the by-products water or alcohol, as shown below for silicate condensation:

$$Si(OR)_3OH + Si(OR)_4 \longleftrightarrow (RO)_3Si-O-Si(OR)_3 + ROH \qquad (6.3)$$

$$2\,Si(OR)_3OH \longleftrightarrow (RO)_3Si-O-Si(OR)_3 + H_2O \qquad (6.4)$$

The reverse of reactions 6.3 and 6.4, namely siloxane bond alcoholysis and siloxane bond hydrolysis, promote bond breaking and reformation processes that, if extensive, permit complete restructuring of the growing polymer.

Since metals of interest for sol–gel film formation (Si, Al, Ti, etc.) have coordination numbers CN greater than 4, complete condensation would lead to compact, particulate oxides. In fact, for electropositive metals like Ti and Zr, it is difficult to avoid particle formation due to the rapid hydrolysis and condensation rates of their alkoxides. To circumvent this difficulty, the metal alkoxide precursors are modified, by chelation with multidentate ligands such as acetylacetone (acac) [32] or alcohol amines [33]. Because these ligands are less sensitive to hydrolysis than alkoxide ligands and add steric bulk around the metal center, they reduce both the effective functionality [32] and the rates of hydrolysis and condensation [32, 33] of the precursor species. This allows for more controllable hydrolysis and condensation, and the formation of nonparticulate solution species.

Because silicon is substantially less electropositive than Ti or Zr, hydrolysis and condensation of silicon alkoxides occur at much lower rates, and the condensation pathway leading to polymer formation can be more easily influenced by steric and chemical factors such as the steric bulk of the alkoxide ligand, the concentration of the acid or base catalyst, the hydrolysis ratio r of H_2O:M in equation 6.2, or sol aging time. However, even for silicon alkoxides, modification of alkoxide properties has proved beneficial in thin film preparation. For example, alkyl substituents have been incorporated as 'templating' ligands in organoalkoxysilane precursors, $R_nSi(OR')_{4-n}$, permitting the preparation of films with controlled pore sizes and pore size distributions. Such modifications may also be used to control the hydrolysis and condensation kinetics of the silicon sol species [34].

The hydrolysis and condensation kinetics of silicate species have been studied by ^{29}Si NMR, and the structure of the silicate sol species has been probed with ^{29}Si NMR, as well as small-angle X-ray scattering (SAXS). There is a considerable body of evidence based on these techniques that indicates, under many synthesis conditions, condensation of silicon alkoxides results in the formation of randomly branched 'polymeric' silicates rather than dense silica particles. Using ^{29}Si NMR, silicate polymers are characterized by the distribution of $Si(OR)_x(OH)_y(OSi)_n$ species, or the 'Q^n' distribution, where $n = 4 - x - y$. The prevalence of Q^1-Q^3 species under many sol–gel synthesis conditions [35–37] clearly indicates that silicates may be quite weakly condensed when compared, for example, to a commercial particulate silica sol [38] (Fig. 6.1a, [39]).

SAXS probes structure on the ~ 0.5–50 nm length scale. The Porod slope P of a plot of log scattered intensity versus log scattering wave vector, $K = (4\pi/\lambda)(\sin\theta/2)$, is related to the mass fractal dimension D and the surface fractal dimension D_S by the following expression [40]:

$$P = D_S - 2D \tag{6.5}$$

For uniform (nonfractal) objects in three-dimensional space, $D = 3$, $D_S = 2$, and P reduces to -4. For mass fractal objects, $D = D_S$, and $P = -D$. For surface fractal objects, $D = 3$ and $P = D_S - 6$. Porod plots of a variety of silica sols prepared from metal alkoxides are compared with the plot of Ludox, a commercial particulate silica sol, in Fig. 6.1b [41]. By simple variation of the hydrolysis ratio r and catalyst addition it is possible to prepare a spectrum of structures ranging from polymers to particles. It is

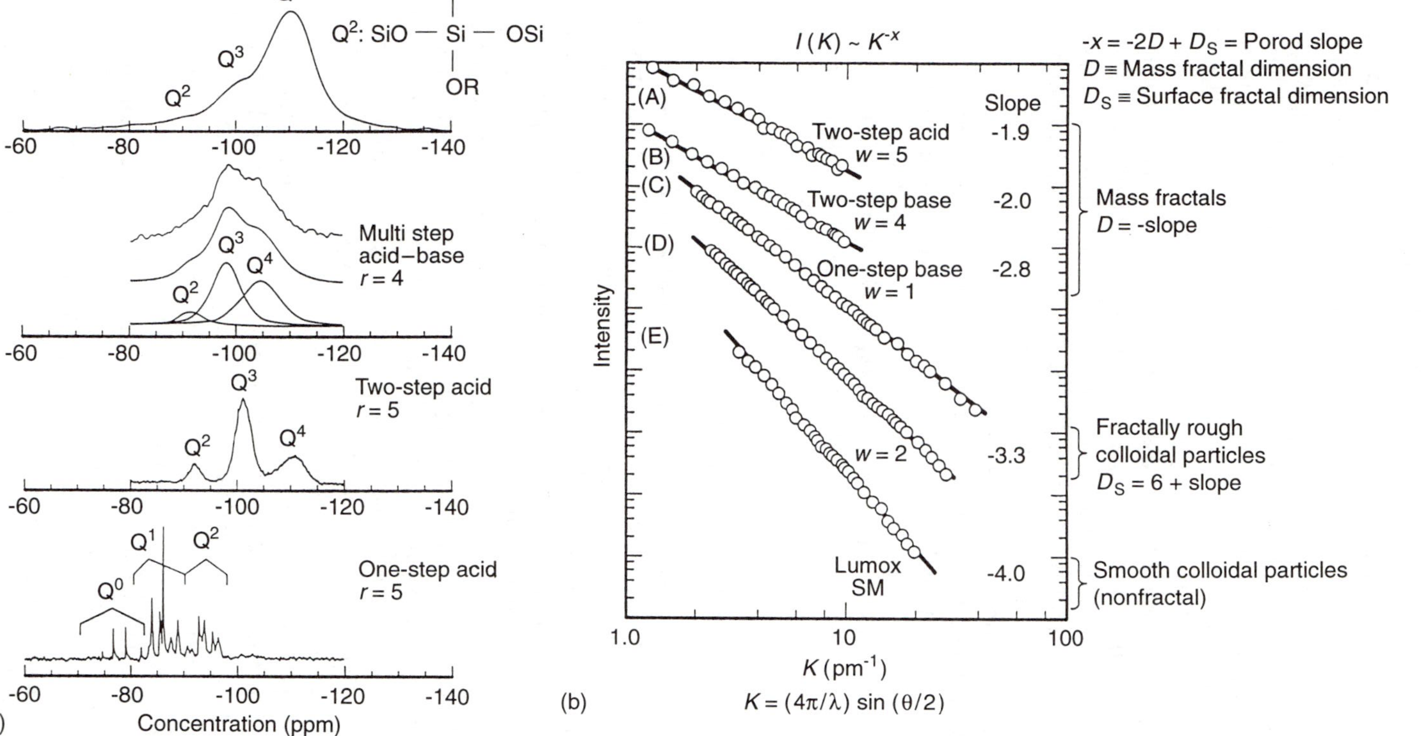

Figure 6.1 (a) Comparison of ^{29}Si NMR spectra of silicate sols prepared from tetraethoxysilane (TEOS) to the spectrum of a commercial silicate sol (Ludox HS40): (a) acid-catalyzed TEOS sol ($H_2O/Si = 2$) after 3 h at room temperature; (b) two-step acid-catalyzed TEOS sol ($H_2O/Si = 5$; (c) multicomponent silicate sol ($H_2O/Si = 5$) aged for 2 weeks at 50°C and ~pH3; and (d) Ludox HS40. (b) Porod plots (log scattered intensity versus log scattering wave vector [$K = (4\pi/\lambda)\sin(\theta/2)$]) obtained by SAXS for a variety of silicate sols: (a) two-step acid-catalyzed TEOS; (b) two-step acid–base catalyzed TEOS; (c) one-step base-catalyzed TEOS ($H_2O/Si = 1$); (d) one-step base-catalyzed TEOS ($H_2O/Si = 2$); and (e) commercial particulate sol (Ludox SM).

generally believed [42] that the prevalence of fractal silicates over a wide range of processing conditions is a consequence of kinetically limited growth mechanisms such as reaction-limited cluster–cluster aggregation (RLCA) [43]. In section 6.7 we show that the fractal properties of sol species may be exploited to tailor the porous microstructure of the deposited films. Structural studies of nonsilicon-based sols are far less numerous. However, Livage and coworkers [32] have shown that by varying the hydrolysis ratio r and the chelation ratio x, e.g. acac/M, it is possible to arrive at a variety of structures ranging from crystalline precipitates (large r, small x) to oxoalkoxide clusters (small r, large x). Schwartz and coworkers have investigated the effects of such precursor variations on thin film processing behavior, such as oligomer aggregation phenomena during film formation, film densification tendencies and film crystallization behavior [44, 45].

Several strategies exist for the preparation of multicomponent precursors: sequential hydrolysis of single component alkoxides, synthesis of mixed metal alkoxides and synthesis of mixed metal oxoalkoxides. Studies to date have focused on control of stoichiometry and homogeneity of the resultant films. Little information exists concerning the structure of hydrolyzed sol species and the structural variations possible.

6.3 DEPOSITION BY DIP-COATING

In dip-coating, a substrate is usually withdrawn vertically from the coating bath at constant speed U_0 (Fig. 6.2) [46]. The moving substrate entrains the liquid in a fluid mechanical boundary layer that divides in two layers above the liquid bath surface, returning the outer layer to the bath (see stagnation point S in Fig. 6.2) [47]. Since the solvent is evaporating and draining, the liquid film has an approximately wedge-like shape that terminates in a well-defined drying line ($x = 0$, in Fig. 6.2a). When the upward-moving flux becomes balanced by that due to evaporation and draining, the process remains spatially steady with respect to the liquid bath surface. In other words, both the shape and position of the depositing film profile remain steady. This situation allows us to use *in situ* optical techniques for characterization, as discussed below.

The hydrodynamic factors in dip-coating (pure liquids, ignoring evaporation) were first calculated correctly by Landau and Levich [48] and recently generalized by Wilson [49]. In an excellent review of this topic, Scriven [47] states that the thickness of the deposited film is related to the position of the streamline dividing the upward- and downward-moving layers. A competition between as many as six forces in the film deposition region governs the film thickness and position of the streamline [47]: (1) viscous drag upward on the liquid by the moving substrate; (2) force of gravity; (3) resultant force of surface tension in the concavely shaped meniscus; (4) inertial force of the boundary layer liquid arriving at the deposition region; (5) surface

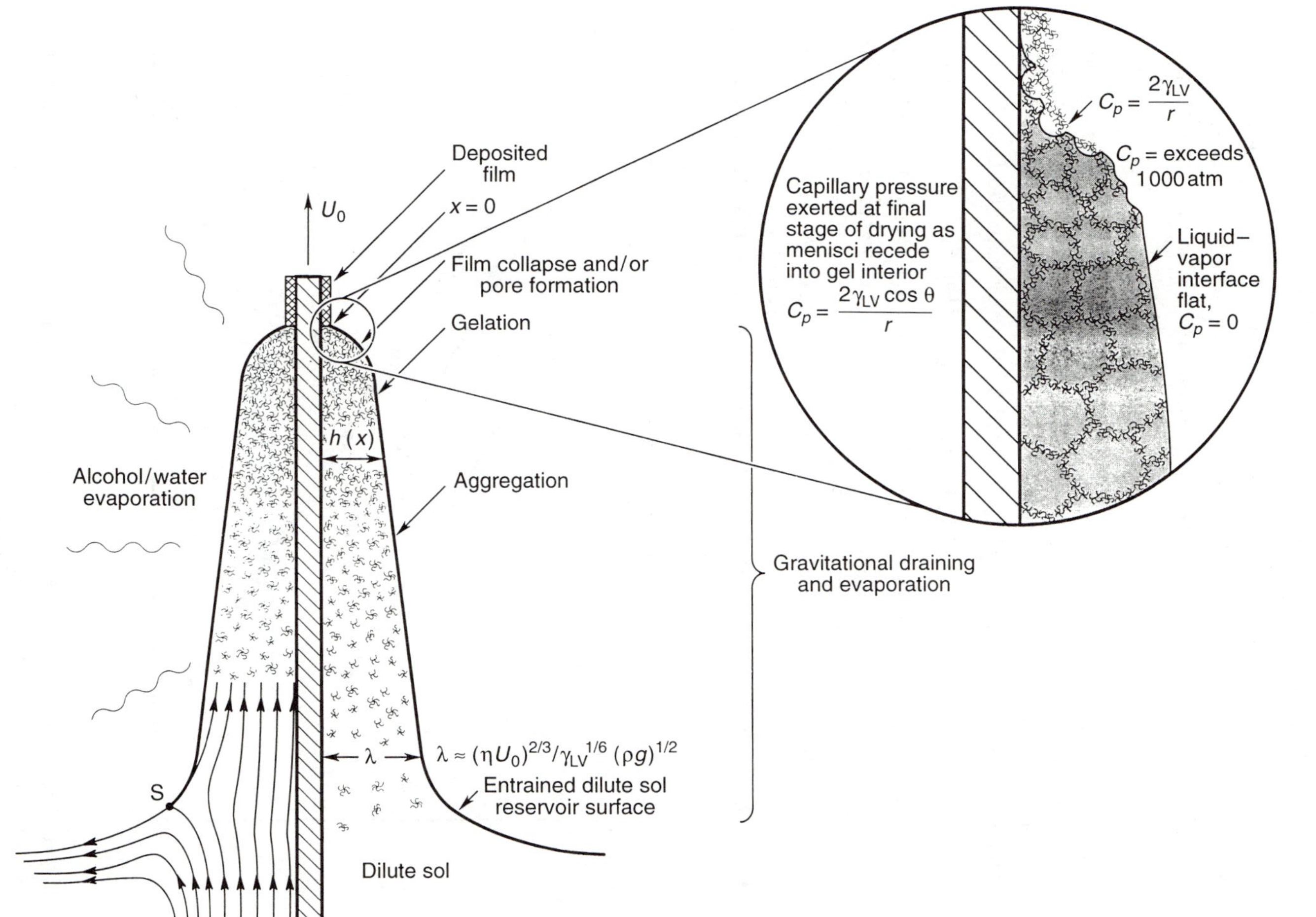

Figure 6.2 The sol–gel dip-coating process controls the pore size of the resultant film through variations in the size, structure and composition of the sol species, through the rates of condensation and evaporation and through capillary pressure.

tension gradient; and (6) the disjoining (or conjoining) pressure (important for films less than 1 μm thick).

When the liquid viscosity η and substrate speed are high enough to lower the curvature of the gravitational meniscus, the deposited film thickness h is that which balances the viscous drag, proportional to ηU_0, and gravity force ρgh:

$$h = c_1 \eta U_0/\rho g)^{1/2} \qquad (6.6)$$

where the constant c_1 is about 0.8 for Newtonian liquids. When the substrate speed and viscosity are low (often the case for sol–gel film deposition), this balance is modulated by the ratio of viscous drag to liquid–vapor surface tension (γ_{LV}) according to the relationship derived by Landau and Levich [48]:

$$h = 0.94(\eta U_0)^{2/3}/\gamma_{LV}^{1/6}(\rho g)^{1/2} \qquad (6.7)$$

Figure 6.3 plots the excess optical thickness $h(n-1)$ versus U_0 for films prepared from a variety of silicate sols in which the precursor structures ranged from rather weakly branched polymers characterized by a mass fractal dimension to highly condensed particles [50]. The slopes are in the range 0.61–0.67, in keeping with equation 6.7. This reasonable correspondence between the thicknesses of the deposited films and a theory developed for

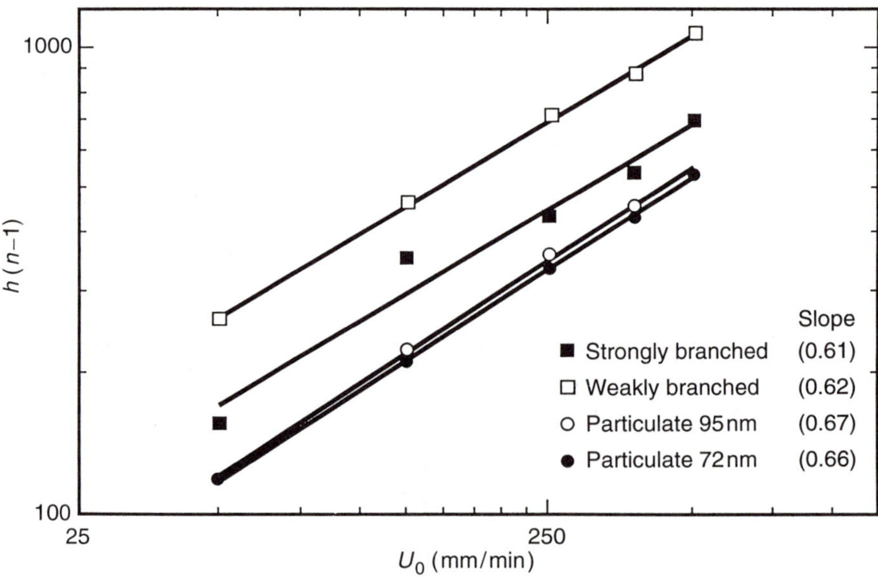

Figure 6.3 Excess optical thickness $h(n-1)$ versus dipping speed U_0. Formulations were a variety of silicate sols that ranged from rather weakly branched fractal polymers to highly condensed particles [50]. The slopes are consistent with the Landau–Levich [48] theory.

gravitational draining of *pure fluids* suggests that the entrainment of the inorganic species has little effect on the hydrodynamics of dip-coating, at least in the early stages of deposition where the entrained sol is quite dilute. Thus some insight into sol–gel film deposition is gained by closer examination of the details of gravitational draining (and evaporation) of pure and binary fluid mixtures often used in sol–gel film deposition.

6.3.1 Film thickness profiles during dip-coating – pure solvents

Previous theories of gravitational draining of pure fluids have not taken into account simultaneous evaporation. Although the thickness of the liquid entrained at the bath surface is not sensitive to evaporation, the film is progressively thinned by evaporation as it is transported by the substrate away from the coating bath. For depositing sols, thinning by evaporation causes a corresponding increase in sol concentration, hence an understanding of simultaneous draining and evaporation is essential to the underlying physics of sol–gel film deposition.

In order to address this problem, Hurd and Brinker [51] developed an imaging ellipsometer that allows acquisition of spatially resolved thickness and refractive index data over the entire area of the depositing film (illustrated schematically in Fig. 6.4). A thickness profile of an ethanol film obtained by imaging ellipsometry is shown in Fig. 6.5 [52]. Instead of the wedge expected for a constant evaporation rate, the film profile is distinctly blunt near the drying line ($x = 0$ in Fig. 6.2 and 6.5a), which indicates more rapid thinning and hence a greater evaporation rate there.

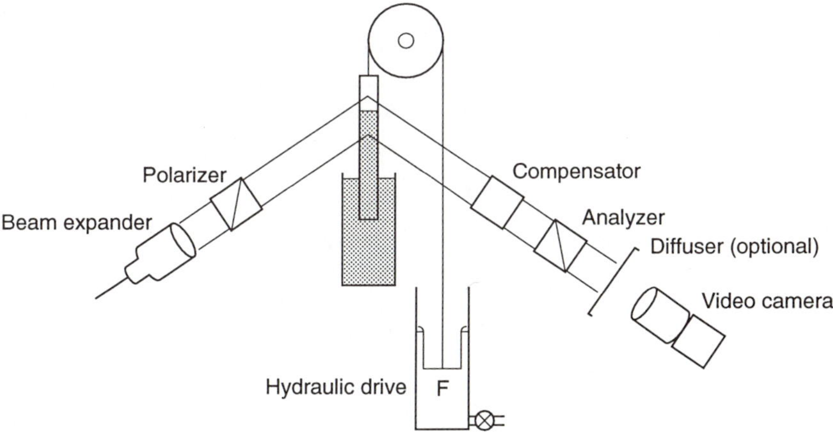

Figure 6.4 Schematic illustration of the imaging ellipsometer developed by Hurd and Brinker [51].

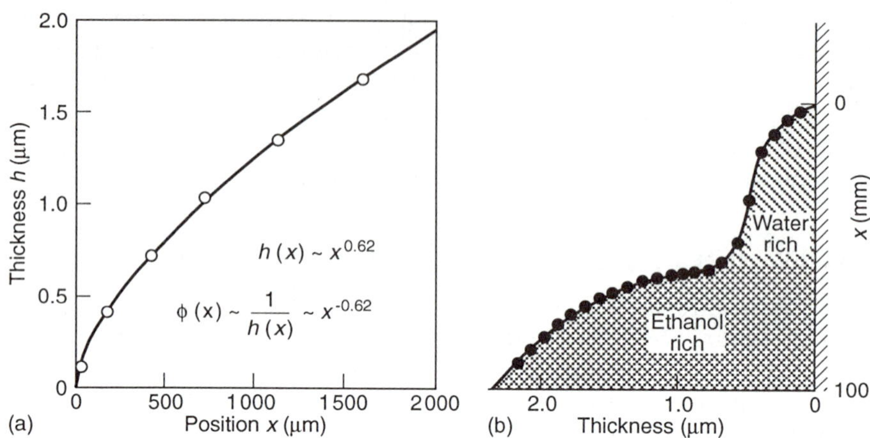

Figure 6.5 (a) Thickness profile of ethanol film from imaging ellipsometry. The parabolic approach to the drying line ($x = 0$) is a consequence of accelerated evaporation, a response by the liquid to cope with a sharp solvent vapor gradient at the terminus of the liquid film. (b) The thickness profile of mixtures often reflects the relative volatilities of the components. The surface tension of the component that survives to the drying line dominates solvent quality and the subsequent capillary collapse. Since the components diffuse freely to the surface, each produces a parabolic step in the profile.

This position-sensitive evaporation rate is a consequence of the film geometry: the blade-like shape of the depositing film in the vicinity of the drying line enhances the rate of diffusion of vapor away from the film surface [52] (at large x, the vapor concentration gradients are normal to the surface, whereas near the drying line $x = 0$, additional gradients are established parallel to the surface). Near any sharp boundaries, the evaporation rate $E(x)$ diverges, but the vaporized mass must remain integrable. For the knife-blade geometry (infinite sheet), E varies with x as follows [52]:

$$E(x) = -D_V a_1 x^{-1/2} \qquad (6.8)$$

where D_V is the diffusion coefficient of the vapor (~ 0.1 cm^2/s) and a_1 is a constant. Since the change in thickness dh is attributable to the local evaporation rate times the area, $E(x)dx$, the divergence in the evaporation rate as $x \to 0$ accounts for the blunt profile shown in Fig. 6.5a, where the data were fit to the form:

$$h(x) \sim \int E(x)dx \sim x^v \qquad (6.9)$$

giving $v = 0.50 \pm 0.01$, as expected from Eq. 6.8.

The singularity strength (exponent) in equation 6.9 is sensitive to the geometry of the film. For coating a fiber, we would expect a logarithmic singularity [53]. As a coated cylinder decreases in radius, the profile should pass smoothly from $x^{1/2}$ toward $\ln x$. Although some experimental evidence exists for this behavior by extrapolating to small cylinder diameters [53] it is a difficult proposition to prove experimentally. Surface tension makes it difficult to coat a fiber fast enough for the drying line to be well separated from the gravitational meniscus at the reservoir surface.

6.3.2 Film fluid profiles during dip-coating – binary solvents

Due to the alkoxide chemistry normally used in sol–gel synthesis, the most common coating sols are composed of two or more miscible liquids, e.g. ethanol–water. Differences in the evaporation rates and surface tensions of the liquid components of the sol alter the composition of the fluid and the shape of the fluid profile in the vicinity of the drying line and they create convective flows within the depositing films. In some cases rib-like instabilities are also observed in a region near the liquid bath surface [53]. Preferential evaporation is important in that it alters both the chemistry and physics of film deposition. For example, preferential evaporation of ethanol from an ethanol–water mixture enriches the fluid composition in water. Not only can this affect the associated chemistry, according to equations 6.2 to 6.4, but it can greatly alter the surface tension of the pore fluid that dictates the magnitude of the capillary tension exerted on the gel during drying and the extent of flows driven by surface tension gradients. Preferential evaporation and its effect on composition and shape of the fluid profile has been studied [54] *in situ* using imaging ellipsometry and fluorescence imaging techniques. Ellipsometric images of alcohol–water films acquired during dip-coating [54] shows two roughly parabolic features (Fig. 6.5b) that correspond to successive drying of the alcohol- and water-rich regions, according to the nonconstant evaporation model (equation 6.9). This suggests that each component has an independent evaporation singularity. If the water-rich phase is denoted as phase 1 and the ethanol-rich phase as phase 2, then the independent profiles are additive [54]:

$$h_1 = a_1 x^{1/2} \qquad\qquad x > 0 \qquad\qquad (6.10)$$

$$h_2 = a_2 (x - x_2)^{1/2} \qquad x > x_2 \qquad\qquad (6.11)$$

$$h_2 = 0 \qquad\qquad\qquad x < x_2 \qquad\qquad (6.12)$$

where $h = h_1 + h_2$ is the total thickness and x_2 is the position of the 'false' drying line created by the substantial depletion of ethanol.

The 'foot' feature in Fig. 6.5b typical of binary solvents is not due to differential volatility alone. Since each component has a different surface tension γ, surface tension gradients are established [53]:

$$d\gamma/dx = (\gamma_1 - \gamma_2)d\phi_1/dx \qquad (x > x_2)$$
$$= 0 \qquad (0 < x < x_2) \tag{6.13}$$

The surface tension is assumed to follow a simple mixing law, $\gamma = \phi_1\gamma_1 + \phi_2\gamma_2$ where ϕ_i is the volume fraction of component i. For ethanol–water mixtures the surface tension does not obey a simple linear mixing law. The surface tension can be approximated by $1/\gamma^{2.7} \approx (\phi_H/\gamma_H^{2.7}) + (\phi_E/\gamma_E^{2.7})$ where the subscripts H and E refer to water and ethanol, respectively. Since at the liquid–vapor boundary, the viscous shear force must balance the force imposed by surface tension gradients, $\eta\,du/dz = d\gamma/dx$ $(z = h)$, liquid flows into the water-rich foot with velocity u:

$$u = 1/\eta\,[d\gamma/dx]z - U_0 \tag{6.14}$$

the so-called Marangoni effect. The foot slowly grows until this flux is balanced by that of evaporation from the expanding free surface.

The surface tension gradient driven flow of liquid through the thin neck created by the preferential evaporation of alcohol can produce quite high shear rates during dip-coating. A striking example is that of toluene and methanol [53]. The surface tension gradient driven flows are strong enough to greatly distort the double parabolic profile, creating a 'pileup' of toluene near the drying line. A crude estimate of the surface tension gradient, $\Delta\gamma/\Delta x \approx (10\,\text{dyne/cm})/0.1\,\text{cm}$ $(100\,\mu\text{N}/0.1\,\text{cm})$, leads to a shear rate $du/dz \approx 10^4\,\text{s}^{-1}$ in the thin region, from equation 6.14 and assuming $\eta = 0.01$ P. As we discuss in section 6.4.3, these shear fields may be sufficiently strong to align or order the entrained inorganic species.

For binary fluid mixtures, fluorescence imaging has been used to determine *in situ* the variation in fluid composition with position x in the depositing film. In this technique, low concentrations of organic optical molecules (10^{-3} to 10^{-4} M) are added to the sol. Spatially resolved fluorescence spectra are acquired during dip-coating from the entrained molecules, which serve as molecular sensors of the local coating environment. Using pyranine, Nishida and coworkers [54] determined the composition of binary alcohol–water mixtures as a function of distance from the drying line. Their results indicated that, for ethanol–water compositions richer in water than 12 vol%, the composition at the drying line is 100% water. Even for the composition corresponding to the boiling azeotrope (5 vol% H_2O) there was substantial enrichment (87 vol% water at the drying line). Since water is a reagent in sol–gel chemistry and strongly influences the surface tension, these results indicate that preferential evaporation will impact the extent of condensation and the magnitude of the capillary pressure as well as induce surface tension gradient driven flows. The consequences of these phenomena on film formation are addressed in section 6.7.

6.4 DEPOSITION BY SPIN-COATING

Spin-coating differs from dip-coating in that the thin films are deposited by centrifugal draining and evaporation. Bornside *et al.* [55] divide spin coating into four stages: deposition, spin-up, spin-off and evaporation, although for sol–gel coating, evaporation normally overlaps the other stages. An excess of liquid is dispensed on the surface during the deposition stage. In the spin-up stage, the liquid flows radially outward, driven by centrifugal force. In the spin-off stage, excess liquid flows to the perimeter and leaves as droplets. As the film thins, the rate of removal of excess liquid by spin-off slows down, because the thinner the film, the greater resistance to flow, and because the concentration of the nonvolatile components increases, raising the viscosity. In the final stage, evaporation takes over as the primary mechanism of thinning. According to Scriven [47], an advantage of spin-coating is that a film of liquid tends to become uniform in thickness during spin-off and it tends to remain uniform provided that the viscosity is not shear-dependent and does not vary over the substrate. This tendency is due to the balance between the two main forces: centrifugal force, which drives flow radially outward, and viscous force (friction), which acts radially inward. The thickness of an initially uniform film during spin-off is described as follows:

$$h(t) = h_0/(1 + 4\rho\omega^2 h_0^2 t/3\eta)^{1/2} \tag{6.15}$$

where h_0 is the initial thickness, t is time, ρ is the density and ω is the angular velocity. Even films that are not initially uniform tend monotonically toward uniformity, sooner or later following equation 6.15. By assuming that the spin-off and evaporation stages are sequential, Bornside showed that the final thickness h_f scales as

$$h_f \sim \omega^{2/3} \tag{6.16}$$

Equation 6.15 pertains to Newtonian liquids that do not exhibit a shear rate dependence of the viscosity during the spin-off stage. If the liquid is shear thinning (often the case for aggregating sols), the lower shear rate experienced near the center of the substrate causes the viscosity to be higher there and the film to be thicker. This problem might be avoided by metering the liquid from a radially moving arm during the deposition/spin-up stage.

6.5 EFFECTS OF ENTRAINED CONDENSED PHASES

The previous subsections have largely ignored the effects of the entrained inorganic species, namely polymers or particles. These species are initially concentrated by evaporation of solvent as they are transported from the coating bath toward the drying line during dipping or during the four stages of spin-coating. They are further concentrated (compacted) at the final stage of the deposition process by the capillary pressure created by liquid–vapor menisci as they recede into the drying film (Fig. 6.2 inset). During dip-coating, the shape of the entrained fluid profile is altered very little by the presence of the condensed phase [52]. Thus by consideration of the conditions that establish the steady-state fluid profiles, we are able to quantify the changing environment of the inorganic species within the thinning film.

Above the stagnation point S, all fluid elements are moving upward, so all the entrained inorganic species that survive past the stagnation point are incorporated in the final deposited film. Steady-state conditions in this region require conservation of nonvolatile mass, thus the solids mass in any horizontal slice of the thinning film must be constant [52]:

$$h(x)\phi_S(x) = \text{constant} \tag{6.17}$$

where ϕ_S is the volume fraction solids. From equation 6.17 we see that ϕ_S varies inversely with h (Fig. 6.6a). Since for a planar substrate, we expect a parabolic thickness profile, ϕ_S should vary as $1/h \approx x^{-1/2}$ in the thinning film. When coating a fiber, we expect $\phi(x) \sim (\ln x)^{-1}$.

The rapid concentration of the entrained inorganic species by evaporation is more evident from consideration of the mean particle (polymer) separation distance $\langle r \rangle$ which varies as the inverse cube root of ϕ_S, $\langle r \rangle \sim x^{1/6}$. As shown in Fig. 6.6b [53], this is a precipitous function: half the distance between particle (polymer) neighbors is traveled in the last 2% of the deposition process (~ 0.1 s). This evaporation-driven transport may be viewed like that induced by centrifugation or electrophoresis: the centrifugal acceleration needed to cause an equivalent rate of crowding is as much as 10^6 g's! Since

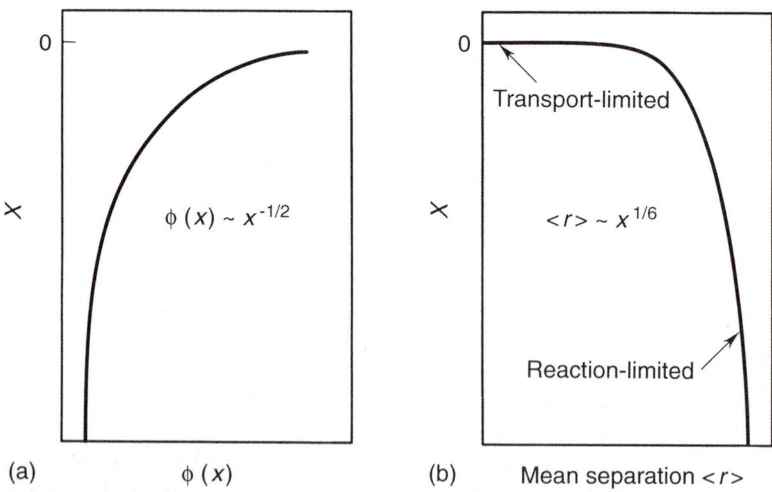

Figure 6.6 Effects of entrained condensed phases: (a) profile–concentration relationship and (b) evaporation-induced crowding.

rheological properties of suspensions are often concentration dependent [56], the increasing concentration of the inorganic species is expected to alter the rheology from Newtonian (dilute conditions) to shear thinning (aggregated systems) or thixotropic (ordered systems). The viscosity also exhibits concentration dependence, at least for high concentrations. Bornside *et al.* [55], in their studies of spin-coating, assumed the following relationship:

$$h = h_0(1 - c_A)^4 + h_s \qquad (6.18)$$

where h is the viscosity of the sol, h_s is the viscosity of the solvent, h_0 is the viscosity of the polymer and c_A is the mass fraction of solvent. From this relationship, we see that the viscosity will change abruptly in the vicinity of the drying line during dip-coating or during the evaporation stage of spin-coating.

The short time of the deposition process (~ 10 s) and, more importantly, the extremely short time in which the inorganic species are in close proximity ($\ll 1$ s) establish important timescales for the dip- and spin-coating processes and distinguish sol–gel thin film formation from bulk (monolithic) gel formation. For bulk systems, the gelling sol is normally maintained at a constant concentration; the gelation, aging and drying stages occur sequentially. Often the time required to prepare a monolithic bulk gel is on the order of days to months. During dip-coating (and spin-coating) the drying stage completely overlaps the gelation and aging stages. The depositing sol (which initially is often less concentrated than sols intended for bulk gels) remains rather dilute until the final stage of the deposition process, when it is rapidly concentrated. Thus, the condensation reactions (equations 6.3 and 6.4) have little time to create a cross-linked gel.

Using drying and reaction theory, Cairncross *et al.* [57] showed that the time scale of drying can change when gelation occurs: (1) slow drying (as in bulk gels) results in gelation before solvent removal, (2) rapid drying in thick films causes a gelled film to form on the surface of the coating, (3) rapid drying in thin films causes gelation to occur after drying is completed, or (4) rapid drying in thin films can create dry ungelled films. Figure 6.7 demonstrates this effect for drying silicate films [57] using the kinetic scheme of Kay and Assink [58] to predict the reaction rates. However, the reaction rates have been shown experimentally to be much more concentration dependent than implied by Kay and Assink's theory. During the rapid concentration of reactants, the condensation reaction rate rises sharply; so the characteristic condensation rate in a rapidly drying coating is difficult to determine.

We anticipate several consequences of the short timescale of the film deposition process. (1) There is little time available for reacting species to 'find' low energy configurations. Thus (for reactive systems) the dominant aggregation process responsible for network formation may change from reaction-limited (near the reservoir surface) to transport-limited near the drying line. The reduced fractal dimensionality of transport-limited aggregates should result in more porous networks. (2) For sols composed of repulsive

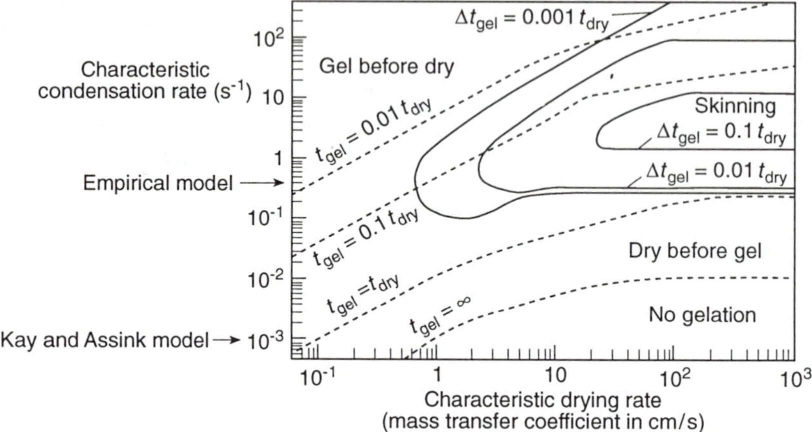

Figure 6.7 Predictions of drying and reaction from Cairncross *et al.* [57]. Varying the reaction rate and drying rate produces several types of drying phenomena. Arrows indicate typical reaction rates using the kinetic scheme of Kay and Assink and an empirical reaction scheme. $-\Delta t_{gel}$ refers to the onset of skin formation and t_{gel} refers to time for complete gelation.

particles, there is little time available for the particles to order as they are concentrated in the thinning film. (3) There is little time available for condensation reactions to occur. Thus gelation may actually occur by a physical process, through the concentration dependence of the viscosity (e.g. equation 6.18), rather than a chemical process. (In some systems this is evident by the fact that the deposited film is quickly resolubilized when immersed in solvent.) (4) Since the gels are most likely weakly condensed compared to bulk gels, they are more easily compacted, first by evaporation and then by the capillary pressure exerted at the final stage of the deposition (drying) process (Fig. 6.2 inset). Furthermore, the maximum capillary pressure is likely to be higher due to the collapse of the network (and hence reduction in pore size) that precedes the final stage of drying. Of course, greater capillary pressure also promotes greater compaction of films compared to bulk gels.

6.6 DRYING OF FILMS

As stated in the previous sections, drying accompanies both dip- and spin-coating processes and largely establishes the shape of the fluid film profile. The rising solids concentration that results from drying and the development of physical or chemical linkages between polymers or particles often leads to the formation of an elastic or viscoelastic gel-like state. During this sol–gel transition, the rheological response of the liquid changes from Newtonian to shear thinning (aggregated systems) or thixotropic (ordered systems) and then to viscoelastic. Eventually gelation extends throughout the film and the material no longer yields, i.e. the film behaves as an elastic solid.

6.6.1 Solidification and capillary pressure

It is at this final stage of the deposition process that the capillary pressure C_p, created by tiny liquid–vapor menisci as they recede into the gel, is maximized (Fig. 6.2 inset). The curvature of the menisci causes the liquid to be in tension (i.e. subambient pressure in the pore liquid) and the network to be in compression. Generally the magnitude of the capillary pressure is estimated by the Young–Laplace equation:

$$C_p = 2\gamma_{LV}\cos(\theta)/r_p \tag{6.19}$$

Here θ is the wetting angle, r_p is the pore size and γ_{LV} is the liquid–vapor surface tension. For wetting pore fluids ($\cos\theta \to 1$), and $C_p = 2\gamma_{LV}/r_p$ represents the maximum capillary pressure obtainable in a pore of size r_p. Because r_p can be of molecular dimensions, the magnitude of $C_{p\,max}$ can be enormous. For ethanol $\gamma_{LV} = 22.75$ dyne/cm ($22.75\,\mu N/mm$) at $20°C$, r_p is estimated to be about 1.3 nm and $C_{p\,max} = 350$ bar (35 MPa). For water $\gamma_{LV} = 72.8$ dyne/cm ($72.8\,\mu N/mm$) at $20°C$, r_p is estimated to be $1.1-1.55$ nm, so $C_{p\,max}$ is $940-1320$ bar ($94-132$ MPa)! These large tensile pressures drive the solvent into metastable states analogous to superheating. Burgess and Everett suggest that the liquid does not boil because nucleation cannot occur in such small pores [59]. Because C_p can approach or possibly exceed 1000 bar (100 MPa), the capillary pressure represents a strong driving force to densify a drying film.

It is the balance between the tensile capillary pressure in the liquid and the compressive elastic stress in the solid which establishes the final density and pore size of the film. The elastic modulus of the network represents its ability to resist collapse. Initially, as the solvent evaporates, the elastic modulus is low, and the gel is able to shrink. As the volume fraction of solids increases and the reactions produce more cross-linking, the modulus E rises. For both wet and dry silica gels, the following relationship between modulus and porosity is observed [60]:

$$E \propto \phi^{3.8} \tag{6.20}$$

As the spacing between polymers decreases, the effective pore size shrinks, and the maximum possible capillary pressure (capillary pressure at which menisci would recede into pores) in the pores rises as approximately $1/\langle r\rangle \sim \phi^{1/3}$. Thus for silica films, the elastic stresses eventually become large enough that the liquid–vapor menisci invade into the pore space.

In a simplified analysis of drying, stress development and pore shrinkage in a thin sol–gel film, Chen and Schunk [61] (see also [62]) have computed predictions of the dynamic evolutions of solid phase stress, liquid phase pressure, and mean pore size for a model sol–gel system. Their representative results (Fig. 6.8) show how the average solid phase stress and the maximum attainable liquid phase pressure stress (defined as porosity times the maximum capillary pressure based on the pore radius) are scaled by the initial network modulus (E_0). The maximum liquid phase pressure rises as the pore radius decreases until the liquid phase and solid phase stresses are equal. We call this point the critical point at which shrinkage stops, thereby establishing the pore size and density. As shown in Fig. 6.8, for sol–gel systems with low enough modulus or large enough capillary pressure, it is possible to collapse the pores completely because the driving force for network shrinkage (the maximum effective pressure force) can always be greater than the resistance to deformation (solid phase stress). Of course, as pores collapse to molecular dimensions, the concept of capillary pressure breaks down, requiring additional physics.

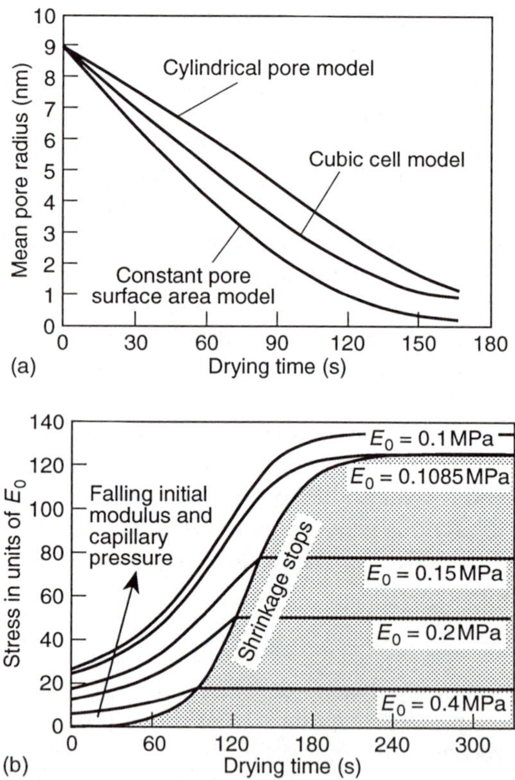

Figure 6.8 Predictions of pore evolution in a shrinking coating: (a) mean pore radius and (b) maximum liquid phase pressure versus drying time. The stress in the solid network rises as the gel shrinks, and the maximum capillary pressure in the liquid rises until it is equal to the solid stress; then menisci recede into the pores and shrinkage stops.

Beyond this so-called critical point, any further solvent loss creates porosity within the film. For systems in which E exhibits a much weaker dependence on ϕ, e.g. as a consequence of organic modification or complexation of the metal centers with multidentate, nonhydrolyzable ligands, we might expect the network to be completely collapsed by the rising capillary pressure. However, the rising viscosity accompanying solvent loss combined with the short timescale of the deposition process may represent a kinetic limitation to achieving a nonporous state. Also, thermolysis of organic ligands after drying can normally create porosity, even though the as-deposited film could be considered nonporous.

6.6.2 Stages of drying

Scherer [63] divides the drying of gels into two stages: a constant rate period (CRP) and a falling rate period (based on [64]). During the constant rate period, mass transfer is limited by convection away from the gel surface, whereas during the falling rate period, mass transfer is limited by transport within the pores of the gel. Extending these concepts to dip-coating, the CRP should exist throughout most of the deposition process because the pores do not form and the liquid–vapor interface remains located at the exterior surface of the thinning film, except at the final stage of drying (Fig. 6.2 inset). A constant evaporation rate implies a wedge-shaped film profile, which is not observed for pure fluids, nor is it observed for inorganic sols. Even for a titanate sol prepared in ethanol, the thickness varies with distance from the drying line as $h(x) \sim x^{0.62}$, so the evaporation rate increases near the drying line. Thus, the film profile, and hence the concentration profile, are largely established by the dependence of the evaporation rate on the geometry of the depositing film.

For sols containing fluid mixtures of differing volatilities, the fluid composition changes with distance x, contributing to further changes in the evaporation rate. The rate of drying in the CRP is usually calculated from an external mass transfer correlation such as [65]

$$\text{mass flux per unit area} = k_{\text{mt}}(\rho_{\text{s}} - \rho_{\infty}) \tag{6.21}$$

where ρ_{s} is the theoretical density of solvent in equilibrium with the surface of the coating, ρ_{∞} is the theoretical density of solvent vapor far removed from the coating surface and k_{mt} is the mass transfer coefficient (m/s). For accurate modeling of sol–gel dip-coating, k_{mt} must be position dependent, or replaced by large-scale two-phase calculations of the fluid mechanics and transport [66].

The critical point, at which liquid–vapor menisci start receding into the pores, corresponds to the beginning of the falling rate period. Depending on the distribution of the liquid in the pores, e.g. funicular or pendular, the drying rate is limited by liquid flow (funicular state – pore filling) or gas phase diffusion (pendular state – discontinuous) [67]. For compliant molecular networks that are collapsed prior to the critical point, drying occurs by Fickian diffusion if the temperature is above the glass transition temperature of the mixture [68]. The onset of a falling rate period near the drying line may account for the differences in the exponents that describe the shape of the pure fluid and the sol film profiles.

6.6.3 Drying stress and cracking

As a film dries, it shrinks in volume. Because the film is normally adhered to the substrate, the volume reduction is accommodated completely by a reduction in film thickness. Croll [69] estimated the stress σ as

$$\sigma = [E/(1 - n)][(\phi_s - \phi_r)/3] \qquad (6.22)$$

where E is Young's modulus (Pa), n is Poisson's ratio, ϕ_s is the volume fraction of solvent at the solidification point, and ϕ_r is the volume fraction of residual solvent in the 'dry' film. For a polymer film, the solidification point was defined as the concentration at which the glass transition temperature equals the temperature of the coating. Thus, stress is proportional to Young's modulus and the difference between the volume fraction of solvent at the solidification point and in the dried coating. Scherer [63, 67] states that the stress in the film is nearly equal to the tension in the liquid ($\sigma \sim C_p$). Despite large stresses induced by capillary pressure, cracking of films often does not occur if the film thickness is below a certain critical thickness $h_c = 0.5$–$1\,\mu m$ [67]. For films that adhere well to the substrate, the critical thickness for crack propagation or the growth of pinholes is given by [70, 71]

$$h_c = (K_{Ic}/\sigma W)^2 \qquad (6.23)$$

where K_{Ic} is the critical stress intensity and W is a function that depends on the ratio of the elastic modulus of the film and substrate (for gel films $W = 1$). In films thinner than h_c, the energy required to extend the crack is greater than the energy relieved by crack growth, so cracking is not observed [67]. When the film thickness exceeds h_c, cracking occurs, and the crack patterns observed experimentally are qualitatively consistent with fractal patterns predicted by computer simulation [72]. Atkinson and Guppy [73] observed that the crack spacing increased with film thickness and attributed this behavior to a mechanism in which partial delamination accompanies crack propagation. Such delamination was observed directly by Garino [74] during the cracking of sol–gel silicate films.

Based on equations 6.22 and 6.23, strategies to avoid cracking include (1) increasing the fracture toughness (K_{Ic}) of the film, (2) reducing the modulus of the film, (3) reducing the volume fraction of solvent at the solidification point, (4) reducing the surface tension of solvents and (5) reducing the film thickness. In organic polymer films, plasticizers are often added to reduce the stiffness of the film and thus avoid cracking [65]. For sol–gel systems, analogous results are obtained by organic modification of alkoxide precursors [75], chelation by multidentate ligands such as β-diketonates [33], or a reduction in the extent of hydrolysis of alkoxide precursors [74].

It should be noted that for particulate films Garino [76] observed that the maximum film thickness obtainable without cracks decreased linearly with a reduction in particle size. Since for unaggregated particulate films, the pore size scales with the particle size, this may be due to an increase in the stress caused by the capillary pressure ($\sigma = P$) or an increase in the volume fraction of solvent at the solidification point, resulting from the manner that the electrostatic double-layer thickness (estimated by Debye–Hückel screening length) varies with particle size [67].

6.7 CONTROL OF FILM MICROSTRUCTURE

From the preceding discussion we expect that the final film structure should depend on the competition between such phenomena as evaporation and capillary pressure, which tend to compact the structure, and condensation reactions and aggregation processes, which tend to stiffen the structure, resisting its tendency to collapse. In this section we document how these factors, along with the structure of the entrained inorganic species, influence film microstructure. We characterize the film microstructure (surface area, percentage porosity and pore size distribution) *in situ* by analyzing N_2 adsorption–desorption isotherms acquired using a surface acoustic wave (SAW) technique developed by Frye and coworkers [13]. The film is deposited on a piezoelectric quartz substrate (Fig. 6.9) which acts as a feedback element

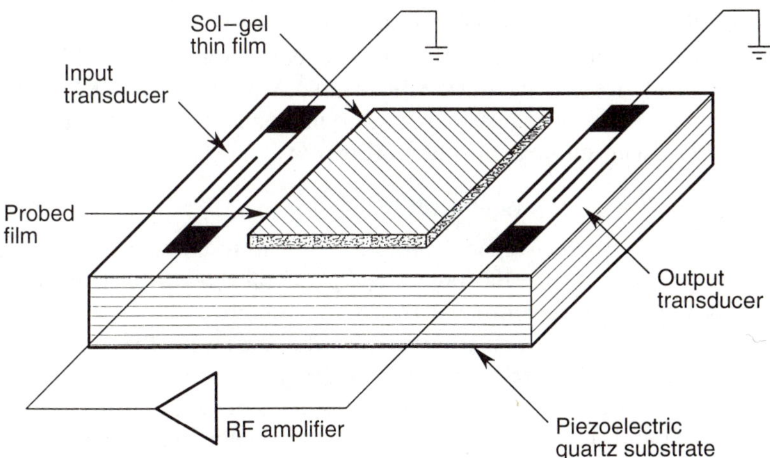

Figure 6.9 Acoustic wave sensor using a piezoelectric quartz substrate.

of an oscillator circuit. The substrate contains two sets of interdigital transducers; the application of an alternating field produces alternating strain, launching a surface acoustic wave. Changes in the mass of the film due to N_2 adsorption or desorption cause a change in the SAW velocity, which is sensed as a change in the resonant frequency of the oscillator circuit. The extreme sensitivity of the SAW device, $\sim 80\,pg/(cm^2\ film)$, permits accurate adsorption data to be obtained easily on a 1 cm^2 area of film less than 100 nm thick.

6.7.1 Influence of precursor structure and reactivity

The mass fractal dimension D relates an object's mass M to its radius [77]:

$$M \sim r^D \tag{6.24}$$

where for mass fractal objects, D is less than the dimension of space. In three-dimensional space, the surface fractal dimension D_s relates an object's area A to its size r [77]:

$$A \propto rD_s \tag{6.25}$$

where in three dimensions $2 < D_s < 3$. Since in three dimensions, $D < 3$, the density of a mass fractal object decreases with distance from its center of mass:

$$\rho \propto 1/r^{(3-D)} \tag{6.26}$$

Because density is inversely related to porosity, this relationship requires that, unlike Euclidian objects, fractal objects become more 'porous' as their size increases.

The possible structures of the inorganic precursors range from weakly branched polymers to highly condensed particles (Table 6.1) [77]. The packing efficiency of fractal objects (i.e. the volume fraction of solids, ϕ) depends on the fractal dimension, size and condensation rate. The fractal dimension and size dictate steric constraints. Mandelbrot [78] has shown that if two structures of radius R are placed independently in the same region of space, the mean number of intersections, $M_{1,2}$ is expressed as

$$M_{1,2} \propto R^{D_1 + D_2 - d} \tag{6.27}$$

where D_1 and D_2 are the respective fractal dimensions and d is the dimension of space ($d = 3$). Thus if each object has a fractal dimension less than 1.5, the probability of intersection decreases indefinitely as R increases. These structures are mutually transparent: during film formation, they should freely interpenetrate as they are forced into close proximity by the increasing concentration. Alternatively, if the fractal dimension of each object exceeds 1.5, the probability of intersection increases algebraically with R. The structures, though porous, are mutually opaque; when concentrated they are unable to interpenetrate, much like an assemblage of tumbleweeds.

Table 6.1 Microstructure of Sol–gel films tailored to applications

Sample aging time[a]	Refractive index	Porosity (%)[b]	Median pore radius (nm)	Surface area (m^2/g)	Applications
Unaged	1.45	0	<0.2	1.2–1.9	Dense protective, optical and electronic films
0–3 days					Microporous films for sensors and membranes
3 days	1.31	16	1.5	146	Mesoporous films for sensors, membranes, catalysts and optics
1 week	1.25	24	1.6	220	
2 weeks	1.21	33	1.9	263	
3 weeks[c]	1.18	52	3.0	245	

[a] Aging of dilute sol ($71\,SiO_2$–$18\,B_2O_3$–$7\,Al_2O_3$–$4\,BaO$) at 50 °C and pH 3 prior to film deposition.
[b] Determined from N_2 adsorption isotherm.
[c] The 3 week sample gelled; it was reliquefied at high shear rates and diluted with ethanol prior to film deposition.

These concepts of mutual opacity or transparency assume that every intersection results in irreversible 'sticking' – conditions chemically equivalent to an infinite condensation rate. In fact, the condensation rates of silicates are quite low (e.g. $10^{-4}\,1/mol\,s$) [79] and very dependent on $[H^+]$ or pH [38, 80]. Thus the probability of sticking at any point of intersection is $\ll 1$, causing film structures to be generally more compact than expected from equation 6.27. For example, deposition of rather weakly branched silicate sols with $D = 1.9$ (sample B in Fig. 6.1b) at pH ~ 2, where the condensation rate is minimized, produces dense films characterized by a type II N_2 adsorption–desorption isotherm (Fig. 6.10b) [81] and surface area equivalent to the geometric area of the film [13]. This indicates that there is no porosity accessible to molecules larger than 0.4 nm, the kinetic diameter of the N_2 molecule. Because the condensation rate is low under these conditions, we expect that, although $D > 1.5$, the precursors can interpenetrate as they are concentrated on the substrate surface by evaporation, leading to densely packed, but compliant, configurations with molecular-scale pores. During the final stage of drying, the capillary pressure is enormous due to the small pores (equation 6.19), causing further compaction of the compliant network.

The differences in timescales of bulk and thin film sol–gel processing are evident from the comparison of adsorption–desorption isotherms of the film and bulk xerogel (dried gel) prepared from identical silicate precursors (Fig. 6.10). Whereas the film is characterized by a type II isotherm and low surface

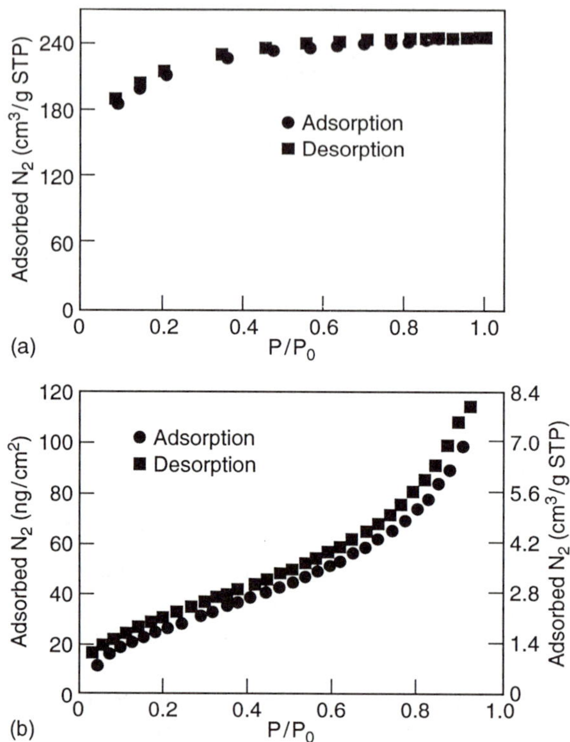

Figure 6.10 Bulk and thin film samples prepared from identical precursors have different microstructures: (a) microporous bulk silica xerogel and (b) nonporous silica film.

area (Fig. 6.10b), indicating no accessible porosity, the bulk xerogel isotherm (Fig. 6.10a) is of type I, indicative of a microporous material, and the N_2 BET (Brunauer–Emmett–Teller) surface area exceeds $800\,m^2/g$. The long duration of the gelation, aging and drying stages inherent in the bulk gel process allows the structure to stiffen at an early point of the drying stage through continued condensation reactions. The result is that the gel is compacted less by evaporation during the initial stage of drying, so the pores are larger and the capillary pressure smaller (equation 6.19) during the final stage of drying. Both increased stiffness and reduced capillary pressure cause the bulk sample to be much more porous than the film.

The interplay between precursor structure and condensation rate is well illustrated by a series of films prepared from more highly branched fractal clusters characterized by $D = 2.4$ and a higher $Q_4/(Q_2 + Q_3)$ ratio (sample C in Fig. 6.1b) [82]. Dip-coating at pH = 3.5, where the condensation rate is high, produces films exhibiting a reciprocal relationship between refractive index (proportional to ϕ) and polymer (cluster) size prior to film deposition (determined by quasi-elastic light scattering) [82]. Since the density of a fractal object decreases as $1/R^{(3-D)}$, this behavior appears consistent with an assemblage of mutually opaque clusters: high values of D and the condensation rate preclude interpenetration of large clusters, so the density (refractive index) of an assemblage of opaque clusters also decreases with R. Smaller clusters are both more dense and, according to equation 6.27, less opaque leading to denser films.

Corresponding SAXS data have been shown to be consistent with this fractal viewpoint [82]. Power law regions indicative of $D = 1.7$ and $D = 2.8$ are delineated by a crossover length comparable to the pore size, 1.9 nm. The conclusion is that the original polymeric clusters aggregate by a diffusion-limited process ($D = 1.7$) as they are rapidly concentrated at the final stage of the deposition step (Fig. 6.6a). The individual clusters are then partially collapsed by the capillary pressure created by the tiny menisci, causing D to increase on short length scales. The crossover occurs at a value of $1/K$ approximately equal to the pore radius determined from the desorption branch of the corresponding N_2 isotherm.

From Table 6.1 [46] we see that all of the important structural properties (percentage porosity, pore size, surface area and refractive index) of this series of films may be controlled by the aging (growth) step prior to film deposition. This allows the films to be tailored to specific applications, such as protective coatings, membranes, sensor surfaces or catalyst supports. Figure 6.11

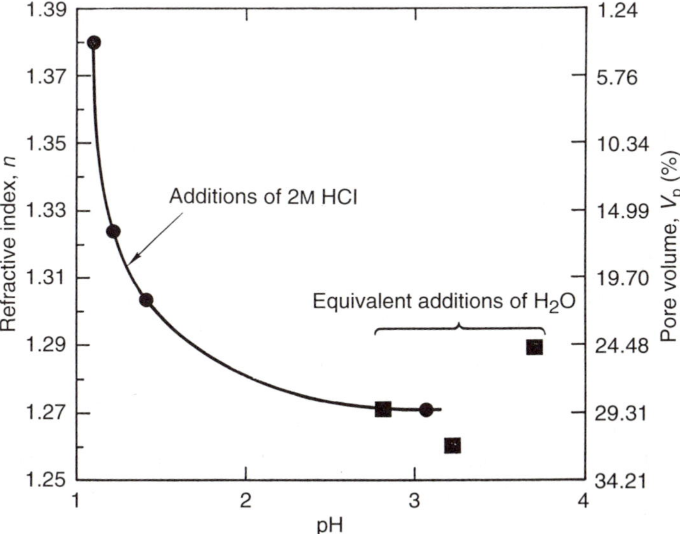

Figure 6.11 Refractive index and pore volume are very sensitive to pH as a result of the reduction in condensation rate with decreasing pH.

illustrates that a progressive reduction in the condensation rate during deposition (via a reduction in the pH of the sol immediately prior to deposition) causes a corresponding increase in the refractive index (increase in ϕ) of films prepared from precursors aged 2 weeks at pH $= 3$ prior to dip coating [50]. Although $D \gg 1.5$, the reduction in the condensation rate promotes interpenetration and retards stiffening, leading to denser films. This system demonstrates that microstructural tailoring may be accomplished by controlling the size and structure of the polymeric precursors or their rates of condensation (or both)!

6.7.2 Effects of capillary pressure and aging during deposition

In sol–gel processing of silicon alkoxides, the hydrolysis step has been carried out with H_2O/Si ratios exceeding 50. Since water is produced by condensation, a H_2O/Si ratio of 2 is theoretically sufficient to achieve complete hydrolysis and condensation of $Si(OR)_4$ to form SiO_2. Even if no condensation occurs, $H_2O/Si > 4$ results in 'excess' water. Thus whenever $Si(OR)_4$ is dissolved in alcohol and hydrolyzed with H_2O/Si ratios exceeding 4, there will always be excess water in the precursor sol.

As discussed in section 6.3, one consequence of excess water is water enrichment in the vicinity of the drying line due to preferential evaporation and surface tension gradient driven flows. This causes a corresponding increase in the capillary pressure (due to the greater surface tension of water) during the final stage of drying as liquid–vapor menisci recede into the film interior (Fig. 6.2a and equation 6.19). Assuming complete wetting ($\theta = 0$ in equation 6.19), the maximum increase in pressure would be a factor of ~ 3.

In order to evaluate the effect of water enrichment on the structure of the deposited film, we prepared silicate sols in which excess water was varied from about 0.5 to 10.5 vol% (assuming complete hydrolysis and condensation). The refractive index of the deposited films increased from 1.342 to 1.431 as the excess water was increased from 0.5 to 6 vol%, corresponding to a reduction in porosity from 22 to 7%. Further increases in the excess water caused a reduction in refractive index (increase in porosity). We believe that this behavior reflects the competition between capillary pressure and aging (stiffening) of the silicate skeleton, which are both enhanced by the increasing water concentrations. These results differ from those of Glaser and Pantano [83], who observed a monotonic increase in refractive index with water concentration for spin-coated silicate films. The difference between dip- and spin-coating is the evaporation rate and correspondingly the time available for aging. Spinning creates a strong forced convection in the vapor above the substrate [47], increasing the evaporation rate. Thus there is little time for aging to occur, and the structure of the film is dominated by the effects of capillarity.

6.7.3 Effect of substrate speed

An increase in the substrate speed U_0 results in an increase in film thickness according to equations 6.9 and 6.10. Since virtually all solvent evaporation occurs at the exterior surface of the entrained sol (i.e. at the sol–vapor interface), thicker films take longer to dry, increasing the aging time for reactive sols and,. as we shall see, the time for ordering of repulsive sols. A second effect of the substrate speed is the shear field induced within the depositing sol. When the effects of shear experienced by a particle overwhelm those due to diffusion, it is expected that mutually repulsive particles could align in close-packed planes oriented parallel to the substrate surface or that asymmetrically shaped particles might align with their long axes parallel to the shear field.

Nitrogen adsorption–desorption isotherms acquired for 55 nm particulate sols by SAW techniques show that the volume fraction of porosity decreases from 36 to 26% as the coating rate is increased from 12.7 to 45.7 cm/min. For packings of monosized spherical sols, this corresponds to a change from random dense packing (consistent with a liquid-like structure observed in previous SAXS experiments [82]) to face-centered cubic (or hexagonal) close packing. Thus we infer that increased coating rates result in particle ordering.

The smooth increase in refractive index with U_0 argues against a discrete phase transition from a liquid-like structure to a crystal-like structure. Experiments on monosized, spherical colloids [84] at constant concentration, however, indicate such a transition when $\phi = 0.3$ and the Peclet number (Pe, a dimensionless parameter reflecting the relative rates of shear and diffusion):

$$Pe = ya^2/D_0 \qquad (6.28)$$

exceeds about 1, where y equals the shear rate (s^{-1}), a equals the particle diameter (cm) and D_0 equals the diffusion coefficient (cm^2/s). For 55 nm diameter particles deposited at 45.7 cm/min from a low viscosity alcohol-based sol, we calculate $Pe = 0.04$ near the liquid bath surface where y is maximized ($\sim 275\,s^{-1}$) for a pure fluid. This value is obviously too low to account for ordering. However, according to the previous discussion of the deposition of sols from binary fluid mixtures, both the shear rate and sol viscosity are expected to be maximized in the vicinity of the drying line due to water enrichment, surface tension gradient driven flows and the concentration dependence of viscosity. These combined factors probably cause Pe to exceed 1 there. We note that, to date, preliminary light-scattering experiments have not provided any direct evidence of ordering within the depositing sol, although we may not have the necessary spatial resolution to acquire data sufficiently close to the drying line, where we anticipate ordering to occur.

A second consequence of increased withdrawal speeds is increased drying time. Increased time may benefit the ordering of repulsive particles because finite time is required for alignment and registration of the particles into a crystal-like structure. Conversely, for the reactive sol, increased drying times caused increased stiffness of the silicate skeleton (aging) prior to the final stage of drying, leading to more porous films. In both cases the coating rate can be used to tune porosity of the deposited film.

6.8 APPLICATIONS

6.8.1 Protection for terrestrial and space mirrors

Mirrors used in solar arrays for both terrestrial and space applications frequently utilize a front surface design to maximize the light incident on the surface. Unlike a conventional second surface mirror (e.g. household mirror) which is protected on the front surface by glass and on the back surface by an opaque paint, front surface mirrors have an exposed reflecting surface which requires rigorous protection, depending upon the choice of reflector material and the operating environment. Terrestrial mirrors require protection against acid rain or other chemical pollutants, particulates and ultraviolet radiation. For space applications in low earth orbits (LEO, defined as altitudes of ~ 200–700 km) materials are exposed to atomic oxygen (AO), ultraviolet radiation, electrons, protons and cosmic rays, in addition to particulates such as micrometeorites. Of these, AO exposure presents the primary challenge to the design of protective films for many space materials.

Atomic oxygen is the most abundant LEO species, e.g. comprising 90% of the total atmosphere at an altitude of 500 km [85]. Spacecraft orbiting at this altitude at a velocity of 8 km/s are exposed to AO with a kinetic energy of about 5 eV and a flux of ~ 1000 atoms/cm^2 s [86, 87]. Reflecting materials, such as silver, copper and osmium, as well as various polymer or epoxy composite structural materials, organic paints and adhesives, and thermal control coatings are aggressively attacked by atomic oxygen. In addition to AO corrosion resistance, design requirements for protective films for space materials must include simple deposition techniques amenable to large areas and complex geometries, high transparency, reasonable abrasion resistance and good stability under vacuum and fluctuating temperature conditions.

A variety of characterization methods and facilities exist to determine AO degradation of candidate space materials. Exposure to RF plasma discharges in air or oxygen (asher test) is commonly used to roughly simulate LEO conditions and give relative indications of material durability. A wide range of environments can be simulated using the RF plasma asher by controlling the RF power, source gas and pressure. Plasma reaction conditions are monitored using erosion of polyimide (Kapton) as a reference material. Kapton asher erosion data is compared with erosion data collected during spaceflight experiments in LEO to estimate comparable space exposure. As an example, in a NASA study by Gulino *et al.* [88], 1 h exposure to a plasma generated using ambient air as the carrier gas resulted in Kapton erosion equivalent to that observed by exposure to a fluence of 1×10^{19} atoms/cm^2 in shuttle experiments. Based on this calibration, a Kapton exposure of 26 h approximated one year in LEO. Alternatively, numerous atomic oxygen test facilities [89] allow materials to be tested under conditions where a number of critical parameters can be controlled, e.g. AO fluence, energy and energy distribution, charged species population and partial pressure, temperature, UV radiation and gas purity and composition. Finally, experiments such as the limited duration candidate exposure, coordinated by the Center on Materials for Space Structures at Case Western Reserve University, Cleveland OH, permit exposure of candidate materials to actual LEO conditions during space shuttle flights. The most significant difference between LEO exposure tests and asher simulations is the kinetic energy of the oxygen atoms – atoms in LEO have a translational energy of 5 eV, whereas atoms generated in the asher have energies of 0.04–0.06 eV. Another major difference is the interaction of AO with other space components such as UV/VUV and micrometeorites. AO facilities which generate a variety of species at different energies and fluxes can more closely simulate actual space environments but at significantly greater cost.

Sol–gel protective films for mirror applications

Sol–gel thin films meet the criteria outlined above for protective films and, as such, are attractive candidates for both terrestrial and space mirror applications. For most solar applications, cost limitations preclude mechanical preparation of the substrate surface therefore any subsequent surface coating must be compatible with large-scale, continuous deposition coating techniques. Sol–gel coatings have been used in terrestrial solar thermal designs to produce high performance, low cost reflective surfaces which can function as both the optical concentrator and as a structural component of collectors. In these mirror designs sol–gel thin films performed a dual role: (1) a sol–gel planarizing layer enhanced the surface smoothness of the metal substrate and improved specular reflectance while preventing corrosion of the stainless steel substrate, and (2) a sol–gel protection layer minimized degradation of the silver reflecting surface [90].

Planarized mirrors were constructed of 400 series 0.005 in. (0.1 mm) thick stainless steel foil, SiO_2 sol–gel planarization layer (400–500 nm total thickness), silver ($\sim$150 nm thickness, sputter-deposited), and a sol–gel protective film (nominal film thickness of 100–200 nm) [12]. Both the protective and planarizing sol–gel films were deposited from SiO_2 sols prepared using a two-step hydrolysis method [77] with H_2O/Si molar ratios (r) ranging from 2.5 for the planarizing layer to 5–10 for the protective layer. As discussed in section 6.7.2, variation of the H_2O/Si ratio r is commonly employed in sol–gel processing to vary the structures of the inorganic polymers. Reflectance properties of the samples were measured with a

Beckman 5270 spectrophotometer equipped with an integrating sphere accessory and with a Device and Services model 15R portable specular reflectometer. Hemispherical (total) reflectance data were referenced to a calibrated mirror standard (NIST) and have an uncertainty of 0.005 reflectance units (where 100% reflectance is 1.000 reflectance units). Specular reflectances were measured at 600 nm at an angular aperature of 26 mrad and have an uncertainty of 0.01 reflectance units. Solar averaged values were calculated using the NIST-corrected data and an air mass zero solar spectral distribution (ASTM E490-73a) because of the extraterrestrial mirror application.

Sol–gel planarization of the substrate resulted in an improvement in the specular reflectance of the mirror from $\sim 36\%$ for an unplanarized mirror to over 92% [12]. Sol–gel protective films of varying compositions were deposited onto planarized and unplanarized mirrors prior to simulated AO testing (plasma asher) at NASA's Lewis Research Center. The plasma was generated by exciting ambient air with 100 W of continuous wave RF power at 13.56 MHz and an operating pressure of ~ 50 mtorr (6.7 Pa). Specular and hemispherical reflectance measurements were made at Sandia National Laboratories following total exposures of 0, 50, 100, 180, and 295 h. An unprotected control sample, i.e. no sol–gel overcoat, was exposed for 1 h. The AO fluence at the various sampling times was determined from empirical data obtained from Kapton erosion under identical conditions. The approximate Kapton equivalent fluences yielded an effective flux of $\sim 3 \times 10^{15}$ atoms/cm^2 s [91]. Based on this calibration, 295 h asher exposure represented > 5 yr exposure at an altitude of 465 km.

The hemispherical reflectance of sol–gel protected mirrors following asher exposure is shown in Fig. 6.12. The unprotected sample shows reflectance degradation from 0.95 to 0.65 after only 1 h exposure. In contrast, sol–gel protective coatings with $r = 5$–10, i.e. comprised of weakly branched polymers which freely interpenetrate upon deposition to form relatively dense films [77], show excellent protection after exposure to conditions which simulate $\sim 30\%$ (5.75 yr) of the AO fluence expected during 20 yr in LEO. However, when protective films were deposited from an aged, multicomponent sol (wt% $= 71SiO_2$–$18B_2O_3$–$7Al_2O_3$–$4BaO$; Table 6.1) and the coated mirrors were subjected to asher exposure, little protection from the corrosive effects of atomic oxygen was observed. This result is not surprising due to the highly branched nature of the polymeric species – fractal polymers of this type do not interpenetrate upon deposition and lead to highly porous films which are unsuitable for protective applications. Interpenetration of polymers depends upon polymer fractal dimension and is discussed in section 6.7.1.

Results of depth profiling Auger analysis (argon sputter) of sol–gel protected mirrors following air plasma exposure are shown in Fig. 6.13a (unplanarized mirror) and Fig. 6.13b (planarized mirror). The unplanarized mirror, i.e. silver sputter-deposited directly on stainless steel followed by a sol–gel SiO_2 protective layer, clearly shows migration of iron through the

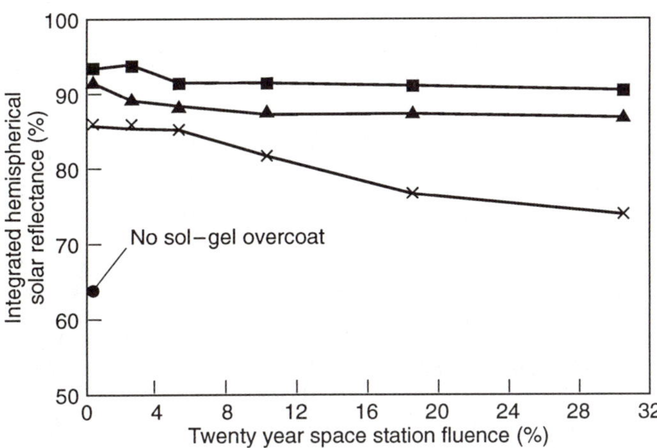

Figure 6.12 Sol–gel overcoats preserve the hemispherical solar reflectance of treated mirrors under simulated atomic oxygen exposure conditions. Sol compositions: (■) SiO_2, $r = 10$; (▲) SiO_2, $r = 5$; (×) aged multicomponent sol, see Table 6.1.

silver layer into the sol–gel protection layer. The consequence of iron diffusion into the silver is severely reduced hemispherical reflectance. In this sample, silver is also seen in the near-surface region of the protection layer. In contrast, the planarized mirror shows minimal movement of iron into the planarization layer and no silver or iron migration into the near-surface region of the sol–gel protective coating.

The Auger profiles show significant differences in diffusion resistance between the planarizing layer and the protection layers. The absence of diffusion through the planarizing layer suggests that this layer, prepared from a sol with $r = 2.5$, is more resistant to diffusion and presumably more dense than the protective films. This result initially appears to conflict with data presented by Brinker and coworkers [50] which show decreasing film porosity as excess water increases from 0.5 to 6% (corresponding to $r = 2.5$ to $r = 7.5$). At even higher water contents ($r > 7.5$), they observed increasing porosity (lower refractive index), attributed to competition between capillary pressure and aging (stiffening) of the silicate skeleton – factors which are both enhanced by increasing water concentration. Our result, i.e. that low r films appear functionally more dense than films with r values ranging from 5–10, can be explained by processing differences between the films. The planarizing layer consisted of three separate coatings which were each fired to 450°C to cause partial densification. However, processing of the single protective layer was limited to low temperature ($<250°C$) to minimize damage to the silver. In Brinker's experiment, very similar refractive indices (i.e. similar film densities) were measured for films deposited from sols with $r = 5$ and $r = 10$. Thus,

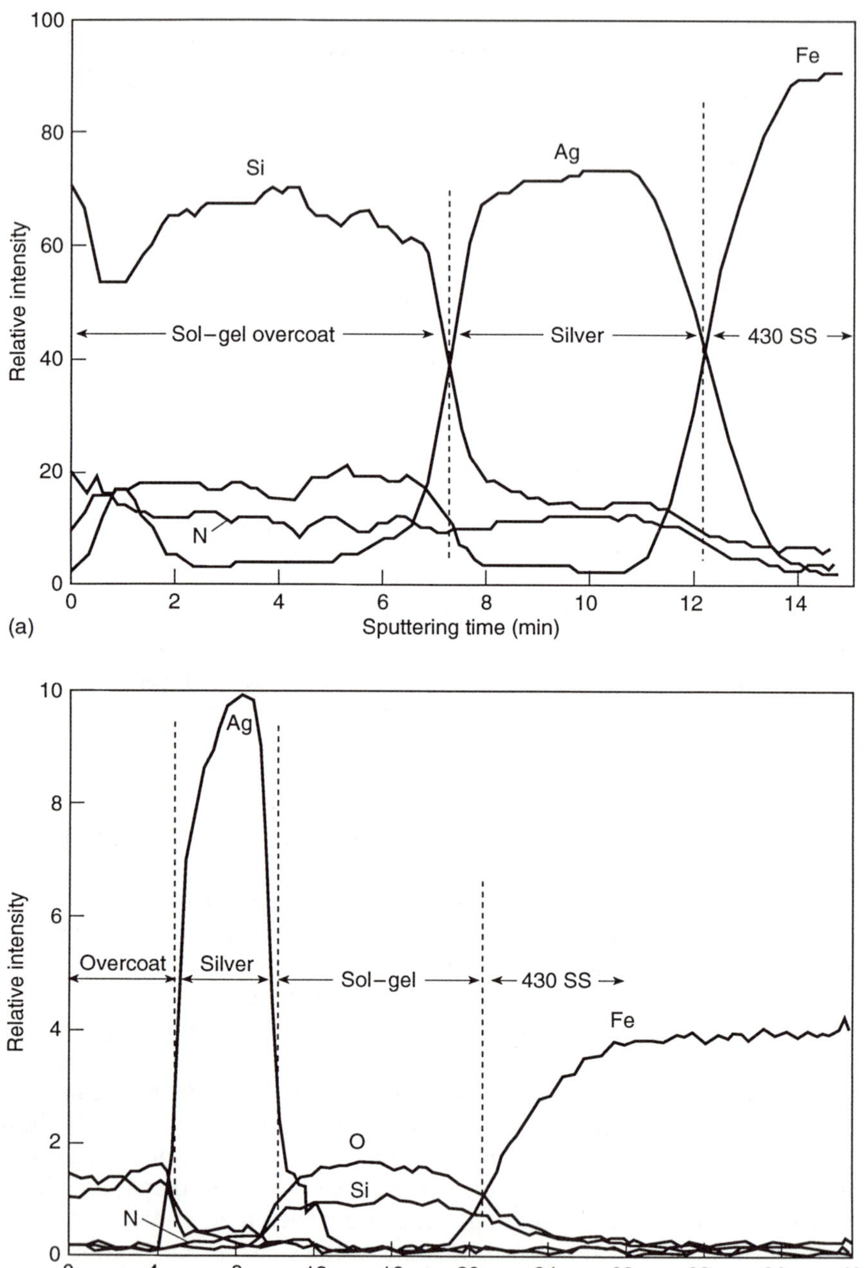
100
80
60
40
20
0
Relative intensity
Fe
Si
Ag
Sol–gel overcoat
Silver
430 SS
N
0   2   4   6   8   10   12   14
Sputtering time (min)
(a)
10
8
6
4
2
0
Relative intensity
Ag
Overcoat
Silver
Sol–gel
430 SS
Fe
O
Si
N
0   4   8   12   16   20   24   28   32   36   40
Sputtering time (min)
(b)

the slight differences in diffusion resistance of the two protective films ($r = 5$ and $r = 10$) which we observe are probably attributable to processing variations rather than to significant differences in film microstructure. The slightly lower resistance to diffusion of silver of the SiO_2 film with $r = 10$ may be due to longer asher exposure time (295 h compared to 180 h for the SiO_2 $r = 5$ film) or to thermally enhanced diffusion during prolonged exposure. As expected, mirrors which were protected by films deposited from the aged, multicomponent sol, discussed earlier, permitted easy migration of Ag and Fe into the protective film (results not shown). These data support the conclusion of the asher experiment, which determined that these highly porous films offer little protection from the corrosive effects of AO.

6.8.2 Space exposure of sol–gel protected mirrors

Sol–gel protective mirror samples similar to those prepared above were included in a materials exposure experiment aboard NASA shuttle mission STS-51 (shuttle *Discovery*, September 12–22, 1993). The limited duration candidate exposure experiments consisted of two separate payload elements, designated LDCE-4 and LDCE-5, and were coordinated by the NASA Center on Materials for Space Structures (CMSS) at Case Western Reserve University, Cleveland OH.

The payload configuration consisted of two aluminum disks, each 49 cm in diameter, onto which sample trays containing 64 candidate samples were mounted. Each tray also contained Kapton HN polyimide samples as exposure controls. The sample trays were mounted inside a canister which was open only during periods when the payload bay was pointed toward the direction of orbital travel (velocity vector). During shuttle operations where contamination may occur, such as water dumps and thruster firings, the canister was closed, thus restricting exposure of samples to ram atomic oxygen only and minimizing contamination.

The sol–gel protective coatings were deposited onto planarized, silvered mirrors. Based on the results of asher tests and previous experiments to evaluate sol–gel insulating films for electronic passivation [92], SiO_2 sols with $r = 5$ and 12.5 were evaluated, in addition to a combination SiO_2–TiO_2 overlayer. Controls included an unprotected silver sample and mirrors protected by commercial proprietary sputter-deposited oxides which contained

---

Figure 6.13 Auger sputter depth profiles compare an unplanarized mirror with a mirror having a dense sol–gel planarizing layer: (a) no planarization, $r = 10$ overcoat, 295 h NASA plasma and (b) sol–gel planarization $r = 5$ overcoat, 175 h NASA plasma. The planarizing layer is an effective barrier to diffusion of iron from the stainless steel support through the silver mirror and into the protective sol–gel overcoat.

Al$_2$O$_3$ or SiO$_2$ as primary components. The sols were prepared as reported previously [93] and were deposited by dip-coating using a computer-controlled, constant atmosphere coating apparatus. Postdeposition treatment consisted of either 250°C fire or 1 h exposure to oxygen plasma to remove residual organic species. Companion control samples were stored in the laboratory under argon immediately after coating and were maintained in an inert atmosphere until final postflight analysis ($\sim$1.5 yr).

Candidate sol–gel protected mirror samples, 3/4 in. (19 mm) or 1 in. (25 mm) diameter disks, were characterized optically (specular and hemispherical reflectance) and transported to NASA under argon. Preflight characterization was carried out in a clean room and the samples were held in clean, controlled humidity conditions, until launch integration at Kennedy Space Center. During the spaceflight, the materials were exposed at an altitude of 296 km for 38 h. The prediction for the ram atomic oxygen fluence was 7×10^{19} atoms/cm^2; this fluence value was consistent with the average weight loss as measured on the Kapton HN control samples. After flight, the closed canisters were cleaned and opened in a class 100 000 clean room for final analysis.

Specular reflectance measurements were made using 26 mrad aperture and vacuum holding device. Samples designated as controls were held in the laboratory under argon atmosphere after coating until final characterization.

Table 6.2 shows the pre- and postflight optical results for control and flight samples. Samples 1, 4 and 5 experienced some silver degradation prior to the experiment because the sol–gel protective coatings were not deposited onto freshly silvered surfaces. This explains the somewhat lower preflight reflectance values for these samples. As expected, the unprotected sample shows severe degradation during the 38 h shuttle exposure. The sputtered protective coatings were deposited immediately following silver deposition

Table 6.2 Preflight and postflight specular reflectance of mirrors

Sample	Protective coat	Preflight specular reflectance (%)	Postflight specular reflectance (%)
1-control	none	89.3	90.5
1-flight	none	90.5	20.3
2-control	sputtered SiO$_2$	94.6	94.3
2-flight	sputtered SiO$_2$	95.1	94.4
3-control	sputtered Al$_2$O$_3$	94.6	95.2
3-flight	sputtered Al$_2$O$_3$	95.6	94.7
4-control	sol–gel SiO$_2$	88.9	87.9
4-flight	sol–gel SiO$_2$	88.5	76.3
5-control	sol–gel SiO$_2$–TiO$_2$	89.2	89.1
5-flight	sol–gel SiO$_2$–TiO$_2$	89.2	89.4

so that the fresh silver surface was never exposed to ambient air. The sputtered coatings exhibit high initial reflectance values which are retained during AO exposure. The 130 nm thick sol–gel SiO_2 coating alone, which is somewhat porous after firing to only 250°C, is less protective than either sputtered oxide coating. However, the combination sol–gel film, consisting of SiO_2 followed by TiO_2, allowed no silver degradation during AO exposure. TiO_2 is highly resistant to AO but the high acid content of most sol–gel prepared TiO_2 formulations contributes to silver degradation and loss of specularity. However, in this case, the SiO_2 layer prevented acid damage to the silver. It is unknown whether the enhanced protection afforded by the combination film is due to the TiO_2 layer or merely to the increased thickness of the two-layer film.

Sol–gel coatings, especially combination SiO_2–TiO_2 layers, are suitable for use as protective coatings for mirrors in space applications. The silver layer, which was severely degraded in unprotected films, was protected in both laboratory simulations of AO exposure and actual space AO exposure. Sol–gel protective films allow the original, high reflectance values to be maintained and provide protection that is comparable to materials such as sputter-deposited oxides. Although protection of sputter-deposited silver appears to be easily accomplished by the use of a sputtered oxide layer, the use of sol–gel films for oxidation-resistant coatings is not limited to silver mirrors. One of the great advantages of sol–gel processing is the ability to deposit coatings of uniform composition and thickness over large areas and relatively complex geometries without using thermal or vacuum deposition processes. Other exposed spacecraft components and structures, especially those utilizing temperature-sensitive materials such as polymers or composites, may be ideal candidates for sol–gel protection.

Our results support the hypothesis that sol–gel derived oxides are quite nonreactive in AO environments. Materials with bond strengths > 4.2 eV (Si–O, Si–C, C–F) appear to be relatively stable in AO environments, whereas materials with bond energies below this value degrade. There is no driving force for further oxidation of oxides, since they are in their highest oxidation states. This is not the case for carbon-based materials, where the thermodynamic stability of the oxidation product, CO_2 (C–O, 8.3 eV), favors further oxidation. Densification of the sol–gel oxide during prolonged AO exposure may also occur as has been noted upon exposure of unheated, porous sol–gel derived oxide films to laboratory oxygen plasma [94]. It is assumed that densification of a sol–gel protective film is advantageous, however some residual porosity may actually be desirable. A microporous underlayer, such as the SiO_2 underneath the TiO_2, may serve to minimize stress buildup during continuing AO-driven densification of the exposed film. Alternatively, a microporous overlayer may perform sacrificially to mitigate the impact of debris on a dense underlying film. Sandwich structures such as these may hold the greatest potential for long-term protective schemes.

6.9 SUMMARY

During sol–gel dip-coating, inorganic precursors are rapidly concentrated on the substrate surface by gravitational draining with concurrent evaporation and condensation reactions. We have presented examples of several factors that influence the structure of the deposited films: (1) size and opacity of fractal precursors; (2) relative rates of condensation and evaporation; (3) capillary pressure; (4) shear resulting from surface tension gradient driven flows; and (5) substrate withdrawal speed. By control of these factors it is possible to tailor the pore size, surface area, pore volume and refractive index of the deposited film for applications ranging from dense protective or passivation layers to films with precisely controlled pore sizes for sensor surfaces and separation membranes.

ACKNOWLEDGMENTS

The authors thank numerous collaborators who have contributed to this work, including Don Stuart, Jeff Zink, Bruce Dunn, and Tim Boyle. We thank Terry Garino for a careful review of the manuscript. This work was performed at Sandia National Laboratories under DOE Contract # DE-ACO4-94AL85000.

REFERENCES

1. Dislich, H. (1988) in *Sol–Gel Technology for Thin Films, Fibers, Preforms, Electronics, and Specialty Shapes* (ed. L.C. Klein) Noyes, Park Ridge NJ, pp. 50–79.
2. Schroeder, H. (1969) in *Physics of Thin Films 5* (ed. G. Haas), Academic Press, New York, pp. 87–141.
3. Nelson, R.L., Ramsay, J.D.F., Woodhead, J.L.J. *et al.* (1981) *Thin Solid Films*, **81**, 329–37.
4. Partlow, D.P. and O'Keeffe, T.W. (1990) *Appl. Opt.*, **29**, 1526–9.
5. Floch, H.G., Priotton, J.-J. and Thomas, I.M. (1989) *Thin Solid Films*, **175**, 173–8.
6. Yoldas, B.E. (1980) *Appl. Opt.*, **19**, 1425–9.
7. Makishima, A., Kubo, N., Wada, K. *et al.* (1986) *J. Amer. Ceram. Soc.*, **69**, C127–9.
8. Yoldas, B.E. and Partlow, D.P. (1985) *Thin Solid Films*, **129**, 1–14.
9. Yoldas, B.E. (1982) *Appl. Opt.*, **21**, 2960–4.
10. Pettit, R.B. and Brinker, C.J. (1982) *SPIE Optical Coatings for Energy Efficiency and Solar Applications 324*, p. 176.
11. Warren, W.L., Lenahan, P.M., Brinker *et al.* (1990) in *Better Ceramics Through Chemistry IV, Mat. Res. Soc. Symp. Proc. 180* (eds B.J.J Zelinski, C.J. Brinker, D.E. Clark and D.R. Ulrich), Materials Research Society, Pittsburgh PA, pp. 413–19.
12. Ashley, C.S., Reed, S.T. and Mahoney, A.R. (1988) in *Better Ceramics Through Chemistry III, Mat. Res. Soc. Symp. Proc. 121* (eds C.J. Brinker, D.E. Clark and D.R. Ulrich), Materials Research Society, Pittsburgh PA, pp. 635–8.
13. Frye, G.C., Ricco, A.J., Martin, S.J. and Brinker, C.J. (1988) in *Better Ceramics Through Chemistry III, Mat. Res. Soc. Symp. Proc. 121* (eds C.J. Brinker, D.E. Clark and D.R. Ulrich), Materials Research Society, Pittsburgh PA, pp. 349–54.
14. Bein, T., Frye, G.C. and Brinker, C.J. (1989) *J. Amer. Chem. Soc.*, **111**, 7640–1.
15. Pettit, R.B., Brinker, C.J. and Ashley, C.S. (1985) *Solar Cells*, **15**, 267–78.

16. Dimos, D., Warren, W.L., Sinclair, M.B. *et al.* (1994) *J. Appl. Phys.*, **76**(7), 4305–15.
17. Hirano, S.I. and Kato, K. (1988) *J. Non-Cryst. Solids*, **100**(5), 38–41.
18. Terabe, K., Iyi, N. and Kimura, S. (1995) *J. Mater. Sci.*, **30**(8), 1993.
19. Brinker, C.J. and Scherer, G.W. (1990) *Sol–Gel Science,* Academic Press, San Diego CA.
20. Prakash, S.S., Brinker, C.J., Hurd, A.J. and Rao, S.M. (1995) *Nature*, **374**, 439–43.
21. Hietala, S.L., Smith, D.M., Hietala, V.M. *et al.* (1990) in *Better Ceramics Through Chemistry IV, Mat. Res. Soc. Symp. Proc. 180,* (eds B.J.J. Zelinski, C.J. Brinker, D.E. Clark and D.R. Ulrich), Materials Research Society, Pittsburgh PA, pp. 433–7.
22. Shane, M. and Mecartney, M.L. (1990) *J. Mater. Sci.*, **25**, 1537–44.
23. Bratton, R.J. and Lau, S.K. (1981) in *Advances in Ceramics,* Vol. 3. (eds A.H. Heuer and L.W. Hobbs), American Ceramics Society, Columbus OH.
24. Atik, M. and Aegerter, M.A. (1992) *J. Non-Cryst. Solids*, **147/148**, 813–19.
25. Kawashima, A., Wang, Y., Akiyama, E. *et al.* (1994) *Zairyo-to-Kankyo*, **43**, 854–9; Kato, K. (1993) *J. Mater. Sci.*, **28**, 4033–6.
26. Affatigato, M., Osborne, D. and Haglund, R.F. (1995) *J. Non-Cryst. Solids*, **181**, 27–38.
27. Beier, W. and Odler, I. (1990) in *Better Ceramics through Chemistry IV, Mat. Res. Soc. Symp. Proc. 180,* (eds B.J.J. Zelinski, C.J. Brinker, D.E. Clark and D.R. Ulrich), Materials Research Society, Pittsburgh PA, pp. 467–72.
28. Holmes-Farley, S.R. and Yanyo, L.C. (1991) *J. Adhesion Sci. Technol.*, **5**(2), 131–51.
29. Schunk, P.R., Brinker, C.J. and Hurd, A.J. (1996) in *Liquid Film Coating: Scientific Principles and their Technological Implications* (eds P.M. Schweizer and S.F. Kistler), Chapman & Hall, New York.
30. Brinker, C.J. and Scherer, G.W. (1990) *Sol–Gel Science,* Academic Press, San Diego CA, Ch. 13.
31. Brinker, C.J. and Scherer, G.W. (1990) *Sol–Gel Science,* Academic Press, San Diego CA, Chs 2 and 3.
32. Livage, J., Henry, M. and Sanchez, C. (1988) *Prog. Solid State Chem.*, **18**, 259–342.
33. Takahashi, Y. and Wada, Y.J. (1990) *Electrochem. Soc.*, **137**, 267–72.
34. Raman, N.K. and Brinker, C.J. (1995) *J. Membrane Sci.*, **105**, 273–9.
35. Kelts, L.W. and Armstrong, N.J. (1988) in *Better Ceramics Through Chemistry III, Mat. Res. Soc. Symp. Proc. 121* (eds C.J. Brinker, D.E. Clark and D.R. Ulrich), Materials Research Society, Pittsburgh PA, pp. 519–22.
36. Lippert, J.L., Melpolder, S.M. and Kelts, L.W. (1988) *J. Non-Cryst. Solids*, **104**, 139–47.
37. Kelts, L.W., Effinger, N.J. and Melpolder, S.M. (1986) *J. Non-Cryst. Solids*, **83**, 353–74.
38. Iler, R.K. (1978) *The Chemistry of Silica,* Wiley, New York.
39. Brinker, C.J., Hurd, A.J., Frye, G.C. *et al.* (1991) *Ceram. Soc. Jpn*, **99**(10), 862–77.
40. Schaefer, D.W. (1988) *MRS Bull.*, **8**, 22–7.
41. Ludox plot from Schaefer, D.W and Keefer, K.D. (1986) in *Fractals in Physics* (eds L. Pietronero and E. Tosatti), North-Holland, Amsterdam, pp. 39–45.
42. Brinker, C.J. and Scherer, G.W. (1990) *Sol-Gel Science*, Academic Press, San Diego CA, Chs 2 and 5.
43. Meakin, P. (1983) *Phys. Rev. Lett.*, **51**, 1119.
44. Schwartz, R.W., Voigt, J.A., Boyle, T.J. *et al.* (1995) *Ceram. Engng Sci. Proc.*, **16**(5), 1045.
45. Schwartz, R.W., Assink, R.A., Dimos, D. *et al.* (1995) in *Ferroelectric Thin Films IV, Mat. Res. Soc. Symp. Proc.* 361, Materials Research Society, Pittsburgh PA p. 377.
46. Brinker, C.J., Hurd, A.J., Frye, G.C. *et al.* (1990) *J. Non-Cryst. Solids*, **1221**, 294–302.
47. Scriven, L.E. (1988) in *Better Ceramics Through Chemistry III, Mat. Res. Soc. Symp. Proc. 121,* (eds C.J. Brinker, D.E. Clark and D.R. Ulrich), Materials Research Society, Pittsburgh PA, pp. 717–29.

48. Landau, L.D. and Levich, B.G. (1942) *Acta Physiochim. URSS*, **17**, 42–54.
49. Wilson, S.D.R. (1982) *J. Engng Math.*, **16**, 209.
50. Brinker, C.J., Frye, G.C., Hurd, A.J. and Ashley, C.S. (1991) *Thin Solid Films*, **201**(1), 97–108.
51. Hurd, A.J. and Brinker, C.J. (1988) *J. Physique*, **49**, 1017–25.
52. Hurd, A.J., Brinker, C.J. (1990) in *Better Ceramics Through Chemistry IV, Mat. Res. Soc. Symp. Proc. 180* (eds B.J.J. Zelinski, C.J. Brinker, D.E. Clark and D.R. Ulrich), Materials Research Society, Pittsburgh PA, pp. 575–81.
53. Hurd, A.J. (1994) in *The Colloid Chemistry of Silica, Adv. Chem. Ser. 234* (ed. H.E. Bergna), American Chemical Society, Washingon D.C., pp. 433–50.
54. Nishida, F., McKernan, J.M., Dunn, B. *et al.* (1995) *J. Amer. Ceram. Soc.* **78**(6), 1640–8.
55. Bornside, D.E., Macosko, C.W. and Scriven, L.E. (1989) *J. Appl. Phys.*, **66**, 5185–93.
56. Sakka, S. and Kamiya, K. (1982) *J. Non-Cryst. Solids*, **48**, 31–46.
57. Cairncross, R.A., Francis, L.F. and Scriven, L.E. (1992) *Drying Technol. J.*, **10**, 893–923; Cairncross, R.A., Francis, L.F. and Scriven, L.E. (1994) *AIChE J.*, **42**, 55–67.
58. Assink, R.A. and Kay, B.D. (1988) *J. Non-Cryst. Solids*, **107**, 35–9.
59. Burgess, C.G.V. and Everett, D.H. (1970) *J. Coll. Interface Sci.*, **33**, 611.
60. Smith, D.M., Scherer, G.W. and Anderson, J.M. (1995) *J. Non-Cryst. Solids*, **188**, 191–206.
61. Chen, K.S. and Schunk, P.R. (1994) Modeling of drying, stress development, and pore structure evolution in membrane thin-film fabrications. Paper presented at the 6th Annual Joint Meeting of the New Mexico Sections of the American Ceramic Society and Materials Research Society, Albuquerque NM.
62. Cairncross, R.A., Chen, K.S., Schunk, P.R. *et al.* (1996) in *Computational Modelling of Materials and Processing*. Proceedings of the American Ceramic Society National Meeting Symposium, American Ceramic Society, Cincinnati OH (in press).
63. Scherer, G.W. (1992) *J. Non-Cryst. Solids*, **147/148**, 363.
64. Sherwood, T.K. (1929) *Ind. Engng Chem.*, **21**, 12–16 and 976–80.
65. Cohen, E.D., Gutoff, E.B. and Lightfoot, E.J. (1990) in *Mechanics of Thin Film Coating*. Proceedings of the International Symposium at the AIChE Spring Meeting, American Institute of Chemical Engineering, Orlando FL.
66. Schunk, P.R. and Rao, R.R. (1994) *Int. J. Numerical Methods in Fluids*, **18**(4), 821.
67. Brinker, C.J. and Scherer, G.W. (1990) *Sol–Gel Science*, Academic Press, San Diego CA, Ch. 8.
68. Yapel, R.A. and Scriven, L.E. (1990) in *Mechanics of Thin Film Coating*. Proceedings of the International Symposium at the AIChE Spring Meeting, American Institute of Chemical Engineering, Orlando FL.
69. Croll, S.G. (1979) *J. Appl. Polym. Sci.*, **23**, 847.
70. Evans, A.J., Dory, M.D. and Hu, M.S. (1988) *J. Mater. Res.*, **3**, 1043–9.
71. Thouless, M.D. (1988) *Acta Metall.*, **36**, 3131–5.
72. Meakin, P. (1991) *Science*, **252**, 226–34.
73. Atkinson, A. and Guppy, R.M. (1991) *J. Mater. Sci.*, **26**, 3869–73.
74. Garino, T.J. (1990) in *Better Ceramics Through Chemistry IV, Mat. Res. Soc. Symp. Proc. 180* (eds B.J.J. Zelinski, C.J. Brinker, D.E. Clark and D.R. Ulrich), Materials Research Society, Pittsburgh PA, pp. 497–502.
75. Schmidt, H., Rinn, G., Nass, R. and Sporn, D. (1988) in *Better Ceramics Through Chemistry III, Mat. Res. Soc. Symp. Proc. 121* (eds C.J. Brinker, D.E. Clark and D.R. Ulrich), Materials Research Society, Pittsburgh PA, pp. 743–54.
76. Garino, T.J. (1988) Ph.D. thesis, MIT, Cambridge MA.
77. Brinker, C.J., Keefer, K.D., Schaefer, D.W. *et al.* (1984) *J. Non-Cryst. Solids*, **63**, 45–59.
78. Mandelbrot, B.B. (1982) *The Fractal Geometry of Nature*, Freeman, San Francisco CA.

79. Bechtold, M.F., Mahler, W. and Schunn, R.A. (1980) *J. Polym. Sci., Polym. Ed,* **18,** 2823.

80. Coltrain, B.K., Melpolder, S.M. and Salva, J.M. (1991) in *Ultrastructure Processing of Advanced Materials.* Proceedings of the IVth International Conference, Tucson AZ, (eds D.R. Uhlmann and D.R. Ulrich), Wiley, New York, pp. 69–76.

81. Gregg, S.J. and Sing, K.S.W. (1982) *Adsorption, Surface Area, and Porosity,* Academic Press, New York.

82. Brinker, C.J., Hurd, A.J. and Ward, K.J. (1988) in *Ultrastructure Processing of Advanced Ceramics* (eds J.D. Mackenzie and D.R. Ulrich), Wiley, New York, p. 223.

83. Glaser, P.M. and Pantano, C.G. (1984) *J. Non-Cryst. Solids,* **63,** 209–21.

84. Ackerson, B.J. (1990) *J. Rheol.,* **34,** 553.

85. Outlaw, R.A. and Brock, F.J. (1977) *Vac. Sci. Technol.,* **14,** 1269.

86. Arnold, G.S. and Peplinski, D.R. (1986) *AIAA J.,* **24,** 673.

87. Leger, L.J., Visentine, J.T., Kuminecz, J.F. and Spiker, I.K. (1984) Paper 84–0415, presented at the American Institute of Aeronautics and Astronautics 23rd Aerospace Sciences Meeting, Reno NV, Jan. 1985.

88. Gulino, D.A. and Diver, T.M. (1988) *NASA Technical Memorandum TM-100834.*

89. Banks, B.A., Rutledge, S.K. and Brady, J.A. (1988) in *Proceedings of NASA's 15th Space Simulation Conference,* Williamsburg VA, pp. 51–65.

90. Reed, S.T. and Ashley, C.S. (1988) in *Better Ceramics Through Chemistry III, Mat. Res. Soc. Symp. Proc. 121* (eds C.J. Brinker, D.E. Clark and D.R. Ulrich), Materials Research Society, Pittsburgh PA, pp. 631–4.

91. de Groh, K.K. (1989) Unpublished data, NASA Lewis Research Center, Cleveland OH.

92. Warren, W.L., Lenahan, P.M., Brinker, C.J. *et al.* (1991) *J. Appl. Phys.,* **69,** 4404–8.

93. Brinker, C.J. and Harrington, M.S. (1981) *Solar Energy Mater.,* **5,** 159–72.

94. Wielonski, R., Melling, P. and Drauglis, E. (1986) in *Proceedings of the 20th Annual Technical Conference of the Society of Vacuum Coaters,* pp. 254–261.

7

Solid phase cladding

S.B. Dunkerton

7.1 INTRODUCTION

The many metallurgical and ceramic coating techniques are amply illustrated in this book. Most of the chapters cover technologies relating to thin coatings, that is, less than 0.3–0.5 mm thick, which leaves a large market where arduous conditions and/or long life requirements dictate the need for thick coatings. These thicker coatings, or claddings, as they will be referred to, can be deposited by an equally wide range of techniques, all emanating from well-established welding and bonding methods.

One area not covered in this chapter, or elsewhere in the book, lies in the use of fusion claddings, where welding processes such as submerged arc, metal inert gas (MIG), tungsten inert gas (TIG), plasma transferred arc (PTA) are typically used. These processes will deposit a range of hard-wearing and/or corrosion-resistant layers with thickness up to nominally 10 mm in a single pass. Readers wishing to find out more are referred to Dunkerton [1] and Houldcroft and John [2].

Solid phase cladding tehniques are more competitive with the subject areas of this book; they can deposit coatings in the nominal range 0.5–5 mm. This chapter aims to introduce the various techniques, highlighting their particular characteristics and providing background to assist the user in the selection of a process for a particular application. The common thread of the processes covered is that all the coatings are deposited by methods that involve no melting of either the coating or the substrate, that is, they are all 'solid phase'

Metallurgical and Ceramic Protective Coatings. Edited by Kurt H. Stern.
Published in 1996 by Chapman & Hall, London. ISBN 0 412 54440 7

claddings. The advantages of this group of techniques are that unusual combinations of metals can be bonded, there is zero or minimal dilution and, because the deposits are 'welded', excellent adhesion is achieved.

Some of the processes described (explosive cladding, roll bonding) are established techniques which are widely used for specific large-scale application areas. Their niche markets are also established and consequently little new development work is ongoing. Developments that do arise tend to be in the transfer of the technology to new application areas, and this is described in the following sections. Other solid phase cladding techniques are more recent and are still in the embryonic stage of industrial exploitaton. This covers such processes as friction surfacing, a variant of friction welding, resistance cladding, a variant of resistance welding, and hot isostatic pressing, an extension of the diffusion bonding process. The discussion on these latter processes will indicate their future potential and highlight possible application areas.

7.2 EXPLOSIVE CLADDING

Explosive cladding is a well-established process within the high energy rate metalworking field of forming, welding, compaction and hardening. The use of explosives for working metals dates back to the late 1800s and, as with many developments, the application to welding resulted from the need to prevent 'accidental' welds when working on explosive forming. In these early studies, it was discovered that the use of an overcharge could result in preforms or blanks being welded to the die cavity, leading to the expensive replacement of the die. These events led to the first serious study of explosives for welding and cladding in the late 1950s. Apart from interest in remote welding/cladding, significant potential for the process was seen in the ability to join metal combinations with widely different melting points, thermal expansion coefficients and hardnesses.

The process of explosive cladding has grown to encompass applications in many diverse industries such as shipbuilding, chemical processing, nuclear, power transmission and, recently, electronics. Its unique advantages within the solid phase cladding field are large surface area cladding, restricted access cladding, no need for large capital equipment and applicability to a wide range of dissimilar metal combinations.

7.2.1 Principle and mechanism

Explosive cladding is a 'cold' welding process in which bonding is achieved by an oblique, high velocity collision between the two plates being welded. The collision results in a metallic jet being ejected at the point of contact between two plates, and this jet removes surface contamination allowing automatically clean surfaces to come into intimate contact and form a weld.

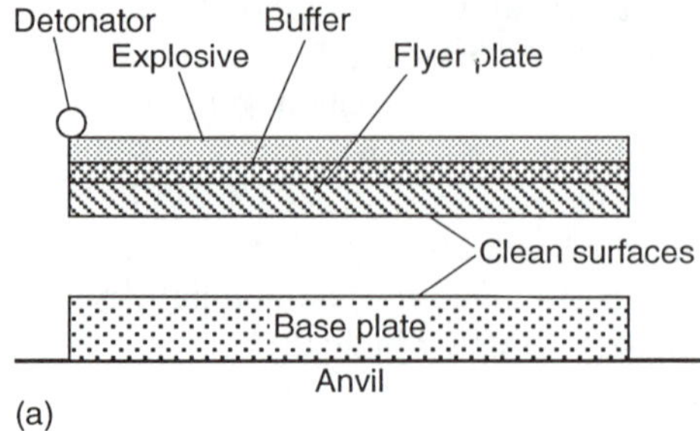

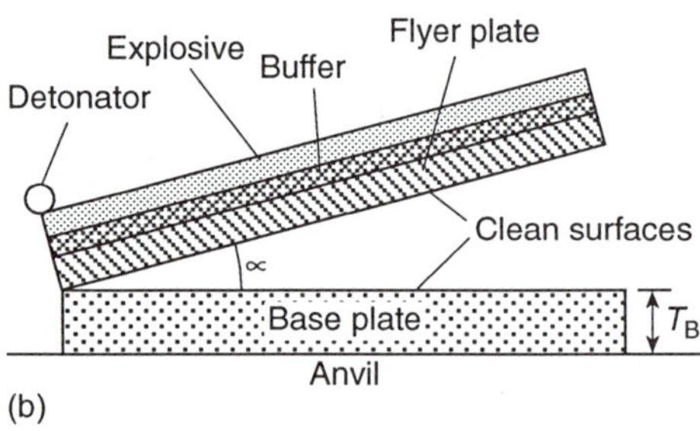

Figure 7.1 Schematic arrangement for explosive cladding: (a) parallel plate and (b) inclined plate.

Two forms of the cladding process are used; Fig. 7.1 shows the arrangements for both parallel-plate cladding and the inclined plate arrangement. Both have the base plate (to be clad) resting on an anvil, with the flyer plate (clad layer) held above the base plate. A buffer layer (of rubber or other suitable material) is laid on the flyer plate, to protect the top surface, and on top of this rests the explosive charge. The explosive is usually denoted from one point on the top plate, such that the flyer plate is progressively forced down against the base plate until the two plates are completely welded.

Bahrani [3] and El-Sobky [4] have provided the theory to the mechanism of explosive cladding and only a brief summary is provided here. When the explosive charge is fired, a detonation wave starts to travel along the explosive charge at the detonation velocity of the explosive. This accelerates the flyer plate by means of shock waves and, as the detonation front traverses the flyer plate, it forces the flyer plate to move towards the base plate with a constant velocity (some 4–5 km/s). The flyer plate velocity is related to the detonation velocity, starting angle and impact angle between the flyer and base plate. On collision, the kinetic energy of the flyer plate is dissipated to produce very high pressure and shear stresses at the collision point. These high pressures cause the materials at the point of impact to behave rather like inviscid fluids. Under these fluid conditions, a jet is ejected from the metal surfaces (believed to contain parts of both the flyer and base plate) which also moves forward at a velocity related to the detonation charge and collision angles. Although this reentrant jet has a small mass, it has a very high velocity, so it provides an effective scouring action.

In setting up the explosive cladding process, the plate arrangement – either parallel or inclined – is critical in controlling angles of collision; therefore, angle of incline or gap should be set carefully. With the parallel gap arrangement, the plates may be separated by small beads. Where metallic beads are used, these can trap the reentrant jet and leave small unbonded areas around the particles. Also, when cladding with thin sheet, the metal particles can be visible from the top clad suface. Polystyrene beads partly solve this problem; another option is to use a curtain of pressurized air. A curtain is particularly useful for thin plate, and it also allows the possibility of using an inert gas as the pressurizing medium, which may be important for highly reactive materials (e.g. titanium). The understanding of cladding mechanisms and experimental results are such that conditions for many metal combinations are now well known or readily established. The other critical areas are the explosive charge and point of detonation. With safety regulations, it is essential that the people responsible for cladding are fully trained in the handling and safety of explosives. Coupled with the need to carry out the explosive operations in remote sites (e.g. quarries, specially constructed bunkers, etc.), this means that most cladding work is carried out by a number of qualified and established explosive metalworking companies throughout the world.

Preparation of the plates involves machining to a flat, smooth finish ($R_a = 2–3\,\mu m$) then simple abrasion with emery paper, wiping with a clean dry cloth and possibly degreasing on site prior to cladding. Where surfaces are rough, the asperities may interfere with jet generation and cause additional heating at the weld interface, with the possibility of excessive melt formation. Therefore, although the explosive charge can be increased to compensate for a rough finish, this may have a deleterious effect on clad quality.

7.2.2 Materials and metallurgy

A wide range of metal combinations have been welded/clad by the explosive process with general agreement that over 300 combinations have been successfully joined. The factors determining weldability include the ductility, melting point, density and thickness of one or both components [5]. With respect to ductility, Du Pont [6] has suggested that an elongation of at least 5% in a standard tensile test or an energy of at least 14 J in a Charpy V-notch impact test is required in the flyer plate to avoid cracking. It is usual to make the more brittle material the backer plate, but dependent on size and material, the brittle flyer plate can be preheated to impart more ductility.

Metals of very low melting point (e.g. solderable alloys) can melt very quickly under impact and this interferes with the transmission of the shock waves. For metals with a large difference in density, welding becomes more difficult and Du Pont has again suggested a limiting factor of $9\,g/cm^3$ difference. Above this factor, there is an increasing loss of symmetry in the field behavior of the two metals at the collision point such that more jetted material becomes trapped at the interface.

A surprisingly wide range of material thicknesses can be clad onto a base plate and Chadwick and Jackson [5] list a range of commercially available clad material thicknesses (Table 7.1) showing that flyer thicknesses of 0.4–50 mm (50 mm for aluminum) have been successfuly welded. More typical thicknesses tend to be in the range 1.5–10 mm.

The combinations of metals which have been joined by explosive welding are numerous. Figure 7.2 attempts to represent a simplified range of combinations, taken from [5] and [7]. In addition, limited trials have shown some success in the joining of thermoplastics by explosive cladding.

Because explosive cladding is a cold process, there is no thermal macrostructure associated with the weld and the features to be observed often include a sinusoidal or 'wavy' interface, localized melting and intermetallic formation

Table 7.1 Various cladding materials and thicknesses [5]

Cladding metal	Thickness for routine cladding (mm)	
	Min	Max
Aluminum	6	50
Aluminum bronze (CA106)	1.5	20
Brasses	1.5	20
Copper (OF, DHP and DLP)	1.5	22
Cupronickel (90/10)	1.5	22
Cupronickel (70/30)	1.5	22
Hastelloy (B, B2, C, C-276, C4 and G)	1.5	13
Incoloy (800 and 825)		
Incoloy (600 and 625)	1.5	20
Monel (400, 404 and K-500)		
Nickel (Gd. 200, 201)	1.5	20
Nickel–silver (Du Pont)	1.5	20
Platinum	0.4	ND[a]
Silicon bronze (Du Pont) (Everdur-1015 and high silicon 655)	1.5	16
Stainless steel (austenitic and ferritic)	1.5	25
Tantalum	0.5[b] or 1.5	>6
Titanium	1.5	20
Zirconium	ND	12

[a] ND = not determined.
[b] If copper interlayer is used.

	Alloy steels	Aluminum alloys	Brass/bronze	Carbon steels	Copper alloys	Copper nickle	Gold	Nickle alloys	Platinum	Silver	Stainless steels	Tantalum	Titanium	Vanadium	Zinc	Zirconium
Alloy steels	X	X		X	X	X		X			X			X		
Aluminum alloys	X	X		X	X	X		X			X		X			
Brass/bronze			X					X								
Carbon steels	X	X		X				X			X	X	X			
Copper alloys	X	X			X			X			X		X			
Copper nickle	X	X				X		X			X					
Gold							X									
Nickle alloys	X	X	X	X	X	X		X		X	X					
Platinum									X							
Silver										X						
Stainless steels	X	X		X	X	X		X			X					
Tantalum	X			X								X				
Titanium	X	X		X	X								X			
Vanadium	X													X		
Zinc															X	
Zirconium																X

Figure 7.2 Combination of metals which may be joined by explosive cladding.

(specific dissimilar metal combinations), possibly with cracking or inadequate bonding. The wavy interface (Fig. 7.3) is almost synonymous with explosive welding, although it is not essential for welding. It is formed as a result of the fluid mechanical behavior of the surfaces during impact. Various theories for wave formation have been postulated and these are reviewed in [4].

Despite the lack of metallurgical change in the materials, there is evidence of localized metallurgical effects as seen by hardness increases and associated increases in tensile strength and reductions in ductility. They are believed to be caused by shock hardening and their level of occurrence will depend on the parent materials and their prior cold work or other hardening mechanism. Indeed, highly worked steels may see local softening because the localized interfacial heat causes some annealing.

Detailed examination of the wavy interface of an explosive weld (Fig. 7.3) shows the localized metal movement in the vortices of the waves. Also, although not clearly evident, it is possible that some melting has occurred in the center of these vortices. In some welds, a more clear cast structure can become evident; this is preferably avoided because of the poor properties associated with it.

Two popular dissimilar metal combinations for explosive cladding are aluminum to steel and titanium to steel. Both of these combinations suffer from intermetallic formation, and indeed these welds can exhibit small entrapped locations of intermetallic. With correct choice of parameters, the

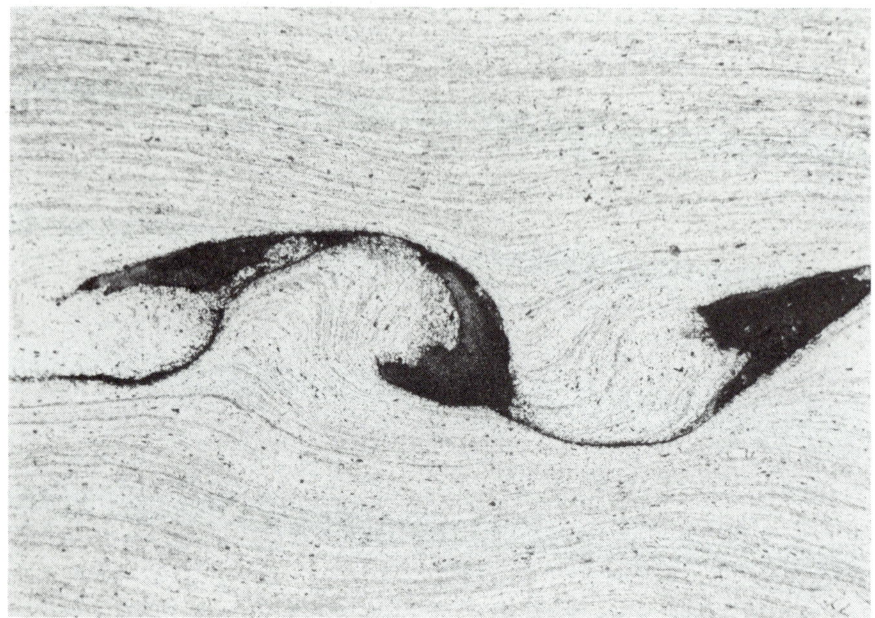

Figure 7.3 Typical microsection of explosive weld showing characteristic wavy interface.

intermetallic formation is minimized and satisfactory welds are achieved. It should be remembered that, as a cold welding process, explosive cladding is probably the most attractive way to join such a difficult combination as titanium to steel. The limitations are primarily associated with the restricted geometries that can be clad.

Where it is difficult to achieve the required level of control during welding, some dissimilar combinations are welded with an appropriate interlayer of another metal. For example, 5000 series aluminum is not readily welded to 300 series stainless steel directly but can be welded via a pure aluminum interlayer [8].

7.2.3 Equipment

There is no equipment, as such, in explosive welding. The key feature is the applicaion of the explosive layer which is often a manual process, and detonation occurs through the manual ignition of suitably located detonators.

The safety features associated with the storing and handling of explosives, together with the envrironmental issues arising from large-scale detonations, mean that explosive cladding is a specialist service activity. If plates are required clad in this way, users will purchase from the explosive cladding companies.

There may be exceptions to this for small-scale cladding, and developments have been made in the provision of hand-operated explosive welding machines which will carry charges of a few grams. Such equipment can spot weld 1 mm thick material and the noise is no greater than that given by a 12 mm bore shotgun. However, the requirements for handling and storing the explosive will remain.

7.2.4 Applications

Flat clad plate

By far the largest tonnage of explosive cladding comes in the form of flat plate. The requirement for clad plate spans many industries and essentially gives the designer and manufacturer the advantage of increased performance at minimal cost. Typically the aims are to increase operating pressures and temperatures and/or to operate in increasingly hostile environments. Explosively clad plate tends to fulfill the market for heavier section plate than is possible for roll bonded plates, and with a wider range of clad materials.

Explosively clad plate may be further processed by hot and/or cold rolling. Reductions of 90% or more are possible even with such dissimilar combinations as titanium on steel. However, in examples such as these, careful control of the temperature must be maintained to restrict the growth of intermetallic compounds at the interface and hence maintain ductility. Limiting temperatures for common clad plates are shown in Table 7.2, and Fig. 7.4 shows the

Table 7.2 Limiting temperatures for clad plate [9]

Combination	Maximum temperature (°C)	
	Fabrication (hot forming)[a]	Service
Stainless–CS[b]	1100	
Monel–CS	1100	
Hastelloy–CS	1200	
CuNi–CS	1000	
Copper–CS	925	540
Titanium–CS	800	400
Copper–aluminum	260	150
Aluminum–CS	315	260

[a] Minimum time at temperature.
[b] CS carbon steel.

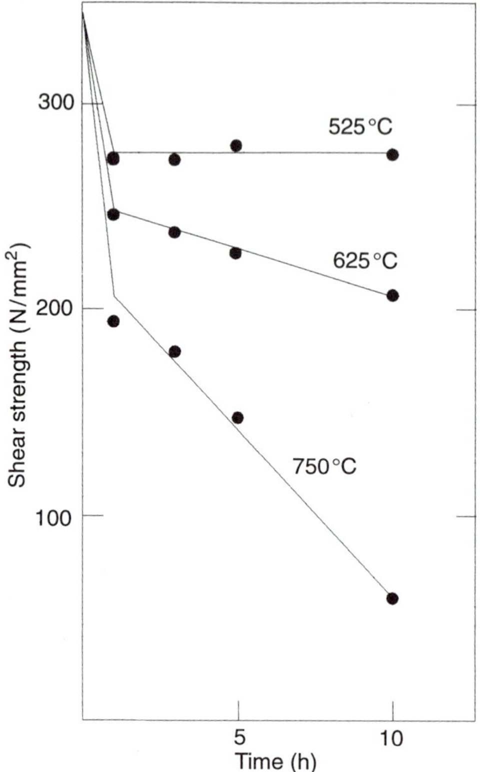

Figure 7.4 Titanium–carbon clad shear strength after postweld heating [9].

reduction in shear strength of titanium–steel clad plate after heating to temperatures up to 750 °C [9].

The explosive cladding process can also be combined with roll bonding to produce a competitively priced plate. An example includes a thin layer of stainless steel explosively clad to mild steel and subsequently roll bonded to structural steel.

Marine

The use of aluminum as a superstructure material in the shipbuilding industry has highlighted the shortcomings of bolted or riveted dissimilar metal combinations. In the presence of seawater, steel and aluminum form a galvanic cell such that corrosion can take place at the interface. Within the ship's structure, crevice corrosion could go unnoticed until replacement is required. The use of explosively welded transition pieces [10] removes the crevice, and although corrosion can still occur, it is moved to the external surfaces where it can be monitored and corrosion products removed. From these transition joints, exploitation has moved to the use of clad plate, providing the opportunity for new designs in ship superstructure.

Chemical process vessels or heat exchangers

Clad plate has been a valuable construction material for chemical plant vessels and heat exchanger tube-plates for many decades, due to the combined cheapness of the carbon steel substrate and the high corrosion resistance of the clad layer (e.g. stainless steel). When these products first became commercially available in the 1960s, there was an improvement in quality and range over previously available plates. Although stainless steel remains the most widely used cladding material, other more exotic materials such as titanium, zirconium and tantalum have found significant application. The size of clad can be considerable, up to $27\,m^2$ in a single firing and individual tube-plates up to 450 mm thick and weighing over 50 tonnes have been clad [11].

For chemical process vessels it is necessary to form the clad plate into a cylindrical vessel. Forming of the plate is not a problem, and conventional processing can be used (with the temperature restrictions mentioned earlier). However, special procedures are required for the welding of clad plate involving a reactive metal, such as titanium.

Electrical

Aluminum, copper and steel are the most common metals used in high current–low voltage conductor sysems, and combinations are required to maximize the advantages of the specific properties of each material. Transition joints are therefore used which are machined from clad plate in aluminum–steel and aluminum–copper. These are subsequently introduced by conventional welding between like materials. This concept is routinely employed by the primary aluminum reduction industry in the fabrication of anode rods. The mechanical properties (shear, tensile, impact strength) of the aluminum–steel explosive weld exceed those of the parent pure aluminum. Similarly, copper–aluminum junctions are used as bus bars in the electrical industry.

Power generation

Considerable emphasis is placed on the reliability of power generation systems and subsystems. Where seawater is used for cooling there has been continuous upgrading of the materials used to construct the plants to resist corrosive attack. The most favorable material for resistance against such corrosion is titanium, and this is now being used, in clad form, in condenser tubing and in the tube sheet.

Cylindrical cladding

Much of the discussion so far has concentrated on the simple plate arrangement for explosive cladding. However, increasingly cladding is being carried out on tubular configurations, furthering the development of bimetallic cylinders. The main application of duplex, bimetallic tubing lies in the field of heat exchangers and transport of chemicals, but marine components have also been required as in, for example, flange bore cladding of mild steel with cupronickel. Generally this process has been limited to short, large diameter cylinders which are more readily welded, followed where necessary by extrusion or drawing to meet the final dimension requirements.

7.2.5 New developments

Onward rolled clad plate

The combined use of roll bonding and exlosive welding was mentioned earlier and in itself is not new. However, certain combinations of metal are difficult to onward roll where the clad layer is significantly stronger than the backing layer as in, for example, high nickel alloys to carbon steel. In these cases, the lower yield strength backer component tends to roll preferentially, causing extension of this layer and an associated inward curling of the composite clad.

This combination of metals is seen to be important for flue gas desulphurization in power stations. The proposed solution [12] is to roll two explosively clad plates simultaneously with a separating agent between the two. Successful results have been achieved with this route, and an added advantage is that the nickel alloy is kept clear of the rolls so that contamination or deformation is minimized.

Clad tube

Cylindrical cladding, as described earlier, has been widely used to produce bimetallic tubing for the chemical industry. Another major application is in the oil and gas industry for the transportation of corrosive media. Here, it is important that longer tubes are available to make the process cost-effective. Considerable development is therefore being addressed in the production of ingots 3 m long, between a carbon steel and an inner corrosive alloy such as stainless steel or nickel alloy. Such ingots are then sectioned and drawn, making three 6 m lengths of clad tube [12]. The advantages of this approach include consistency and accuracy of the final tube dimensions when compared to a direct clad product, and the additional length of tube minimizes the not inconsiderable onward joining costs. Bond integrity can also be more uniform because of the elimination of end effects.

Hybrid metal packages

Many of the earlier applications refer to the explosive cladding process being applied to structural components of relatively large section. A recent innovation demonstrates the versatility of hermetic packaging within the electronics industry.

Hermetic metal packaging of electronic devices, particularly high power hybrids, usually involves brazing. This process can be prone to porosity and stress senstivity; stress sensitivity results from joining materials of different thermal expansion coefficients. The materials involved in electronic packaging include cold rolled steel, stainless steel, molybdenum, aluminum, copper, Glidcop (dispersion-hardened copper) and Kovar (iron–nickel–cobalt alloy). Kovar is a particularly popular material because its coefficient of thermal expansion is close to that of the ceramic substrates inside the package, and the glass seals used to insulate the package leads.

The recent development involves the use of explosively bonded metal packages where Kovar or stainless steel is joined to a hard copper base. The

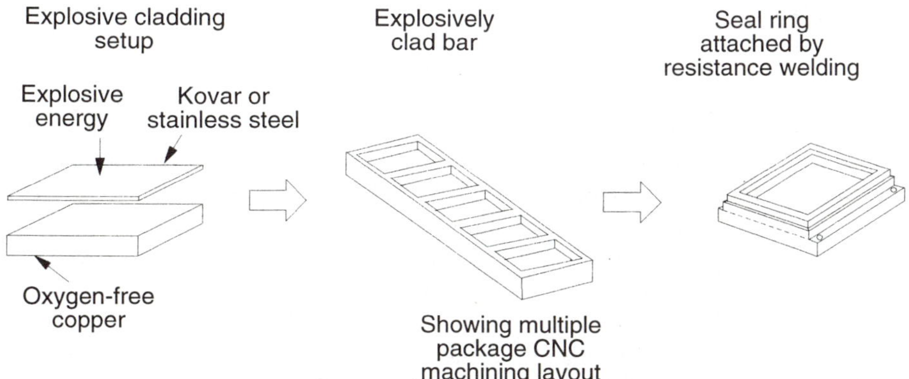

Figure 7.5 Typical explosive weld procedure for manufacture of an electronic hybrid package [13].

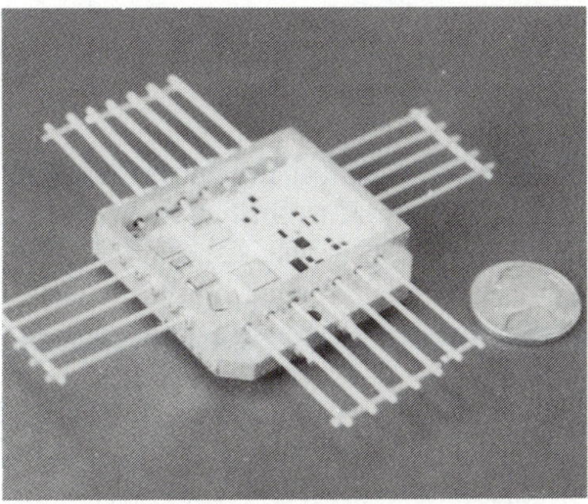

Figure 7.6 Explosively clad electronic hybrid package [14].

whole is subsequently machined (by CNC) to produce the cavity within which the electronic substrates are mounted, Fig. 7.5 [13]. A typical completed package is shown in Fig. 7.6. The bonds produced by this technique provide a strong and hermetic seal, with the potential for high yields and reliability.

A further interesting feature is the ability to bond to the new composite metal packages comprising aluminum reinforced with silicon carbide. These materials have been developed specifically for their low thermal expansion coefficient (close match to silicon) but pose problems when it comes to hermetic sealing because of the difficulty of welding or brazing them. Explosive cladding is essentially a cold welding process, so it enables these metal composites to be joined to metals such as stainless steel [14].

7.3 ROLL BONDING

Hot roll bonding is said to account for more than 90% of the clad plate production worldwide [15]. Corrosion-resistant clad steel is being increasingly employed in the oil and gas industries because of the move towards more corrosive fields. The initial high cost of clad plate is more than offset by its ability to transport corrosive fluids, by the savings in the use of inhibitors and by reduced inspection and maintenance.

A summary of clad product production methods from [15] is reproduced in Table 7.3. It can be seen that hot roll bonding is used for plate thicknesses of 6–200 mm, where the clad layer can be from 1.5 mm to 40% of the total thickness. Also, a significant advantage with hot rolling compared to other methods is the ability to produce tube lengths up to 14 m.

7.3.1 Principle and mechanism

The plate surfaces are prepared for bonding by mechanical abrasion or machining (grinding) followed by degreasing. The cleaned surfaces are then brought into contact and are heated and forced through sequential rollers to reduce the cross-sectional thickness and to encourage metallurgical bonding between the materials. The reduction in thickness leads to an expansion of the materials at the surface and the breakup of surface oxides. The superclean surfaces, brought into intimate contact, are bonded together by atomic diffusion across the interfaces (i.e. diffusion bonding).

In practice, two clad plates can be rolled simultaneously, effectively forming a sandwich. The plates are assembled such that the corrosion-resistant layers are next to each other but are separated by a stop-off compound such as chromia or zirconia. This has the advantage of preventing contact of the alloy with the rolls, hence stopping contamination. An additional advantage is the reduction in distortion caused because the two dissimilar metals elongate at different rates. 'Sandwich' rolling compensates for this effect and leads to much flatter plate production. Prior to rolling the clad plates can

Table 7.3 Summary of clad product production methods and dimensional availability [15][a]

Product	Wall thickness	Width or diameter	Max length
Roll bonded plate	6–200 mm Cladding 1.5 mm to 40% of total thickness	1000–4400 mm	14 m
Explosion bonded plate	Cladding 1.5–25 mm Minimum base 3 × cladding thickness No limit to max thickness of base	50–3500 mm	5 m
With hot rolling		1000–4400 mm	14 m
Overlay welded plate	Base metal > 5 mm Clad layer > 2.5 mm	Limited only by access of equipment	Limited only by access of equipment
Longitudinally welded pipe	6–32 mm	219–1016 mm	8 m or 12.8 m depending on supplier
Liner pipe (thermal, shrink fit)	7–24 mm Total wall liner 2–20 mm	60–400 mm	9.6 m or 12 m depending on diameter
Liner pipe (explosive mechanical joint)	Outer pipe > 5 mm Liner < 5 mm (depending on diameter)	50–500 mm	12 m
Seamless pipe Extruded Plug/mandrel mill Seamless pipe	6–25 mm	60–400 mm	12.8 m
Explosion metallurgical joint	6–20 mm	200–250 mm	3 m or 5 m
After cold rolling	2–20 mm	50–200 mm	6–12 m
Centricast pipe	10–90 mm Clad layer minimum 3 mm	100–400 mm	4–6 m
HIP clad pipe or fittings	> 5 mm Clad layer minimum 2 mm	25–100 mm	2 m
Weld overlay fittings	Base metal > 5 mm Clad layer > 2.5 mm	25 mm minimum	For small diameters limited by torch length, e.g. 1 m for diameter 50 mm No limit on large diameter

[a] Not all combinations of wall thickness, diameter and length may be possible; readily available sizes are quoted; exceptional sizes may be possible.

be welded together to prevent separation. Generally rolling takes place in air, but to prevent or minimize oxidation, an inert atmosphere such as argon may be used.

7.3.2 Materials and metallurgy

Typical alloys comercially clad to carbon steel are shown in Table 7.4. With certain of these combinations it is possible for brittle phases to be formed at the interface during roll bonding, due to the formation of intermetallics or complex carbides. This is usually prevented in production by plating the substrate with a layer of nickel or iron to act as a buffer.

Because of the wide use of clad plate within the offshore industry, various standards have been developed to control the quality of the product:

ASTM A 264   Specification for stainless chromium–nickel steel clad plate, sheet and strip
ASTM A 265   Specification for nickel and nickel-base alloy clad steel plate
ASTM A 578   Straight beam ultrasonic examination of plain and clad steel plates for special applications

Plate rolling is normally followed by a heat treatment to restore the cladding to the solution-annealed condition and return the backing steel to its appropriate condition (e.g. quenched and tempered, normalized). Selection of a heat treatment to suit both cladding and backing materials can be problematic. Where low annealing temperatures are involved ($<1000\,°C$), simple heat treatments can be used; but as this temperature increases, to over $1150\,°C$, there can be problems with grain growth and loss of toughness in the backing steel. From the point of view of corrosion properties, Charles, Jobard and Catelin [16] have examined a series of corrosion tests on clad plate subjected to simulated heat treatments. For plates clad with both stainless steels (austenitic and duplex) and nickel alloys (625 and 825), satisfactory corrosion resistance was found with the conclusion that hot rolled plate provided a cost-effective option for use in oil and gas process industries.

As well as the stainless steels and alloys referred to earlier, hot roll cladding is also applicable to other dissimilar combinations, including metals such as copper, titanium and zirconium. The use of the process will depend upon the forming properties of the materials and the formation of brittle compounds at joint interfaces. As indicated earlier, the formation of brittle compounds may be alleviated by the use of suitable buffer interlayers but with added cost in the more complex combinations (e.g. titanium to steel).

Other materials which are conventionally roll bonded are aluminum and its alloys, to themselves or to other nonferrous metals. Solid phase bonding of aluminum alloys is difficult because of their inherently stable oxide, but the conditions of heat and deformation that prevail in hot roll bonding are sufficient to break up the surface oxides and allow sound bonds to be made.

Table 7.4 Typical cladding alloys[a]

	UNS designation	Nominal composition						
		C_{max}	Cr	Ni	Mo	Cu	N	Other
Alloy 405	S40500	0.08	11.5–14.5					Al 0.15
Alloy 410	S41000	0.15	11.5–13.5					
Alloy 304L	S30403	0.030	18–20	8–10				
Alloy 316L	S31603	0.030	16–18	11–13	2.5–3			
22 Cr duplex[a]	S31803	0.030	22	5	3		0.15	
25 Cr duplex[a]	S32550	0.030	25	6.5	3.5	1.6	0.23	
Alloy 904L[a]	N08904	0.020	20	25	4.5	1.5		
Alloy 28[a]	N08028	0.020	27	31	3.5	1		
Alloy 825[a]	N08825	0.020	21	42	3	2		
Alloy 625[a]	N06625	0.030	21.5	58 min	9			Nb 3.6
Alloy C22[a]	N06022	0.01	20–22.5	bal	12.5–14.5			Fe 2.6 W 3
Alloy C276[a]	N10276	0.01	15.5	bal	16			Fe 4–7 W 3.75
Alloy 400	N04400			63 min		31		Fe 2.5
90–10 CuNi	C70600			10		bal		Fe 1.5 Mn
70–30 CuNi	C71500			30		bal		Fe 0.6

[a] Other alloys may be clad and also pure metals such as nickel, copper, titanium (and alloys) and zirconium.
[b] Cladding alloys of particular interest in oil and gas production.

Typical examples included Al to Al–Si alloy and aluminum alloys clad to zinc, copper and tin.

7.3.3 Equipment

There is no specialist equipment for roll bonding; it requires the exended use of conventional hot rolling machinery as used in primary metal fabrication plant. Therefore the main manufacturers of roll bonded plate are, for example, the steel producers and the aluminum sheet/plate producers. Sheet or plate is usually preheated and then passed through a series of heated rolls, each leading to some further reduction in plate thickness. When dissimilar metals are passed through the rolls, for bonding, additional care is necessary because of the differential deformation characteristics of the two metals. This problem is avoided in severe cases by the sandwich or simultaneous rolling mentioned earlier.

7.3.4 Applications

The main application of hot rolled plate is in the oil and gas or chemical processing industries, where typically a carbon or low alloy steel is clad with a corrosion-resistant alloy. The plate is often formed into pipe and longitudinally welded using submerged arc, TIG or MIG welding. It is important to ensure a sound root and provide a continuous corrosion-resistant layer at least as thick as the internal cladding.

Another prime application is in heat exchangers, particularly aluminum plate/fin heat exchangers for industrial and aerospace use. These exchangers are manufactured from many alternate layers of aluminum plate and finned aluminum sheet, stacked and then dip or vacuum brazed. To ease assembly, the braze alloy is placed by roll bonding the required alloy (usually Al–Si base) to the 3003 aluminum plate. Such plates are required in small assemblies up to the very large exchangers which can measure 1.5×6 m.

Similar bonded sheets but with zinc, copper or tin bonded to aluminum have been used for diffusion bonding studies and applications, where the cladding is used to aid the subsequent aluminum–aluminum bond.

7.4 FRICTION SURFACING

Friction energy has long been used as a means of providing the heat needed to weld a wide variety of materials. The method has today gained a prominent position worldwide as a commercial joining process in industries as diverse as aerospace, automotive and offshore oil exploration.

The solid phase process of friction welding has been further modified to allow the deposition of surface layers onto a substrate. This method of metal transfer is now known as friction surfacing. Although this technology is regarded as novel, the original concept was first reported in 1941 by Klopstock and Neelands [17]. Since that time and until the early 1980s there is little reference to the technique. However, the potential of the process has grown with the increasing interest in surface engineering, and publications on the process can be found on a worldwide basis.

7.4.1 Principle and mechanism

Friction surfacing is another solid phase cladding technique; it uses a combination of heat and deformation to clean surfaces and metallurgically bond metals together. In its simplest arrangement (Fig. 7.7) a rotating consumable bar is brought into contact, under a low load, with a stationary substrate. At initial contact, the rotating bar is preferentially heated to form a plasticized layer by the frictional motion. Once this plasticized zone has developed sufficiently (usually in 1–3 s), the substrate plate is traversed tangentially to the bar and the plasticized metal is transferred to the plate.

Bonding occurs by the combination of self-cleaning between the two materials and the application of heat and pressure to encourage diffusion across the interface, thereby forming a solid phase metallurgical bond. The process relies on producing precisely the right temperature and shear conditions at the interface between the rotating bar and substrate via the plasticized layer.

A typical deposit laid down by friction surfacing is shown in Fig. 7.8. In this particlar case, the deposit is of stainless steel onto low alloy steel. A flat, uniformly rippled and oxidized surface finish can be clearly distinguished. Figure 7.9 is a photomicrograph through this deposit, which can be considered representative of most deposits made by this method. Closer examination indicates a small zone of lack of adhesion at the outer edges (cold laps) and the presence of a narrow, crescent-shaped heat affected zone (HAZ) in the substrate.

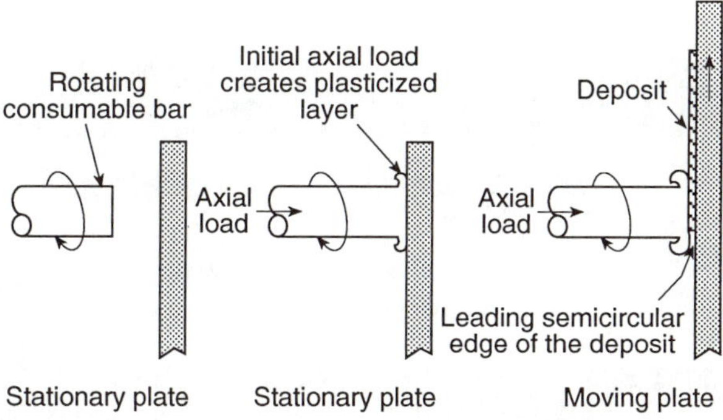

Figure 7.7 Schematic arrangement for friction surfacing.

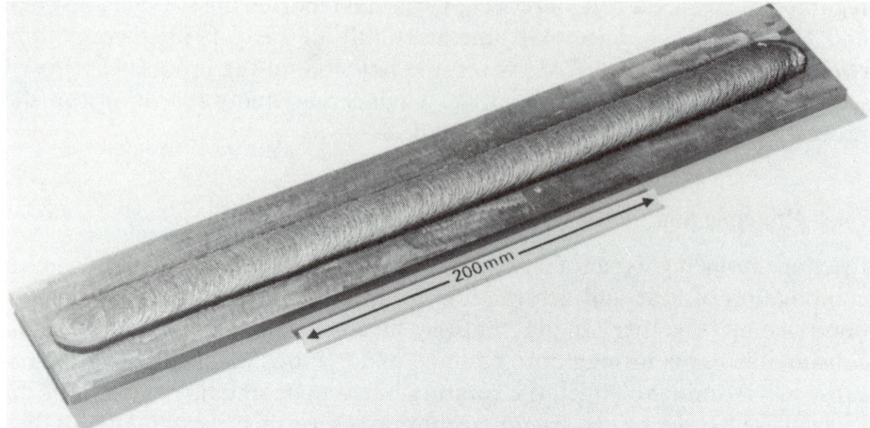

Figure 7.8 Typical friction surface deposit: stainless steel on low alloy steel.

Surfacing can also be made onto a circular substrate (i.e. shaft or tube) in which case a uniform HAZ can be formed and the small overlap zone (start/stop position) does not detrimentally affect the previously formed bond between the clad layer and substrate (Fig. 7.10). Detailed inspection of the friction surface deposit, under optimum surface conditions, shows some slight difference in appearance between the two side edges. With the forward movement of the substrate, the velocities from the angular rotation of the consumable and from the substrate are added together algebraically. They intensify each other on the advancing side of the rotating consumable and are subtracted from each other on the retreating side (Fig. 7.11). In some respects, a comparison can be made with a helicopter in motion, but under normal process conditions the contribution of the linear traverse is less than 1%. This would not be expected to lead to any significant asymmetry in heating or forming the plasticized layers. However, this lack of symmetry is

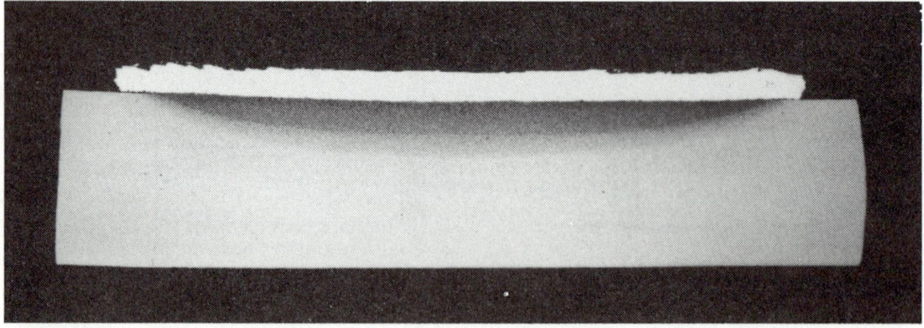

Figure 7.9 Cross section through a friction surface deposit: stainless steel on low alloy steel.

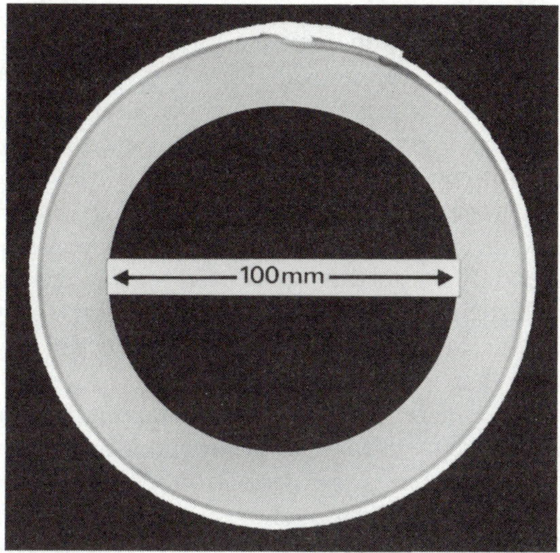

Figure 7.10 Cross section of stainless steel clad onto a tubular component.

always more noticeable at the lower rotational velocities, presumably because the relative velocity of the substrate is greater; moreover, with the lower rotational speed, an increased bulk of plasticized material is generated [18].

The deposition zone of the substrate receives a scouring action, the result of relative motion while under an axially applied load, via the plasticized

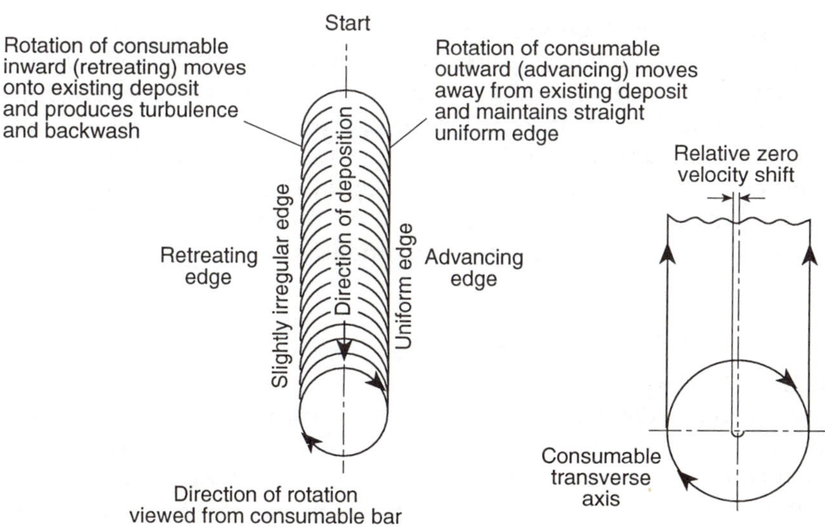

Figure 7.11 Deposition characteristics of friction surfacing.

layer. This scouring action is greater for the central portion of the deposition zone than the edges, which only experience a proportionally brief period of surface action from the outer consumable edge. This leads to the so-called cold lap; an additional factor is the unequal distribution of frictional energy at the free edge of the consumable because this edge is unable to transmit the same degree of pressure.

Friction surfacing is regarded as an efficient process which relies on friction energy to produce the right temperature conditions precisely at the interface between deposit and substrate; this is produced by the mechanism of frictional contact, via the plasticized layer. The consumption of the rotating bar occurs because of an unequal temperature distribution between the comparatively small consumable bar and the bulk substrate. A relative increase in temperature occurs preferentially at the consumable interface, which then offers lower resistance to the shear mechanisms taking place in the consumable such that metal from the bar is transferred to the 'colder' substrate. Friction surfacing enhances this effect by providing a lateral motion which continually introduces new ambient temperature substrate to the rotating consumable bar.

After the initial dynamic contact by a rigid consumable and after traverse has been initiated, the scouring action necessary to disperse the continuous intrusion of the substrate oxide barrier is continued, not by the contact face of the consumable, but by the plasticized layer produced from the consumable. While it is recognized that this plasticized layer has lower mechanical strength and correspondingly produces less of a scouring effect than during the initial contact phase (provided the axially applied force is maintained), it has been shown practically that the oxide dispersal will continue during processing and will result in sound bonds.

The relative lateral movement between consumable and substrate exposes the continually generated semicircular leading edge of the deposit to the atmosphere. This may introduce oxides into the deposit (Fig. 7.7) and these may influence the quality of the deposit, especially for the more reactive materials (e.g. titanium), which may require the use of a suitable gas shield. Dispersal of the existing oxide into the deposit is unlikely to be modified to any great extent by process conditons. Nevertheless, sound deposits of good mechanical strength and adhesion are produced.

7.4.2 Materials and metallurgy

Compared to explosive cladding and roll bonding, friction surfacing is a relatively recent commercial process and therefore fewer material combinations have been examined, although the process does have potential to weld many dissimilar combinations. With the standard process of friction surfacing, the limitations on materials include availability of bar stock – now less of a limitation (see later) – and the consumable must generally have a lower melting point and must not be excessively harder than the substrate. The melting point and hardness resrictions are to prevent melting of the substrate or drilling into the substrate rather than surfacing onto it. Metallurgical compatibility may also be a factor, primarily to avoid combinations forming excessively brittle intermetallic phases. However, the intermixing between the two materials is minimal, occurring over only a few microns or tens of microns, so the process is much more forgiving than any of the fusion technologies. The materials which have been examined in greatest detail are described below.

Mild steel to mild steel

The earliest trials covered mild steel to mild steel because of the ease of welding and the metallurgical simplicity when evaluating a new process. These preliminary tests quickly demonstrated the viability of friction surfacing and provided guidelines to the welding variables required to lay down high integrity deposits of nominally 2 mm thickness at deposition rates approaching 4.5 kg/h. It was determined that deposition was tolerant to a relatively wide change in variables, e.g. speed variation from 900 to 1500 rpm; friction surfacing forces from 28 to 45 kN; and substrate traverse rates of 3–7 mm/s.

Mild steel deposited on 1.5 Ni–1 Cr–0.25 Mo low alloy steel

Using welding variables identical to those developed for mild steel to mild steel has resulted in sound deposits for mild steel deposited on 1.5 Ni–1 Cr–0.25 Mo low alloy steel; good interface adhesion was proved by longitudinal bend testing. Despite deposition onto a hardened alloy steel, the substrate shows no signs of cracking and the substrate microstructure is transformed to martensite. The latter raises Vickers hardnesses in the martensite region in excess of 550. Therefore, no preheat of the low alloy steel substrate is required, even with a carbon content of 0.4%. In direct contrast, preheat would most certainly be needed to prevent cracking if the more conventional fusion cladding processes were to be used with this substrate.

Austenitic stainless steel deposited on mild steel

Mild steel is readily surfaced with austenitic stainless steel, even though the stainless steel has higher hot-strength characteristics. However, the rotational speed of the consumable can considerably influence the deposition characteristics. At high speeds (2190 rpm), the deposit layer is reduced in thickness and exhibits a partly glazed surface; this, together with lower friction loads, will produce a narrowing of the deposit width. At much lower rotational speeds (330 rpm), a marked increase in deposit thickness is seen but is accompanied by a reduction in effective bond width. Optimum surfacing conditions are found at 550 rpm rotational speed, 50 kN force and 5 mm/s traverse speed, giving a deposit thickness of 1.3 mm, for a consumable diameter of 25 mm.

Traverse and longitudinal sections through the clad layer formed under these conditions have confirmed sound, basic adhesion with virtually no penetration of the deposit into the substrate. Close examination of the bond interface has shown the presence of a 10 μm deep decarburized zone of ferrite in the HAZ of the substrate. Energy dispersive X-ray analysis has confirmed low dilution of the process within 20 μm of the stainless steel interface (measuring Cr levels).

Through-thickness tensile tests on such deposits have shown strengths of 570 N/mm² with failure typically occurring in the mild steel parent metal away from the interface and HAZ. Bend testing by longitudinal and side-unrestrained bend tests has shown good bond integrity – 180° bonds are achieved with no visible cracks.

Austenitic stainless steel deposited on austenitic stainless steel

Austenitic stainless steel is also readily surfaced with austenitic stainless steel. It has been found necessary to reduce the traverse rate to 4 mm/s to achieve a sound deposit of 1.5 mm thickness without changing the machine settings from those optimized for the stainless steel–mild steel combinations (described above).

Simple longitudinal and side bend tests to angles of aproximately 180° have demonstrated the integrity of the deposit, and through-thickness tensile tests have given strengths of 540 N/mm² with failure away from the deposit interface.

Transverse sections have again shown the typical cold laps at the outer edges of the deposit, however, the clad layer is free from flaws at the bond line and a fine-grained, hot-worked microstructure with no voids is obtained.

Nickel alloy 625 deposited on mild steel

The primary change in welding conditions to achieve satisfactory deposition for nickel alloy 625 deposited on mild steel involves a reduction in the relative rubbing velocity. This can be obtained by reducing the rotation speed (to 410 rpm for a consumable diameter of 20 mm). Deposits laid down under these conditions have been shown to be capable of withstanding tensile stresses of approximaely 460 N/mm² without failure at the clad interface. A side bend test can produce a 170° bend, but a transverse bend test can produce cracking in the deposited alloy 625, although it remains well adhered to each side of the substrate.

Metallurgically, this metal combination again shows the low dilution typical of friction surfaced deposits. Both nickel and iron line scans across the interface, made using energy dispersive X-ray analysis, have revealed a very sharp transition (within 1 μm) across the interface (Fig. 7.12). With

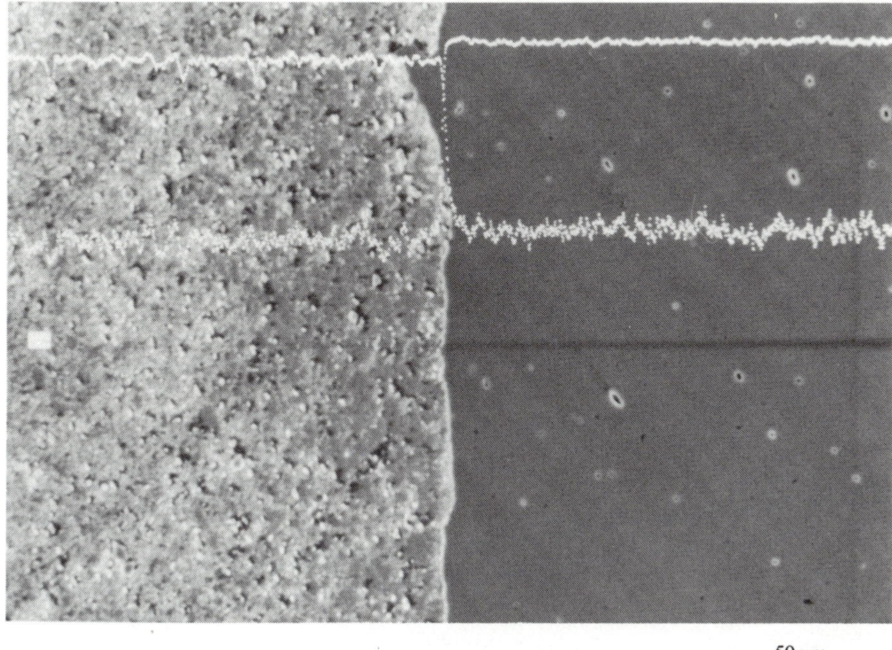

Figure 7.12 Cross section of a friction surface deposit of nickel alloy 625 on mild steel showing abrupt transition across interface by nickel and iron element scans.

virtually no iron dilution, it is not unreasonable to assume that the corrosion characterisics of the alloy 625 would be unimpaired, even with a relatively thin clad layer of 1.30 mm. Hardness measurements have revealed some small hardening of the alloy 625 layer, whereas the HAZ of the substrate is softened compared to the original cold drawn base metal plate.

Hastelloy (CW-12M-1) deposited on stainless steel

Hastelloy is a typical hardfacing deposit on stainless steel, where a uniform microstructure and low dilution are required. With friction surfacing, a low rotational speed has proved necessary to provide an acceptable visual appearance and usable deposit thickness. For a consumable diameter of 20 mm, a rotational speed of 330 rpm, a force of 56 kN and a transverse speed of 4 mm/s has produced coatings with acceptable characteristics.

Mechanical test results have proved acceptable, although longitudinal side bend tests result in small cracks in the clad layer. This is not unusual in the hard layer, but importantly there is no evidence that cracks propagate into the substrate under these static test conditions.

Vickers hardness of the as-deposited Hastelloy increased to greater than 420 when compared to the original consumable of 207. Dilution was limited to $1-10\,\mu m$ from the Hastelloy to the stainless steel.

Stellite grade 6 deposited on stainless steel

Trials with Stellite grade 6 on stainless steel have again shown the need for relatively low surface velocities (rotational speeds) for the higher hot-strength materials (in this case, a cobalt-based hardfacing alloy). At conditions of 330 rpm rotation speed, 39 kN force and 5 mm/s transverse speed, deposits are made with an average thickness of 0.7 mm from a 20 mm diameter consumable. Similar to the Hastelloy deposit, bend tests result in a series of cracks in the Stellite deposit because of its inherent lack of ductility. However, these again do not extend into the stainless steel (under static testing) and the separated sections of the deposit remain firmly attached to the substrate. The adhesion of the deposit is confirmed by through-thickness tensile testing, which has given strengths of 620 N/mm^2 with failure in the base metal substrate.

As with the previous Hastelloy deposit, significant Vickers hardness increases in the as-deposited Stellite were achieved at around 580, compared to the undeposited consumable hardness level of 403.

Aluminum alloy deposited on aluminum alloy (2014A) – Al–4 Cu alloy

Aluminum is readily deposited, but with parameters quite different than those used for the above combinations. Higher rotational speeds are used, typically 600–900 rpm for a 25 mm consumable bar.

Aluminum is deposited at somewhat greater thicknesses than for the higher hot-strength materials, and typical deposits can be approximately 3.0 mm thick. Through-thickness tensile testing shows good adhesion (237 N/mm^2 strength) but failure can occur through cohesion in the aluminum deposit. Transverse macrosections taken from typical aluminum deposits have revealed the typical shallow crescent-shaped HAZ and the usual presence of cold laps at the outer edges of the deposit. The remaining region of the deposit is metallurgically sound. Because a heat treatable copper was deposited in this example, in its hardened state, there occurred some softening of the deposit and the HAZ of the substrate. This was to Vickers hardness 77 compared to 116 for the deposit, and 111 compared to 128 for the substrate.

Widely dissimilar combinations

Surfacing of such widely dissimilar combinations as aluminum or titanium onto a steel substrate is not readily achieved and this is believed to be due to a combination of factors, including efficiency of cleaning, dissimilar melting points and incompatibility (e.g. brittle intermetallics). Similarly, surfacing of hard materials such as steel, onto soft substrates, such as aluminum, is unlikely to be successful because the soft material will deform preferentially.

7.4.3 Equipment

Initial friction surfacing trials were undertaken on conventional friction welding equipment, to which a traverse stage was added to provide the lateral movement between the rotating rod and substrate. As the process has developed, it has been shown that conventional machine tools (e.g. milling machines) can be converted to friction surfacing machines. The main elements are a rotating head (for consumable), a direct drive through which the force is applied and an X–Y table or rotating chuck for movement of the substrate (either plate or shaft).

For larger-scale production use, specialized equipment may need to be manufactured to provide, for example, continuous consumable feed or concurrent machining to dress the deposited layer. Friction surfacing is therefore available as a manufacturing process for use in-house, or as subcontract through specialist (and presently limited) service suppliers.

7.4.4 Applications

The potential applications for friction surfacing are seen to lie primarily in the cladding of localized areas. Typical examples include the anticorrosion surfacing of slide valve plates; the surfacing of the annular contact face of composite pipe flanges; the inlay cladding of strategic materials positioned to suit bearings and seal contact areas on shafts; and the cladding of the exposed regions of shafts that see service in arduous environments. Other applications are reviewed below.

Cutting edges

Bedford [19] has proposed friction surfacing as an option for manufacturing long-life cutting edges on knives, scissors and similar tools. They may be used for cutting paper and wood as well as plastics. Practical difficulties occur where it is difficult to quantify at which point existing tools lose their cutting edge. This may be where the paper is no longer cut cleanly, or the surface of wood is degraded. Any requirement to change or redress tools can lead to a downtime for production equipment and consequent loss of production. Work on the friction surfacing of specific tool steels to carbon steel substrates has shown that blades can be successfully produced with good cutting characteristics that offer an alternative to conventional carbide cutting tips.

Reclamation

A major drive to the increasing interest in surface engineering is to avoid or reduce the problems of worn components. But worn components still occur and often they need to be repaired by building up material to match the original. Tyayer [20] has examined friction surfacing for repairing worn shafts and has concluded that good quality deposits can be achieved.

Shell banding

The defence industry requires a layer of soft material towards the base of artillery shells. Known as a driving band, it improves the shell's rifling characteristics to provide smoothness and accuracy at firing. Typically the soft material is a gilding metal (copper-based alloy) or soft iron and it needs to be deposited onto steel. Alternative friction welding technologies have been examined for this application (e.g. radial friction welding), but they are limited because they require specialized and expensive equipment. Nicholas and Thomas [18] have proposed friction surfacing as a lower cost option, and early work has shown that excellent bond integrity can be achieved. The friction-surfaced driving band must withstand the propellant thrust and any forces experienced while cutting through the rifling of the gun barrel. The band must also react to torque insofar as the rifling imparts spin to the shell. Early trials with this technique have shown promising results.

7.4.5 New developments

Metal matrix composite deposition

Metal matrix composites (MMCs) offer considerable potential as high temperature or wear-resistant materials, and they are finding increasing application in the automotive, aerospace and electronics industries. Not only important as bulk materials, they offer significant advantages if deposited at strategic locations on a surface to provide specific properties where they are most needed. Friction surfacing of consumable bars of MMC is readily accomplished and does not differ from conventional friction surfacing. However, it is not always possible to find the required composite in an appropriate form. In this case, it is possible and potentially economic, to manufacture and deposit the composite in one operation.

The first application of this technique is described by Thomas and Nicholas [21]. The process uses a solid but hollow consumable made from the eventual matrix material. The hollow area is filled with appropriate ceramic powders. During the surfacing process, these powders become widely dispersed within the plasticized metal zone, ultimately providing a deposit of MMC. Alloy steel containing zirconia or alumina, and aluminum reinforced with alumina have been manufactured by such a technique. Although still under development, it is believed that volume fractions of ceramic reinforcement can be obtained up to 25–30%.

More recently, this work has been extended to include the manufacture of MMCs by starting only with powders (i.e. both matrix and reinforcement). The powder is premixed, or previously alloyed, and is contained within a metal or alloy tube sealed at one end with a plug and at the other with a bar held in the chuck of the surfacing machine (Fig. 7.13). Initial trials on this process [22] involved 304L austenitic stainless steel powder, on its own and mixed with chromium boride. It was shown in this work that the maximum volume fraction of chromium boride was 10% and of particular interest was the finding that the containment tube was not deposited, but softened and extruded to form an external flash.

Further studies used an MMC powder of aluminum with approximately 50% volume fraction of silicon carbide (approximate particle size 35 µm).

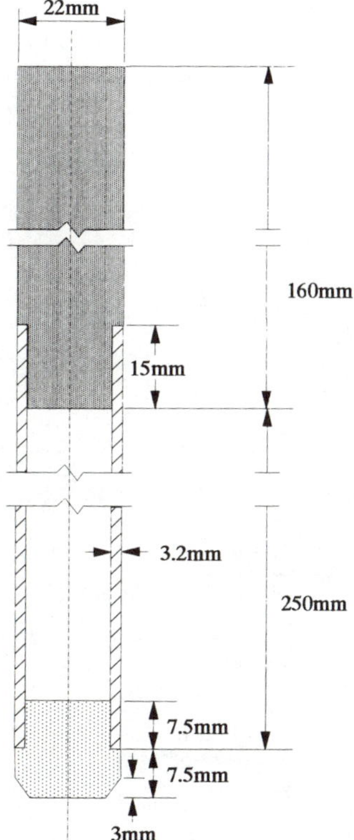

Figure 7.13 Configuration used to manufacture *in situ* metal matrix composite deposits during friction surfacing (not to scale).

This powder was used as supplied and diluted with additional aluminum alloy to give approximate silicon carbide proportions of 25, 30, 40 and 50%. The deposits were sound and there was no adverse reaction between the matrix and composite (Fig. 7.14). Further development is intended to improve the surface appearance and to examine deposition into grooves for better edge adhesion.

Applications of this process are expected in the deposition of long-life cutting edges and surfaces, and the deposition of wear-resistant inserts in engineering components such as pistons. It has the potential to be adapted to deposit a wide range of materials, including mixtures not readily deposited by other means.

In-process forging or forming

It has proved possible to reduce machining, and save a significantly greater proportion of the deposited material during the manufacture of blades and knives, if the deposit is suitably shaped as it is being laid. Figure 7.15 shows a shallow chamfer machined on the cutting side of a grooved substrate. The profile of the groove helps to form the deposit material to a shape similar to that required by the cutting edge. The outer edge of the substrate in this case can be replaced by a secondary material which can be sacrificial,

Figure 7.14 Microstructure of an aluminum–silicon carbide deposit produced by friction surfacing from powder material (50% SiC).

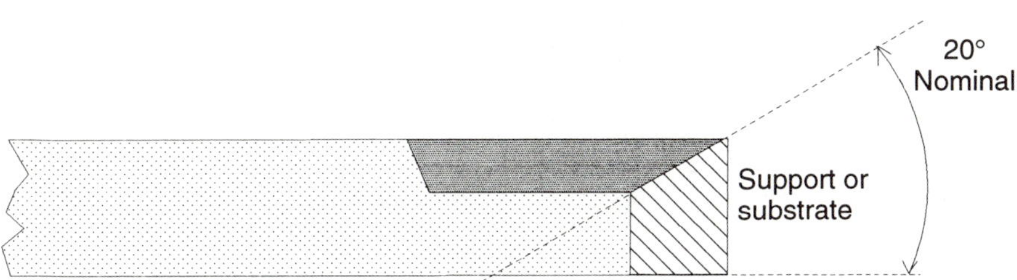

Figure 7.15 Friction surfacing into a shaped recess. The deposit is shaped using a shaped substrate or a support. Shaped deposits reduce machining and save on hard facing materials.

disposable, or reusable. Preferably the support material, such as molybdenum or Monel, is not metallurgically bonded to the deposit and is therefore reusable.

In yet another arrangement, a secondary mould is provided such that, during friction surfacing deposition, the substrate is made to deform and bend into the mould to produce a grooved substrate (Fig. 7.16). Friction surfacing with in-process forging or forming improves the bond quality at the edges of the deposit. And it becomes more economic to manufacture blades and industrial knives because the substrate itself does not require premachining.

Friction surfacing with angled substrate

A substrate may also be traversed across a rotating consumable bar using an angled support as shown in Fig. 7.17. The applied load is a function of

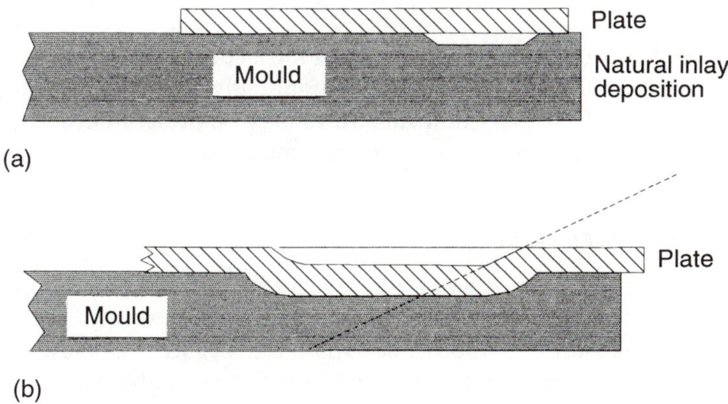

Figure 7.16 Friction surfacing with in-process forging/forming: (a) before deposition and (b) after deposition.

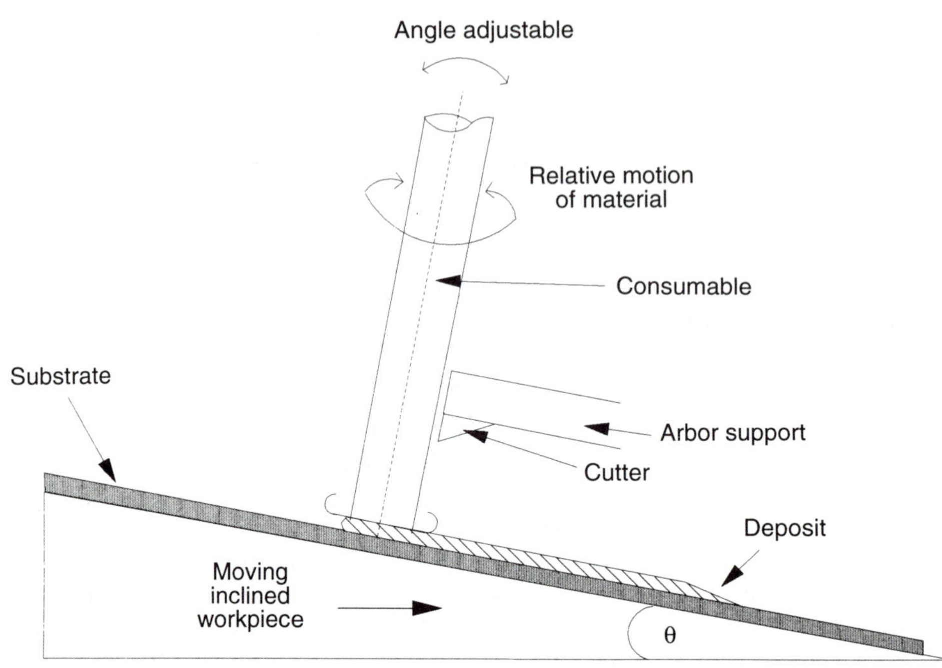

Figure 7.17 Schematic arrangement for friction surfacing with an angular feed substrate.

the rate of advance. The use of an angled substrate is therefore considered an advantage in that the relative advancement between the consumable and substrate is predictably maintained. The optimum condition for surfacing is thus directly controlled by the geometry instead of the applied surfacing conditions.

An angled substrate eliminates the need for a consumable feed system and is particularly suitable for a straight or nominally flat plate, or plates with a grooved recess. Relatively thin substrates may be preset in order to reduce distortion caused by deposition of the surface layer.

Friction coextrusion cladding

Friction coextrusion is a solid phase cladding technique, whereby one material, under a radially compressive load, forms a sheath of nominally uniform thickness around another core material. Figure 7.18 shows the basic principle of the process, in which a composite consumable bar is subjected to relative rotational movement while being pressed against a die, so that it becomes extruded through the die orifice. Friction extrusion takes place at temperatures below the melting point of the outer (sheath) and inner (core)

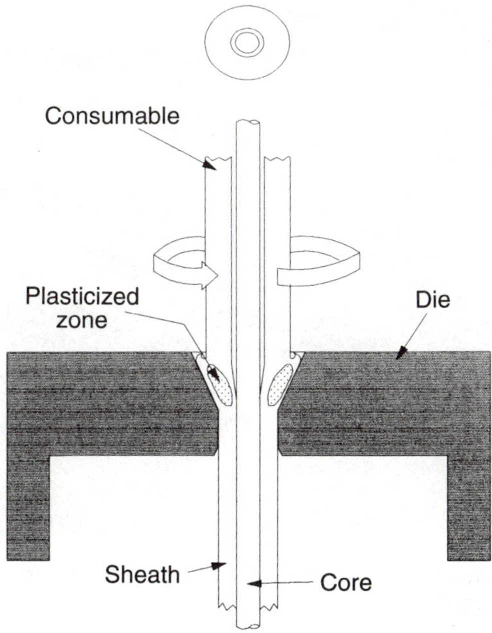

Figure 7.18 Schematic arrangement for friction extrusion cladding.

of the composite bar. Usually the core remains virtually unchanged and does not form part of the frictional shear face or plasticized zone.

The core may serve to structurally strengthen low strength material or the outer material may serve to protect the inner material from corrosion. Figure 7.19 shows a mild steel rod clad with aluminum alloy. Small diameter, multicore extrusions can also be produced, with the layers symmetrically distributed about the center.

A feature of friction coextrusion cladding is that the sheath material is extruded at a different rate than the core material. The axial differential in feed rate between the sheath and core materials assists in the rupture of oxide layers, which provides for good bond integrity.

7.5 RESISTANCE CLADDING

Resistance welding has been a well-established process for many years, used primarily in the automotive industry for spot welding of sheet metal components. It is a well-controlled process with scope for automation and, as such, provides an additional alternative for cladding operations. As with friction surfacing, resistance cladding is relatively new and offers greatest

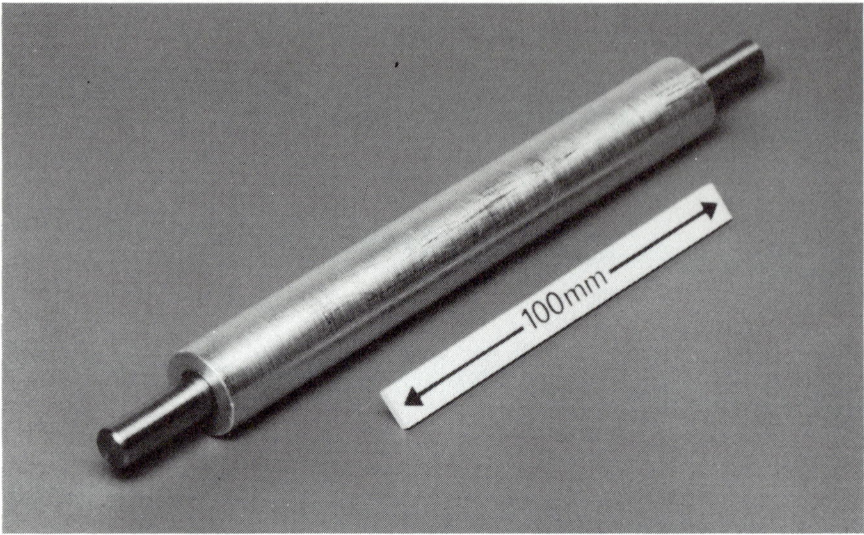

Figure 7.19 A small roller component produced by friction coextrusion cladding aluminum on mild steel.

potential for localized, discrete cladding, particularly for material combinations which are difficult to fusion weld.

7.5.1 Principle and mechanism

Resistance welding processes rely on the application of an applied forging pressure and heat generated by the passage of an electric current through the workpieces via contact electrodes. The electrodes and component geometry are designed to concentrate heat at the interface using increased interfacial resistance.

Resistance cladding is affected by the use of resistance seam welding, whereby electrodes take the form of wheels or rollers which traverse the plates to be welded (Fig. 7.20). The wheels are pulsed intermittently with current to produce a series of overlapping or discrete spots. Current pulsing may be carried out with the electrode wheels, either stationary or rotating, and the spot overlap would normally be 25–50%. Resistance seam welding is mainly restricted to the deposition of coatings onto flat plates. Flat bottomed electrodes can be used to discretely clad the end of a bar or tube.

The principal advantages of resistance cladding are its relatively low cost and its well-proven technology capable of producing joints quickly and reliably. Its disadvantages include a lack of flexibility in joint design, some deformation (approximately 10%), electrode contamination and wear and limited materials applicability.

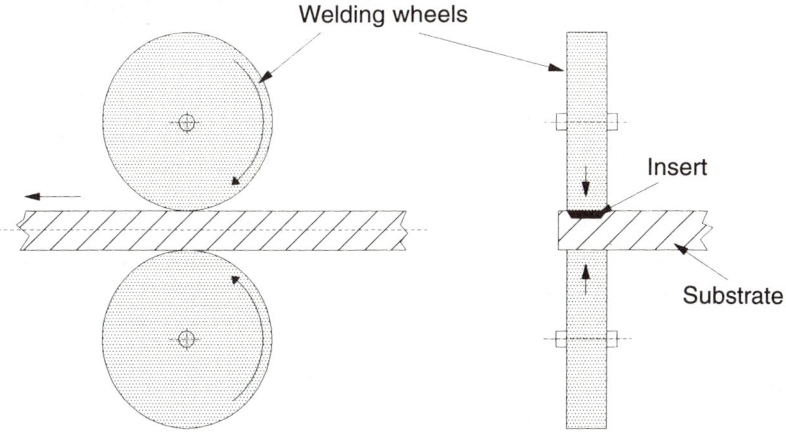

Figure 7.20 Schematic arrangement for resistance cladding.

7.5.2 Materials and weldability

Resistance welding is more traditionally used for joining like materials, and most welds have been made in sheet steel (spot welding), but the process is also applicable to aluminum alloys, nickel alloys and copper alloys.

When welding, the amount of heat generated depends primarily upon the welding current used and the resistance of the work. The resistance usually consists of:

(1) The surface resistance between the two pieces to be welded
(2) The specific resistance of the pieces
(3) The resistance between the electrodes and the work surface

When similar materials are spot welded, then the highest resistance is between the interfaces of the two sheets, provided the correct welding conditions are used. When welding dissimilar metals, the maximum resistance may not exist at the interface; instead it may be located within one of the metals if it has a higher specific resistance than the other metal. Nevertheless, a range of dissimilar combinations can be welded with suitable selection of welding conditions. Examples of successful dissimilar combinations are stainless steel to mild steel and Inconel to stainless steel and mild steel.

7.5.3 Equipment

Resistance welding/cladding equipment is commercially available; usually there is a means to supply electrical power and a means to apply mechanical force through a suitable electrode arrangement. Standard electrodes will be available for traditional welding and some cladding, but specialized electrodes may be required, depending on the materials and geometries being joined.

In recent years, most equipment development has concentrated on power supplies and monitoring. The power supplies provide and control the welding current in either AC or DC modes. The choice of power supply is largely governed by the rate of heat input required for the weld and the degree of control called for. AC is normally used where a relatively slow heat input is demanded; a constant voltage DC often contains facilities to control or ramp the rise and decay of the welding current. As the thickness of the materials to be joined increases, more complex current waveforms may be required to impart the required heat input to the weld.

Welding currents range from a few hundred to several thousand amps, depending on the application; typical values are given by

$$I = 7000t$$

where t is the thickness in millimeters and I is the welding current in amps.

Welding times can be very short; single pulses are often on the order of tenths of a second; overall times depend on the number of cycles and the traversing rates required for cladding. The pressure applied through the electrodes ensures intimate interfacial contact, supresses molten metal expulsion and consolidates the weld. Typical pressures are around $70\,\mathrm{N/mm^2}$.

7.5.4 Applications

Applications for resistance cladding are limited at present as the capabilities of the process are still being established. However, a number of friction surfacing applications could equally be achieved by a resistance approach. The choice may then come down to equipment costs, equipment availability and production rates. One example is an industrial blade, resistance clad into a preshaped tool (Fig. 7.21).

7.6 HOT ISOSTATIC PRESSING

Hot isostatic pressing (HIP), a variant of diffusion bonding, is capable of joining materials in the solid state with minimal deformation. HIP is a well-established commercial process, particularly for consolidating castings, where it has been used for over 25 years. Its high temperatures and pressures close and seal any internal porosity. Although casting consolidation probably remains the largest industrial application of HIP, the consolidation of powders (for powder metallurgy materials) and diffusion bonding are two growing areas.

Large production applications are possible with HIP but the need for specialist and expensive equipment has initially restricted its application to

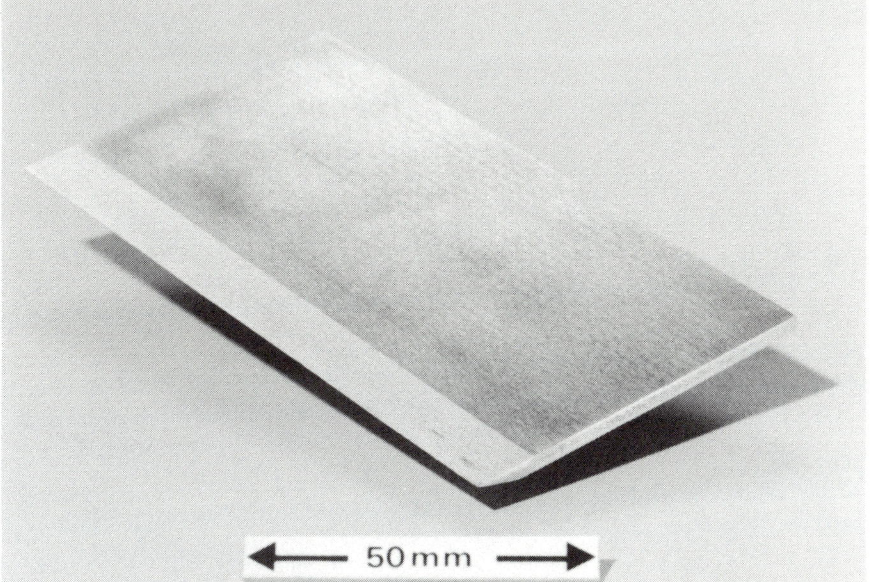

Figure 7.21 Industrial blade with resistance strip cladding.

the aerospace and nuclear industries. Larger commercial furnaces that are now available could allow further exploitation of the process into more cost-sensitive general engineering applications.

7.6.1 Principle and mechanism

Diffusion bonding involves the holding of premachined parts under load at elevated temperatures and in a protective atmosphere. In the uniaxial mode, loads are relatively low (5–10 MPa) sufficient to cause microdeformation at surface asperites but not so large as to deform the components macroscopically. HIP pressures can be higher, up to 100–150 MPa. Temperatures are typically 0.5–0.8 T_m (where T_m = melting point in kelvins). Times at temperature can be 30 min or greater; the overall cycle time is generally quite long, depending on the load in the furnace. Most HIP operations are carried out in atmospheres of high purity argon.

The bonding mechanism is one of microscopic yielding and atomic diffusion. When the two essentially 'flat' surfaces are brought together, there is contact at microasperites. Figure 7.22 shows the voids present at the interface and indicates a surface layer of oxides or containment which can inhibit bonding. As the bonding temperature and pressure are increased, surface yielding helps to break up the surface oxides and reduces the void area so that intimate contact is achieved. Bond development then continues

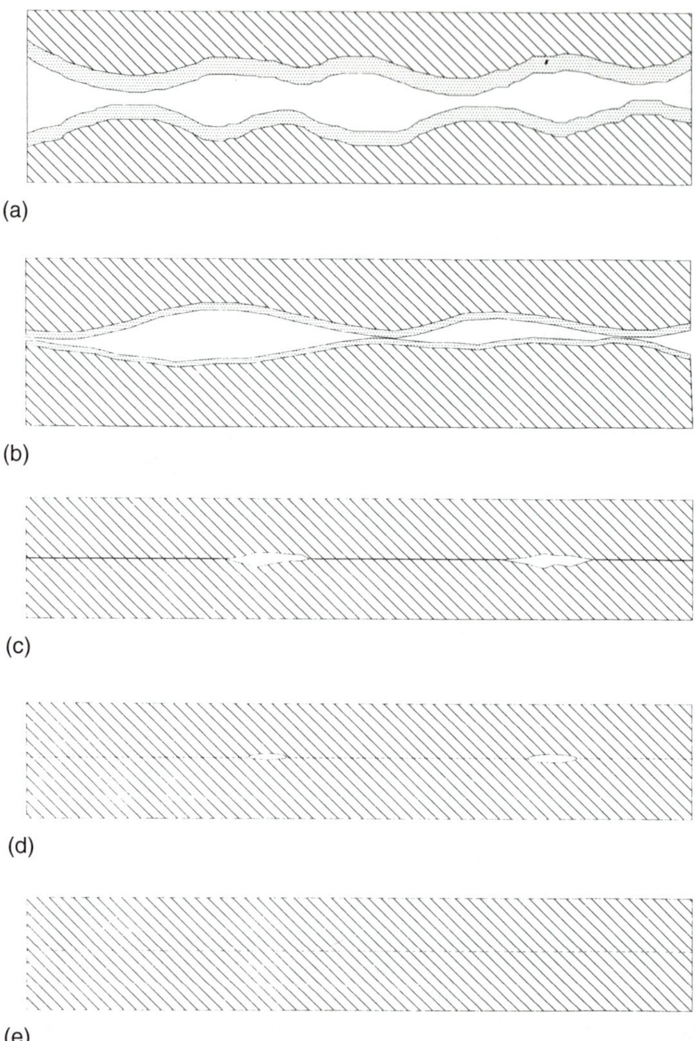

(a)

(b)

(c)

(d)

(e)

Figure 7.22 The mechanism of diffusion bonding. (a) Initial point contact and containment layer. (b) Some point yielding and creep produces a thinner oxide layer with large voids. (c) Final yielding and creep causes voids to remain within a very thin oxide layer. (d) Continued vacancy diffusion eliminates the oxide layer and leaves few small voids. (e) Bonding is complete.

by diffusion-controlled mechanisms, including grain boundary diffusion and power law creep.

In uniaxial diffusion bonding, surface finishes of better than $R_a = 0.4\,\mu m$

have been recommended. However, the higher temperatures and pressures of HIPping allow rougher surfaces to be bonded, possibly up to $R_a = 1-2\,\mu m$.

One unique feature of HIPping for cladding is the need to encapsulate the parts to prevent ingress of the gas medium along the bond interface. Gas ingress would prevent the pressurization medium being applied across the interface. The encapsulation traditionally involves canning in a metal enclosure which is subsequently sealed (by TIG welding) and evacuated. Alternatively a glass sealing route can be used. If the metal components allow, the interfaces can be directly sealed by a welding technique; electron beam welding offers greatest cleanliness and therefore greatest success because it takes place within a vacuum.

7.6.2 Materials and metallurgy

Diffusion bonding (hot isostatic pressing) of similar materials leaves a joint which is generally not discernible by microscopic examination; this is because of grain growth across the interface. However, dissimilar metal joints are generally characterized by a sharp transition zone which may include a reaction layer dependent on the compatibility of the two materials.

A wide range of materials can be bonded by this route, including alloys with different melting points, densities and other physical properties. The limitations come from the compatibility of the materials being joined (that is, do they form undesirable brittle zones) and their respective thermal expansion coefficients. Solutions to these and similar problems may lie in joint design and the use of appropriate compliant or diffusion barrier interlayers. Figure 7.2, explosive welding combinations, also provides a general guide to combinations for HIP and diffusion bonding. A greater range is likely by HIP cladding, but combinations such as titanium to alloy steels and stainless steels will be more problematic because of intermetallic formation. In this specific case, joints are formed with diffusion interlayers; an example is the use of vanadium, copper and nickel [23].

The bonding of ferrous alloys has been well demonstrated; examples include high speed steels to carbon steels and pure iron to a range of carbon steels [24]. Taylor and Pollard [24] observed a narow diffusion zone across the interface, approximately 1 mm wide. These materials were uniaxially bonded but similar results would be expected by hot isostatic pressing.

Lopez *et al.* [25] have evaluated the HIPping of Inconel 625 to a low alloy steel. Although the materials can be bonded, once again, a decarburized zone is formed in the steel side of the joints. This is accompanied by a region of carbide precipitation in the nickel alloy. Elemental analysis of the joints has proved the diffusion of Cr and Ni into the alloy steel with a diffusion zone of $10-20\,\mu m$, depending on the bonding temperature.

In other work, the HIP bonding of a strain-hardening manganese steel to a low alloy steel has been examined [26]. This combination has a thermal mismatch; the manganese steel has a higher thermal coefficient of expansion than the low alloy steel. Direct bonding can therefore lead to cracking during cooling. This is solved by the introduction of a nickel interlayer to absorb thermal strains. The experimental work on 30 mm diameter components demonstrated the need for an interlayer at least 130 µm thick to provide crack-free joints.

In contrast to the other cladding processes described in this chapter, HIP cladding is one technique capable of bonding nonmetals onto metals. Nonmetal claddings, such as ceramics, can open up greater opportunities for wear and/or corrosion resistance. Specific difficulties arise from thermal expansion mismatch, but they can be accommodated by using appropriate metal interlayers or, more recently, by using functionally graded interlayers. Functionally graded interlayers have a ceramic or near-ceramic composition at one surface that grades into a metallic composition at the other surface. They can be manufactured by HIPping (powder consolidation) or by spraying.

7.6.3 Equipment

Hot isostatic pressing furnaces are high capital cost plant but they can be analogous to heat treatment and brazing furnaces, where larger-scale batch production justifies investment. An alternative access may be through an HIP subcontractor.

The furnaces themselves comprise a pressure vessel to withstand the operating pressures of up to nominally 150 MPa and a heating element of, for example, carbon or molybdenum to provide the temperature. Gas consumption is minimized by recycling. The furnaces are generally microprocessor controlled to give the required heating and cooling rates.

Because the furnaces operate at relatively high pressures, certain safety precautions have to be taken (for example, containment in a secure area or gas sensors to check for leaks). However, such furnaces are operated in commercial environments and are used, for example, in the aerospace industries for consolidation of castings and diffusion bonding.

Small-scale equipment is available for laboratory work, which may have a working hot zone of 50–75 mm diameter by 100 mm long. At the other end of the scale, mega-HIP furnaces are established which have a working zone of 1.15 m diameter by 2.2 m height [27].

7.6.4 Applications

Despite the availability of large HIP furnaces, it is expected that most cladding applications will rest with the surfacing of relatively small discrete components which can be processed in batches.

No widespread use of HIP cladding has been reported so far, but its capabilities have been shown by Walker *et al.* [27]. In one example, a valve

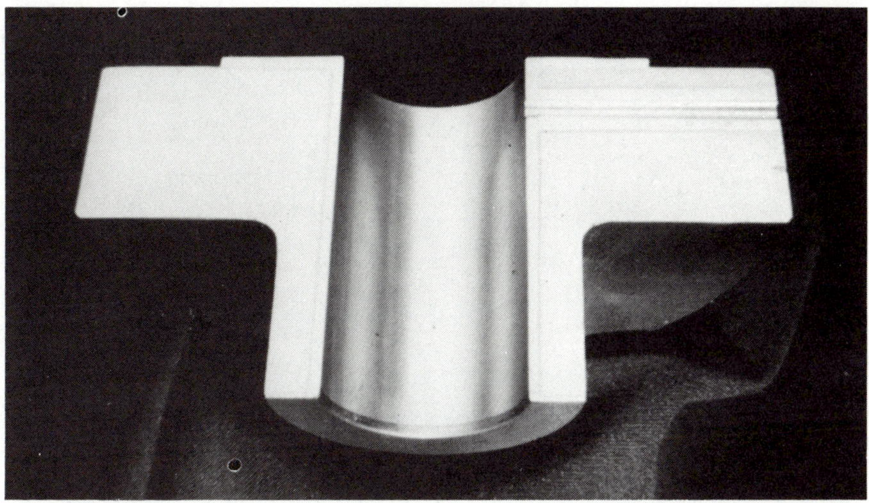

Figure 7.23 Half section of an alloy steel valve component clad on all internal surfaces with IN625 bar tube and consolidated powder [27].

has its internal surfaces clad with Inconel 625 by hot isostatic pressing (Fig. 7.23). The larger bore is clad by placing an Inconel tube inside the bore, with a close fit to the valve body. The smaller bore is clad using a powder starting material which is packed inside the bore. All exposed interfaces are sealed to allow isostatic pressing. Within the process, the Inconel tube is bonded to the inside of the valve and the powder is consolidated and bonded simultaneously. After HIPping, the assembly is machined to bore out the smaller bore and to remove all packaging or sealing.

A similar application has been pursued by McLeish [28] to deposit a hardfacing alloy onto the inside of small bore valves (Fig. 7.24). In this case, there is a serious constraint on access which limits alternative cladding processes. Nickel and cobalt hardfacing alloys have been successfully clad onto the stainless steel valve.

7.7 SUMMARY

A diverse range of processes are available for depositing thick claddings, based on a range of welding/bonding techniques. Some, such as explosive cladding and roll bonding, are well established, whereas others are still in development and are opening up new opportunities for localized cladding, sometimes of new and unusual materials. The difficulty for the user is always the choice of process. Wider even than this chapter, it may extend to sprayed and welded coatings. The answer is rarely simple; it will depend on many technical and commercial requirements. The choice may not be restricted to

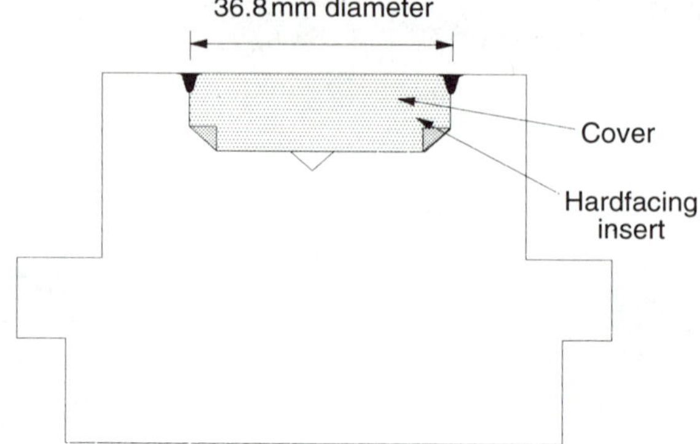

Figure 7.24 Modified design for HIP bonding of 15 mm stop valve [28].

those coatings described in this chapter but may extend to sprayed and welded coatings.

Explosive cladding and roll bonding are particularly relevant to large plate or pipe, from which can be machined smaller items (for example, titanium or aluminum to steel transition pieces). More localized claddings, where the environments are particularly arduous and/or thicker coatings are required (0.03 mm) open up other possibilities, particularly friction surfaced and HIPped claddings.

The future is likely to see more applications develop in these newer processes because of their versatility and their applicability to a wide range of materials. Another area being studied at present is the development of appropriate control or inspection methods to provide for greater reliability. Such methods will include in-process monitoring as well as postinspection such as ultrasonics or thermography.

REFERENCES

1. Dunkerton, S.B. (1992) Advances in thick overlay coatings. Paper presented at New Developments in Surface Engineering, National Centre of Tribology, Risley, UK.
2. Houldcroft, P. and John, R. (1988) *Welding and Cutting*, Woodhead-Faulkner, Cambridge.
3. Bahrani, A.S. (1978) Explosive cladding. *Surf. J.*, **9**(1) 2–9.
4. El-Sobky, H. (1983) Mechanics of explosive welding, in *Explosive Welding, Forming and Compaction* (ed. T.Z. Blazynski), Applied Science Publishers, London, pp. 189–217.
5. Chadwick, M.D. and Jackson, P.W. (1983) Explosive welding in planar geometries, in *Explosive Welding, Forming and Compaction* (ed. T.Z. Blazynski), Applied Science Publishers, London, pp. 219–87.
6. Du Pont (1983) UK Patent 1,168, 246.
7. Birchfield, J.R. (1982) Big boom bonds – metals to resist environments. *Welding Design and Fabrication*, June 1982, pp. 78–83.
8. Patterson, R.A. (1982) Explosion bonding: aluminium–magnesium alloys bonded

to austenitic stainless steel, in *High Energy Rate Fabrication* (eds M.A. Myers and J.W. Schroeder), American Society of Welding Engineers, New York, pp. 15–37.

9. Cleland, D.B. (1983) Basic considerations for commercial processes, in *Explosive Welding, Forming and Compaction* (ed. T.Z. Blazynski), Applied Science Publishers, London, pp. 159–88.

10. Johnson, T.E. and Pocalyko, A. (1982) Explosive welding for the 80s, in *High Energy Rate Fabrication* (eds M.A. Myers and J.W. Schroeder), American Society of Welding Engineers, New York, pp. 63–82.

11. Stone, M.J. (1968) The properties and applications of explosion-bonded clads, in *Select Conference on Explosive Welding*, The Welding Institute, Hove UK, pp. 55–62.

12. Hardwick, R., Brown, D.W. and Nowell, D.G. (1992) *Recent Developments in Explosive Metalworking at ICI Explosives*, ICI Publications, Stevenston, Ayrshire.

13. Mansell, H. (1990) Hybrid metal packages by explosive bonding. *Hybrid Circuit Technology*, Sept. 1990, pp. 67–9.

14. EFTEK Explosion-clad materials for power hybrid and microwave packaging. Brochure from Explosive Fabricators Inc.

15. Smith, L.M. (1992) Clad steel – an engineering option. Paper presented at the 24th Annual Offshore Technology Conference, Houston, Texas, OTC 6911 pp. 343–55.

16. Charles, J. *et al.* (1989) Clad plates: an economical solution for severe corrosive environments. Paper 9 presented at Corrosion '89 Conference, St. Louis, Missouri.

17. Klopstock, H. and Neelands, A.R. (1941) An improved method of joining or welding metals. UK Patent Specification 572789, application date 17 Oct. 1941.

18. Nicholas, E.D. and Thomas, W.M. (1986) Metal deposition by friction welding. Paper presented at the 67th AWS Annual Meeting, Atlanta GA.

19. Bedford, G.M. (1990) Friction surfacing for wear applications. *Metals and Materials*, Nov. 1990, pp. 702–5.

20. Tyayer, K.A. (1959) Friction welding in the reconditioning of worn components. *Svarochnoe Proizvodstov*, **1**(10), 23–4.

21. Thomas, W.M. and Nicholas, E.D. (1988) Surfacing method. European Patent Application 0337 691, filed 11 April 1988.

22. Abson, D. (1983) MMC deposits from powder mixtures by friction surfacing. *TWI Bulletin*, Jan/Feb 1993, pp. 12–13.

23. Baker, T.S. and Partridge, P.G. (1987) Tensile, shear and impact fracture of titanium/stainless steel diffusion bonded joints. Paper presented at International Conference on Diffusion Bonding, Cranfield 7–8 July.

24. Taylor, D.S. and Pollard, G. (1978) The diffusion bonding of steels, in *Advances in Welding Processes* (ed. J.C. Needham), The Welding Institute, Abington Vol. 1, pp. 4–9. Proceedings of the 4th International Conference on Advances in Welding Processes, Harrogate, May 1978.

25. Lopez, B., Gutierrez, L. and Urcola, I.J. (1991) Study of the microstructure obtained after diffusion bonding Inconel 625 to low alloy steel by hot uniaxial pressing or HIPing, in *Diffusion Bonding 2* (ed. D.J. Stephenson), Elsevier, Oxford, pp. 37–48.

26. Atkinson, H.V., Crarre, M.W. and Walker, R.M. (1991) HIP diffusion bonding of austenite to ferritic steels, in *Diffusion Bonding 2* (ed. D.J. Stephenson), Elsevier, Oxford, pp. 49–58.

27. Walker, R.M., Roberts, D.J. and Rickinson, B.A. (1991) Component assembly using HIP diffusion bonding, in *Diffusion Bonding 2* (ed. D.J. Stephenson) Elsevier, Oxford, pp. 261–9.

28. McLeish, J.A. (1989) Small bore valve hardfacing by hot isostatic pressing. Paper C373/012 presented at IMechE Conference.

8

Thermal barrier coatings

R.L. Jones

8.1 INTRODUCTION

The fundamental driving force for the development of thermal barrier coatings has been the continuing quest for ever higher temperatures in gas turbines. Over the period 1940–1970, the temperature capability of superalloys for gas turbine blades, which is defined in terms of 100 h life at 20 000 psi (138 MPa) stress, was increased from approximately 1400 °F (760 °C) to 1900 °F (1040 °C) [1]. This was a costly process which required the development of numerous generations of new superalloys. It was a necessary and successful step, however, in producing the highly reliable and efficient gas turbine engines that we know today. Between 1940 and 1970, higher temperature superalloys, along with improved engine design, allowed specific fuel consumption to be reduced by more than half, thrust-to-weight ratios to be tripled, and time between overhaul to be increased from less than 100 h to over 12 000 h [1]. A recent review of the effect of materials on current and future gas turbine performance has been given by Kool [2].

Boosting temperatures in gas turbines has become difficult in recent years because conventional cobalt- and nickel-based superalloys have been developed to very near their maximum temperature capability (which is ultimately limited by incipient melting). Advances in engine efficiency since the 1970s have been achieved primarily by air cooling wherein air from the compressor section of the engine is pumped through air cooling passageways cast into

Metallurgical and Ceramic Protective Coatings. Edited by Kurt H. Stern.
Published in 1996 by Chapman & Hall, London. ISBN 0 412 54440 7

the blade and vane airfoils. The science of air cooling is well developed, and high performance engines now operate for thousands of hours at gas temperatures that may be 200–300 °F (90–150 °C) higher than the incipient melting point of the airfoil superalloy. Another modern approach to improvement of the high temperature capability of superalloys has been by *directional solidification* (DS) and *single-crystal casting* of airfoils. In these procedures, the casting mold is cooled in such a way, using chill plates, oriented seed crystals or other methods of crystal growth selection, so that the molten superalloy solidifies with the crystal grains all oriented parallel to the longitudinal axis of the airfoil, or with the airfoil as a single crystal [3]. This eliminates grain boundaries perpendicular to the direction of stress, and in some cases allows grain boundary strengtheners to be omitted, which improves the high temperature capability (and often oxidation resistance) of the superalloy.

But these approaches are not without penalty. Cooling the airfoils with compressor air increases the complexity and cost of the engine, and energy is required to compress and pump air through the cooling system. Also, modern air-cooled blades and vanes, which involve superalloys having 12–15 critical alloying elements, exotic vacuum casting techniques, intricate cast-in air cooling passageways and sometimes specialized DS and single-crystal solidification schemes, have become increasingly difficult and expensive to manufacture. This is reflected in higher component prices, with first stage vanes for some engines now having a new cost of as much as $10 000 each.

A third means to increase engine temperature capability, which in fact has been proposed for many years, is to apply a thin coating of a heat-insulating ceramic on the airfoil surface. This idea was principally brought to the front, at least in the United States, by NASA and Air Force research in the late 1950s and early 1960s on thermal control coatings for space vehicles and rocket engines. The engine in the NASA X-15 aircraft was described, for example, as being protected by 'a thermal insulating, 10-mil-thick coating of stabilized zirconia ... over a 5-mil-thick nichrome undercoat' [4]. This composition and structure is, interestingly, not all that different from the thermal insulating coatings, now called thermal barrier coatings (TBCs), used in present-day gas turbine engines.

From this beginning, NASA has taken a leading role in TBC research and development, and continues today as a major contributor in the advancement of aviation engine TBCs. NASA also acts as the technical monitor for Department of Energy (DOE) programs in TBC technology for industrial and ground transportation engines. Many aspects of the history of NASA's involvement and accomplishments in thermal barrier coating development have been recounted in articles by R. Miller of the NASA Lewis Research Center [5,6]. These papers have provided the references for most of the publications relating to NASA research on TBCs that are reported in the present review.

The appeal of thermal barrier coatings for gas turbine manufacturers is made strikingly clear by the graph in Fig. 8.1 [7]. As indicated in Fig. 8.1, thermal barrier coatings have the potential for increasing the temperature capability of gas turbines by 100–200 °F (90–150 °C) above that now achievable with our most advanced superalloys. Such an increase in turbine gas temperature could, for example, provide an improvement of up to 20% in fuel economy [8]. Moreover, this 100–200 °F (90–150 °C) increment is equivalent to at least three generations of new superalloy development that, considering the high temperatures involved, would be difficult, or perhaps impossible, for Co- or Ni-based superalloys. To realize this promised improvement, however, it will be necessary to develop highly reliable TBCs that can be used to their full thermal advantage on critical components such as blades in man-rated engines. Although TBCs are beginning to be widely applied in many types of engines, all commercial applications to date have had to be designed so as to be noncatastrophic in case of TBC failure.

8.2 MATERIAL REQUIREMENTS

It was recognized early on that to be useful as a high temperature thermal barrier coating a material must meet these fundamental requirements: (1) high melting point; (2) low density; (3) high surface emissivity; (4) high thermal shock resistance; (5) low vapor pressure; and (6) resistance to oxidation or

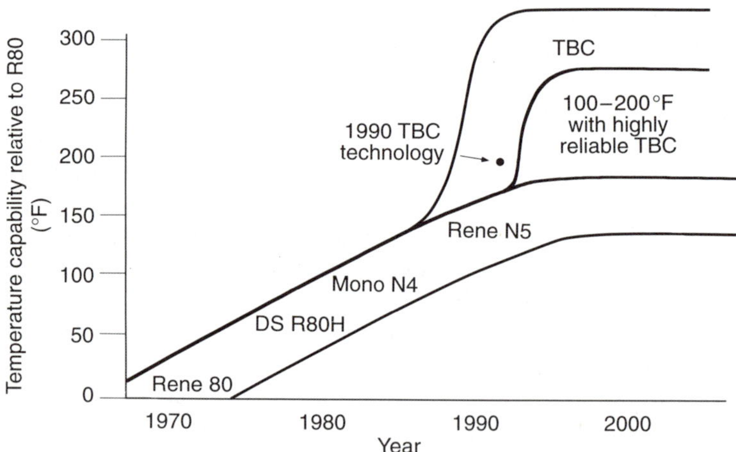

Figure 8.1 Increase in temperature capability of superalloys with time compared to the quantum-step gain potentially obtainable by development of highly reliable thermal barrier coatings (TBCs) for gas turbine blade use. Provided by B. Nagaraj *et al.* [7], and used with permission of the General Electric Aircraft Engine Company of Cincinnati OH.

chemical environment [9]. For engine use, to these must be added (7) low thermal conductivity; (8) high coefficient of thermal expansion; and (9) resistance to gaseous and particulate erosion.

When these requirements are all considered, very few ceramics are found that meet the criteria. As determined originally by NASA, zirconium dioxide (ZrO_2) in its stabilized form appears to be the optimum choice as a thermal barrier coating material. Zirconium dioxide scores especially well for requirements 1, 5, 6, 7 and 8. Stabilized ZrO_2 is outstanding in terms of low thermal conductivity, and has a thermal conductivity that is, for example, approximately 100 times lower than that of Al_2O_3 [10]. Stabilized ZrO_2 also has the highest coefficient of thermal expansion (CTE) of all of the common ceramic oxides (except for MgO), with a CTE of about 11×10^{-6} $°C^{-1}$ at 1000 °C as compared to 17×10^{-6} $°C^{-1}$ for an average Ni-based superalloy [10]. And despite having a high coefficient of thermal expansion, partially stabilized ZrO_2 is regarded as having excellent thermal shock resistance because of microcracking and transformation toughening [10].

Over the years, a number of other ceramics have been suggested or tested for use as thermal barrier coatings and/or anticorrosion coatings in engines. These include $3Al_2O_3 \cdot 2SiO_2$ (mullite), Ca_2SiO_4, $MgAl_2O_4$, $CaTiO_3$, $ZrSiO_4$ and $ZrTiO_4$ [11–13]. However, none of these materials has been developed to the state of being a rival for stabilized ZrO_2. Interest in Ca_2SiO_4 declined, for instance, when it was found to react readily with combustion gas SO_3 to form $CaSO_4$ and SiO_2 [14]. The search for new TBC materials is always an open subject, however, and ceramics from the $NaZr_2(PO_4)_3$ (NZP) system have been recently proposed for possible use as thermal barrier coatings [15]. Because of their low coefficient of thermal expansion, NZPs are being investigated for use as cast-in-place diesel engine port liners [16].

8.3 STABILIZATION OF ZIRCONIA

Although ZrO_2 has many properties well suited for thermal barrier use, it suffers from one deficiency –the need to be *stabilized*. As shown in the ZrO_2–Y_2O_3 phase diagram in Fig. 8.2, pure ZrO_2 transforms from the monoclinic to tetragonal phase at about 1180 °C, and from the tetragonal to cubic phase at about 2370 °C. If the ZrO_2 contains inpurities, these transformations may occur at lower temperatures. The harmful effect of the zirconia phase transformations is that a contraction in volume, initially reported as 9% but now thought to be closer to 4%, results from the monoclinic–tetragonal transition. This change in volume causes pure ZrO_2 to lose its physical integrity, and to crack and crumble, when thermally cycled through the monoclinic–tetragonal transformation temperature. Use of ZrO_2 in engines and other cyclic heat applications therefore requires that the ZrO_2 be stabilized in its high temperature tetragonal (or cubic) form down to ambient temperatures.

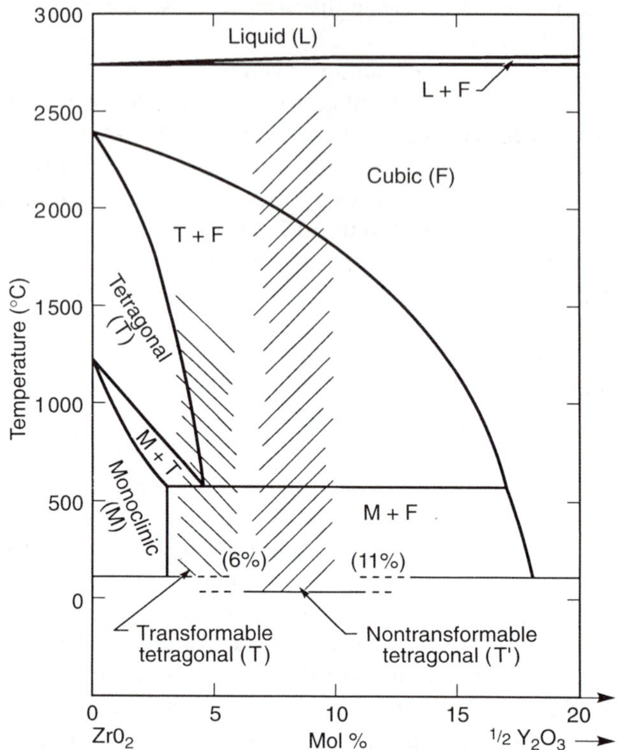

Figure 8.2 Phase diagram of the ZrO_2–$YO_{1.5}$ system indicating the regions of phase predominance for the metastable tetragonal t and tetragonal t′ zirconia polymorphs. After Scott [22].

It was discovered many years ago that the addition of CaO, MgO or other oxides could stabilize ZrO_2 in its high temperature or cubic structures. The oxides currently thought to stabilize ZrO_2 include CaO, MgO, Y_2O_3, CeO_2, Sc_2O_3, In_2O_3 and the majority of the rare earth oxides. Use of these stabilizing oxides results in phase relationships that are generally, but not quantitatively, similar to those illustrated for the ZrO_2–Y_2O_3 system in Fig. 8.2. Because diffusion is very slow in ZrO_2, even at quite high emperatures, phase equilibria are difficult to obtain, and the question is often raised as to the accuracy of even some of the better known stabilized ZrO_2 phase diagrams. The original phase diagrams for CaO–ZrO_2, for example, indicated that CaO stabilized ZrO_2 in the cubic form to below at least 300 °C; more recent work shows however that the cubic structure of CaO–ZrO_2 is actually not stable below 1140 °C [17]. Even recent phase diagrams of the highly studied Y_2O_3–ZrO_2 system have significant differences because of the problem of achieving equilibrium [18]. In some cases, debate may wage for decades, as with TiO_2, before it is finally concluded that the oxide does not give true stabilization of ZrO_2. On the other hand, absolute thermodynamic equilibrium is not always required, and metastable crystal states can be of major engineering importance, as discussed below.

Despite much study, the mechanism of stabilization is still not certain. Many factors are involved, including thermal treatment (annealing temperature, time at temperature, quenching rates) and grain size. In searching for stabilizing oxides, one may invoke a set of working rules; (1) the radius of the oxide cation should be close to that of Zr^{4+} (0.080 nm) so that it can fit into the crystal lattice without serious distortion; (2) the oxide cation should have a stable $2+$ or $3+$ oxidation state; and (3) the crystal structure of the stabilizing oxide itself should be cubic or tetragonal. However these rules may be violated. For instance, stabilizing oxides with cation oxidation states below $4+$ generate oxygen vacancies within the zirconia oxide lattice. These oxide vacancies are thought to be critical in producing stabilization; yet CeO_2, where the cation is Ce^{4+} and no oxygen vacancies should be produced, is still an effective ZrO_2 stabilizer. (In rebuttal to this last question, it is sometimes argued that a certain number of the cerium cations in fact exist in the zirconia matrix in the Ce^{3+} state.)

Although quantitative comparison is difficult, the various stabilizing oxides are known to have different levels of effect, or possibly even different stabilizing mechanisms. For example, Sheu *et al.* investigated the cubic–tetragonal t′ transformation (see below) for several stabilized zirconias, including In_2O_3- and Sc_2O_3-stabilized ZrO_2, and determined that the upper composition limit at which the tetragonal phase could be observed was $\sim 18\,mol\%$ for $YO_{1.5}$ and certain rare earth oxides ($RO_{1.5}$), but $23\,mol\%$ for $ScO_{1.5}$ and $25\,mol\%$ for $InO_{1.5}$ [19]. This led Sheu *et al.* to conclude that In_2O_3 and Sc_2O_3 were less effective stabilizers than Y_2O_3 or the rare earth oxides. The stabilizing ability of the various oxides has also been recently reviewed by Sasaki *et al.*, who provide a plot of the lowering of the monoclinic–tetragonal transformation temperature versus the M^{3+} cationic radius. This plot indicates that the stabilizing efficiency increases with cationic radius from Sc^{3+} up to Y^{3+}, but then decreases down to La^{3+}, which has lower stabilizing efficiency than Sc^{3+} [20].

8.4 PHASE TRANSFORMATION REACTIONS OF STABILIZED ZIRCONIA

If a mixture of 2–$3\,mol\%$ Y_2O_3 (4–$6\,mol\%$ $YO_{1.5}$) with ZrO_2 is sintered at 1400–$1600\,°C$ and cooled at appropriate rates, the ceramic obtained is not the equilibrium monoclinic and cubic mixture predicted by Fig. 8.2, but rather a metastable tetragonal phase termed the t phase, which may contain some cubic phase or be itself contained within a cubic phase matrix depending on the exact Y_2O_3 content and preparation conditions [21]. This phase is referred to as the tetragonal zirconia polycrystal (TZP) phase, or to distinguish it from similar phases formed in the $MgO-ZrO_2$ or CeO_2-ZrO_2 systems, the yttria-tetragonal zirconia polycrystal (Y-TZP) phase. If, on the other hand, a mixture of 4–5 mol% Y_2O_3 with ZrO_2 is heated to 2200–2400 °C or above (i.e. into the high temperature cubic phase region) and quenched rapidly, a metastable tetragonal phase termed the t' phase is produced. This phase has significantly different properties than the Y-TZP phase. It is sometimes described as partially stabilized zirconia (Y-PSZ or PSZ) to differentiate it from the fully stabilized cubic zirconia (Y-FSZ or FSZ) produced at compositions above 18 mol% $YO_{1.5}$. The regions of occurrence of these two metastable tetragonal phases are indicated in the $ZrO_2-Y_2O_3$ phase diagram in Fig. 8.2, which is originally due to Scott who first established the existence of the low temperature metastable tetragonal ZrO_2 phase [22].

The TZP (t) zirconias have very high strength and high fracture toughness, with strength of over 1 GPa and fracture toughness greater than 15 MPa m$^{1/2}$ being achieved [21, 23]. They are among the strongest of ceramics, especially in terms of fracture toughness. Accordingly there is wide interest in TZP for use as extrusion dies, cutting instruments, machine tools, wear parts in engines and machinery, etc. The fracture toughness of TZP is thought to be the result of a *transformation toughening* mechanism wherein the TZP (t) phase, when under stress such as at a crack tip, transforms to the monoclinic ZrO_2 phase. Energy is dissipated, and the fracture toughness of the ceramic is therefore increased by the t tetragonal–monoclinic transformation. In contrast, the tetragonal t' phase is not transformed by mechanical stress, and is thus often called the *nontransformable* tetragonal phase. Despite its promise, a number of potential flaws are known for TZP (t) zirconia: (1) a low temperature degradation when exposed to water vapor at 200–300 °C, which can lead to complete loss of strength; (2) a return to equilibrium monoclinic–tetragonal phase behavior, and hence loss of properties, at temperatures as low as 800 °C; and (3) the need for the TZP (t) grain size to be kept below 1 μm.

The ceramic literature contains a large number of publications concerning the formation of the t and t' tetragonal zirconia phases, their structures and properties, the t tetragonal–monoclinic transformation toughening mechanism, and the relationship between the t and t' crystal morphologies. Many different mechanisms have been proposed, for instance, for the t tetragonal–monoclinic transformation toughening process. For the present review on thermal barrier coatings, it is sufficient to examine only a few of the more important papers.

Garvie in 1978 advanced the theory that the metastable tetragonal t form had a lower surface free energy than the monoclinic, and determined that the tetragonal (t) phase should be stable for crystallite sizes below about 0.1 μm [24]. The existence of a critical grain size, above which transformation from the tetragonal t to the monoclinic phase will occur spontaneously, has since been confirmed by others [25]. In a later paper, Heuer *et al.* critically surveyed the proposed mechanisms for the cubic–metastable tetragonal t, the cubic–metastable tetragonal t' and the t tetragonal–monoclinic phase transformations as they occur in the Y_2O_3–ZrO_2 system [26]. From the Heuer *et al.* review, as well as other work, it appears to be now largely accepted by ceramicists that (1) the transformation of the cubic to the tetragonal t phase is diffusion controlled, with the intrinsic energy of t phase formation being released by the development of interacting microstructures (termed complexes or, if larger, colonies) between the resultant t phase grains; (2) the rapidly quenched tetragonal t' phase is formed by a diffusionless transformation, and is characterized by antiphase domain boundaries (APBs) which result from the reduction in symmetry during the phase transformation; and (3) the t tetragonal–monoclinic transformation toughening process is martensitic, with the grain size and Y_2O_3 content being the primary factors determining the start (M_s) temperature of the martensitic transformation.

In the past there has been considerable question concerning the crystal structures of the various stable and metastable tetragonal phases found with stabilized zirconias. For example, in discussing the Sc_2O_3–ZrO_2 phase diagram, Ruh *et al.* speak of one tetragonal phase as being 'the high-temperature structure supported by Teufer in which neighboring oxygen atoms do not lie in planes parallel to the basal plane' as opposed to a second stabilized tetragonal phase which 'is the distorted fluorite structure normally observed in all stabilized tetragonal zirconias ... [where] the neighboring oxygen atoms do lie in planes parallel to the basal plane' [27]. Recent neutron diffraction studies of fine particle ($\sim 0.5\,\mu m$) metastable zirconia by Igawa *et al.* [28] indicated that the oxygen atoms were shifted in the $\langle 100 \rangle$ direction, with the authors concluding that metastable zirconia has basically the same crystal structure as doped (i.e. stabilized) zirconia at room temperature and tetragonal ZrO_2 at high temperature.

However, the tetragonal t and t' phases, as curently identified and perceived, are now generally thought of as being crystallographically similar (i.e. having the same unit cell arrangement, although the dimensions may vary), but morphologically different (i.e., having different grain microstructure features such as antiphase boundaries, twins and domains) [26, 29]. The t' phase is spoken of as distinguishable, for instance, on the basis of having 'the characteristic herringbone structure and three tetragonal variants,' and exhibiting 'a plate-like domain structure with antiphase boundaries (APBs)' [29]. The regions of t and t' phase predominance should probably not be considered as separated (as sketched in Fig. 8.2), but rather as blending smoothly together. On the other hand, a difference in composition exists between the t and t' phases. This difference may result from the original preparation, where lower Y_2O_3 concentrations are used for TZP (t) than for PSZ (t') production, or it may ensue from thermal decomposition of the t' phase (see below). Being a rapidly quenched single phase, the t' phase has the same composition as the original Y_2O_3–ZrO_2 mixture; in certain cases, however, such as in high temperature aging of plasma-sprayed TBCs [30], the metastable t′ phase may undergo thermal decomposition and phase separation, with its Y_2O_3 content then being proportioned between the higher Y_2O_3 composition cubic phase and lower Y_2O_3 composition tetragonal t phase. The t and t′ phases are often referred to as the low yttria and high yttria tetragonal phases, respectively, to denote the difference in Y_2O_3 composition.

The low temperature degradation of Y-TZP by 200–300 °C water vapor has been studied extensively. It is commonly concluded to arise from the reaction of H_2O with Y_2O_3 to form $Y(OH)_3$ on the Y-TZP grain surface. The formation of $Y(OH)_3$ either causes an accumulation of strain at the Y-TZP surface which in itself ultimately triggers transformation to the monoclinic phase [18], or is accompanied by the generation of tetragonal zirconia *embryos* which, upon growing past a certain critical size, induce transformation to the monoclinic phase at the Y-TZP surface [31]. Doping Y-TZP with CeO_2 inhibits its degradation by 200–300 °C water vapor, presumably because a CeO_2 coating is formed on the Y-TZP grains which protects the *active points* of the tetragonal (t) solid solution from H_2O attack [31].

Many of the more successful stabilized zirconia thermal barrier coatings (see below) contain about 8 wt% Y_2O_3 (i.e., 4.5 mol% Y_2O_3 or 9.0 mol% $YO_{1.5}$), and they are produced by plasma spraying or e-beam physical vapor deposition which can simulate rapid quenching. These TBCs therefore tend to consist, at least initially, of the tetragonal t′ phase, and the properties of this phase are thus of critical interest in TBC science. Although the PSZ (t′) phase, being nontransformable, is not strengthened by the tetragonal–monoclinic martenistic transformation, it is regarded by ceramicists as having excellent strength and crack toughness in its own right. Virkar and Matsumoto propose that the t′ phase is toughened by a ferroelastic domain-switching mechanism that remains active at temperatures up to 1400 °C [32]. This gives a toughening capability to substantially higher temperatures than for the TZP (t) phase, where tetragonal–monoclinic transformation toughening loses effect at 800–900 °C. Moreover, Noma *et al.* have reported the preparation of tetragonal t′ zirconia that withstood annealing for 48 h at 1700 °C without significant transformation [33]. The t′ zirconia phase has also shown exceptional resistance to degradation by 200–300 °C H_2O even for grains of 100–200 μm in size [23, 29].

Tetragonal t′ zirconia, although perhaps having slightly less strength and crack toughness, thus appears superior to the tetragonal t phase in resistance to high-temperature thermal decomposition, in resistance to low temperature degradation by H_2O vapor and in tolerance to large grain size or grain growth. The predominance of the t′ phase in thermal barrier coatings would be expected to be highly beneficial. However, Miller *et al.* have found that the t′ phase, when in the form of a plasma-sprayed TBC, suffers phase separation by diffusion to the tetragonal t and cubic phases when aged at 1200 °C [30]. Similarly, a study by VanValzah and Eaton has shown that quenching (at about 1000 °C/min) tends to retain the t′ phase in plasma-sprayed ZrO_2–8 wt% Y_2O_3, but that the t′ phase was not stable when annealed at 1288–1482 °C [34]. Decomposition of the t′ phase in ZrO_2–8 wt% Y_2O_3 to the cubic (high yttria) and tetragonal t (low yttria) structures after aging at 1200–1600 °C has been observed also by Brandon and Taylor [35]. In a related vein, Shankar *et al.* report that the tetragonal content is decreased, and the cubic content increased, in a ZrO_2–7.8 mol% $YO_{1.5}$ plasma-sprayed coating after 100 h aging at 1150 °C, but the monoclinic content ($<0.5\%$) is not changed [36].

An apparent contradiction exists as to the high temperature stability of the t′ phase, which is reported on the one hand as being stable even up to 1700 °C [33], but to undergo thermal decomposition in the TBCs at temperatures as low as 1200 °C [30]. The explanation may lie in the different methods of preparation. The high stability ceramic specimens prepared by Noma *et al.* were produced by arc furnace melting followed by ultrarapid quenching (100 000 °C/min) to form solid films of 30–40 μm thickness [33]. High levels of chemical homogeneity and favorable phase microstructures are likely to have been achieved. On the other hand, plasma spraying is a less well-controlled process, where the structure and performance of plasma-sprayed TBCs can be altered significantly by just a change in the plasma spray powder [37]. The phase composition and chemical homogeneity of the plasma spray powder are critical, and proper attention has not always been given to these factors. Kvernes and coworkers, who stress the need for high quality powders, have shown that large variation in Y_2O_3 content can occur between individual powder particles of Y_2O_3–stabilized ZrO_2 [38]. The chemical homogeneity and phase microstructure of plasma-sprayed TBCs are virtually certain to be inferior to those of the t′ zirconia phase carefully prepared by Noma *et al.* Thus the exceptional thermal stability of the Noma *et al.* t′ zirconia may be an indication that a large improvement in the high temperature capability of zirconia TBCs may yet be possible by further optimization of the chemical homogeneity and microstructure of the TBC tetragonal t′ phase.

8.5 IDENTIFICATION OF PHASES IN STABILIZED ZIRCONIA

X-ray diffraction (XRD) has been mainly relied on for the identification of the monoclinic, tetragonal (t or t′), and cubic phases in stabilized zirconia. One of the more recent papers in this area has been by Schmid, who includes references to most of the earlier work [39]. Quantitative analysis of mixtures of the monoclinic and cubic phase, or of the monoclinic and tetragonal phase, is based on the ratios of the integrated $(1\,1\,\bar{1})_m$, $(1\,1\,1)_m$ and $(1\,1\,1)_{c,t}$ diffraction peaks. The analysis is straightforward because there is no peak overlap, although the monoclinic phase shows anomalously high X-ray absorption, which must be taken into account. On the other hand, analysis of mixtures containing the tetragonal and cubic phase are difficult because the XRD patterns are very similar and they are differentiated only by the characteristic peak splittings of the tetragonal phase such as $(0\,0\,2)_t$, $(2\,0\,0)_t$, $(1\,1\,3)_t$, $(3\,1\,1)_t$, $(0\,0\,4)_t$, $(4\,0\,0)_t$. Moreover, these additional peaks overlap with both the cubic and monoclinic peaks, and deconvolution of the overlapped peaks is necessary for analysis for the tetragonal phase. Miller *et al.* used a mechanical means of peak resolution to separate the tetragonal and cubic peak components in the $(4\,0\,0)$ region, and so were able to determine the percentage of tetragonal phase present [40]. Illustrative examples of peak deconvolution are shown by Taylor *et al.*, who demonstrated the 1400 °C transformation of the high yttria t′ phase to a mixture of the low yttria t phase and cubic phase by deconvolution of the $(4\,0\,0)$ peaks [41]. A BASIC language computer program for deconvoluting overlapped diffraction peaks in the $Y_2O_3–ZrO_2$ system has been developed by Froning and Jayaraman [42]. For the case where the zirconia particles are very small and substantial peak broadening occurs, synchrotron source XRD has been invoked to distinguish between the cubic and tetragonal phases [43]. However, quantitative XRD analysis of stabilized zirconia when the tetragonal (in both t and t′ forms), monoclinic and cubic phases are all present remains difficult, and is not a routine task even with present-day computer-controlled diffractometers and peak deconvolution computer programs. Later papers have recommended Rietveld analysis of the diffraction data as being a superior method for determining zirconia polymorph phase abundance [44, 45].

Neutron diffraction has also been applied for study of the zirconia phase structures by Shankar *et al.*, who have reported the detection of an ordering of the oxygen anions in the tetragonal phase [36].

In addition to neutron and X-ray diffraction, there has been some application of Raman spectroscopy to identify ZrO_2 phases. The Raman spectra of the monoclinic, tetragonal and cubic zirconia phases are characterized by 18 Raman bands, 5 Raman bands, and 1 Raman band, respectively. This difference in spectra has been used by Srinivasan *et al.*, to distinguish between monoclinic and tetragonal zirconia materials produced by different precipitation techniques for the precursor hydrous zirconia oxide [46]. Raman spectroscopy was also employed by Hamilton and Nagelberg to monitor *in situ* the formation of the monoclinic ZrO_2 phase resulting from the molten sodium vanadate corrosion of yttria-stabilized zirconia at 700 °C [47].

8.6 ZIRCONIA THERMAL BARRIER COATINGS

The evolution of the structure and production of zirconia TBCs, which are inextricably tied together, has been treated in reviews by several authors [5, 6, 48–52]. Most modern thermal barrier coatings consist of an inner metallic layer, or bond coat, and an outer layer of stabilized zirconia. For gas turbine use, the metal bond coat is normally 0.003–0.005 in. (75–125 μm) thick and the zirconia layer 0.010–0.015 in. (250–375 µm) thick. Thermal barrier coatings intended for truck or off-road diesel engines, or for large marine diesel engines, may be much thicker, with a zirconia depth of 0.060–0.250 in. (1.5–6.25 mm); they are sometimes termed *thick* thermal barrier coatings or TTBCs.

In the past, three-layer TBCs have been used, for example, in magnesia–zirconia coatings on gas turbine combustors. Scott and Restall describe combustor coatings which consisted of an NiCr bond coat, a mixed $NiCr + ZrO_2$–24 wt% MgO intermediate layer meant to reduce interfacial stress, and an outer thermal barrier of pure ZrO_2–24 wt% MgO [53]. These coatings gave five times longer combustor life, but ultimately failed by oxidation of the NiCr in the intermediate layer and evaporative loss of MgO (above about 950 °C). Coatings of ZrO_2–7 wt% Y_2O_3 over an NiCrAlY bond coat were subsequently found to be superior to the three-layer magnesia–zirconia TBC. An alternative approach to three-layer TBC systems is to divide the metallic bond coat into an upper oxidation-resistant alloy and a lower 'compliant' alloy next to the substrate, with conventional stabilized zirconia applied on top of this 'composite' bond coat. Such a coating has been investigated by Movchan *et al.*, who report improved performance [54]. In general, three-layer TBCs with graded stress-compliant metal–ceramic intermediate layers have been superseded, particularly for gas turbines, because of oxidation of the metal in the intermediate layer.

8.6.1 Plasma-sprayed thermal barrier coatings

The two principal means at present for applying TBCs are plasma spraying and electron beam physical vapor deposition (EB-PVD). Each method produces its own largely distinctive structure in the resultant TBC, as will be discussed further below. Sputter deposition has been studied to some extent as an alternative [55], but has shown no advantage over plasma spraying or EB-PVD. A number of other possible processes for the deposition of TBCs are listed by Miller [5].

Plasma spray deposition

Derived from research on low thrust plasma arc engines [6], plasma spray deposition is widely used in the coatings industry. The basis of the process, as it pertains to thermal barrier coatings, has been described by Miller [5]. A more detailed review of plasma spraying is presented by Hocking *et al.* [56]. Highly automated, computer-controlled plasma spray systems are now available and coming into use [57]. A schematic representation of the plasma spray gun and plasma spraying process is shown in Fig.8.3. A DC arc is struck (using typically on the order of 30 V and 300–800 A) between the cathode and anode to generate a plasma in the carrier gas, which is usually a mixture of Ar plus H_2 or H_2 and N_2. The temperature of the plasma gas

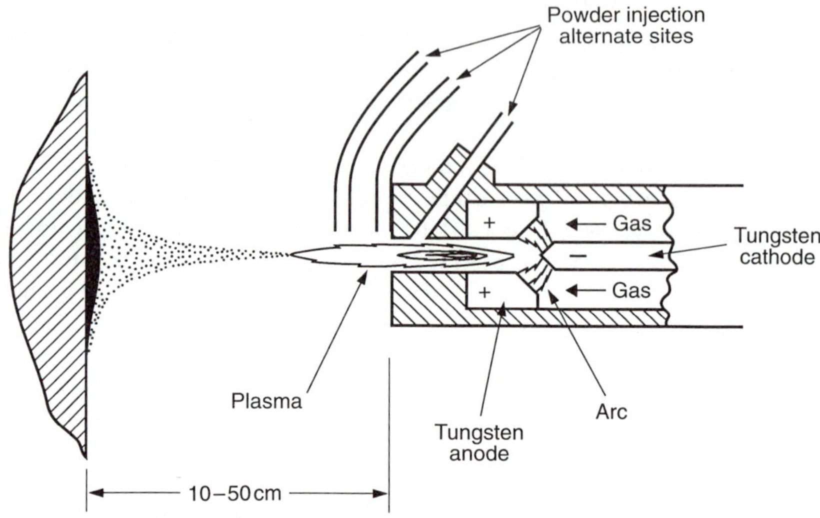

Figure 8.3 Schematic illustration of a plasma spray gun and the plasma spraying process.

at the anode nozzle may be 6000–12000 °C, and its velocity on the order of 200–600 m/s. The plasma spray powder, with particle sizes of 20–100 μm, is injected near the nozzle exit and melts rapidly upon being entrained in the exiting plasma gas. The molten particles then impact upon the substrate, held at standoff distances of about 10–50 cm, to form the desired coating. By injecting the powders further downstream in the cooler sections of the plasma gas, it is possible to plasma spray relatively low melting materials including even plastics [56].

Although plasma spraying is simple in concept, it involves many complex factors, and extensive work has been required to develop it as a reliable means for TBC deposition. As listed by Miller [5], the factors that must be controlled involve (1) the torch design (nozzle diameter, powder injection location, injection angle); (2) the spraying operation (arc power, gas composition, plasma stream temperature and velocity, powder injection rate, powder distribution and dwell time in plasma stream, standoff distance, workpiece surface preparation); and (3) the plasma powder material (size range, shape, crystal phase distribution, chemical homogeneity). Taylor *et al.* have also reviewed the complexities involved in plasma spraying, and cite several references where attempts have been made to model plasma spraying [58].

Properties of plasma-sprayed zirconia layer

A schematic cross section of a typical plasma-sprayed thermal barrier coating is given in Fig. 8.4. Porosity is introduced as the molten droplets splat down,

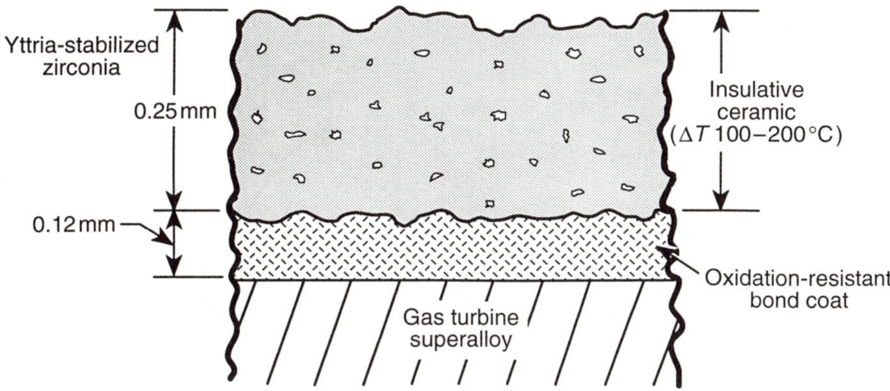

Figure 8.4 Cross section showing the dimensions and features of a typical plasma-sprayed zirconia thermal barrier coating for aviation use. The zirconia layer is comprised of splatted-down particles with porosity of 5–15%.

and the porosity (or inversely, the density) of the zirconia layer is critical in determining the heat insulation, thermal shock, thermal cycle tolerance and erosion resistance of the TBC. It can be varied within limits by the plasma spraying technique, as shown by Taylor *et al.* [58] and Joshi and Srivastava [59]. In general, the more porous zirconia layers favor better heat insulation and spalling resistance, whereas the less porous layers give improved erosion resistance. Thermal cycle life can be quite sensitive to the zirconia density, with low densities being as detrimental as high densities, as shown by Stecura in Fig. 8.5 [60]. The relationship between the mechanism of formation, microstructure and properties of plasma-sprayed coatings has been reviewed by McPherson [61, 62]. Sintering at temperatures as low as 1100 °C has been found to increase the strength and modulus of plasma-sprayed zirconia; the sintering effect increases with temperature and silica impurity content of the zirconia [63]. The sintering effect was interpreted as resulting from the transformation of the zirconia from 'a mechanically interlocked network to a chemically bonded system.' However, recent research by Miller *et al.* indicates that silica impurities strongly decrease the cyclic life of ZrO_2–7 wt% Y_2O_3 TBCs [64].

Lack of reproducibility has been a traditional problem in the plasma spraying of zirconia TBCs. The characteristics of the individual plasma spray powders were long suspected as being a primary cause of TBC variability, but little documentation existed to establish the fact. This problem was addressed by Rigney *et al.* [37], who critically examined 12 different plasma spray powders, all nominally ZrO_2–8 wt% Y_2O_3, to determine how the physical, crystallographic and chemical makeup of the powders influenced the thermal cycle life and erosion resistance of coatings plasma sprayed from

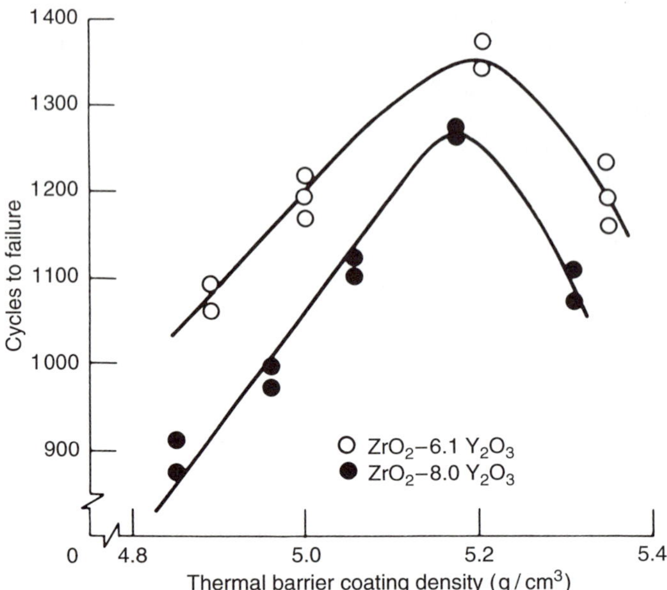

Figure 8.5 Effect of the density of the ZrO$_2$ layer on the thermal cycle life of ZrO$_2$–Y$_2$O$_3$ thermal barrier coatings as determined by Stecura [60].

the 12 different powders. The powders were prepared by five different processes; (1) spray drying, (2) spray drying and sintering, (3) sintering and crushing, (4) casting and crushing and (5) casting, crushing and fusing. A wide variation in performance was found, with spalling life for hourly thermal cycling to 1093 °C ranging from 40 h to >1000 h. The most critical factors were concluded to be the particle size and its uniformity, the chemical homogeneity of the powder and the method of manufacture, with spray drying being the least favored. Wide scatter also was reported in European tests using six different ZrO$_2$–8 wt% Y$_2$O$_3$ powders, as described in the review by Rhys-Jones and Toriz [52], where thermal cycle life varied from only 10 cycles to over 1000 cycles for the different powders. The influence of powder microstructure on the structure and properties of plasma-sprayed ZrO$_2$–8 wt% Y$_2$O$_3$ TBCs has also been recently reported by Pieraggi *et al.* [65].

Spalling life in the Rigney *et al.* results could not be directly correlated with the monoclinic content of the TBC. Although the shortest-lived coating (40 h) had the highest monoclinic content (64%), and the longest-lived coating (>1000 h) was 97% tetragonal, several coatings with 20–25% monoclinic content survived longer than others that were 100% tetragonal. Also, all of the coatings either retained the tetragonal (presumably t') structure, or tended to develop the tetragonal structure if not initially tetragonal, during thermal cycling. This is at variance with the findings of thermal decomposition of the t′ phase during high temperature aging [30, 34, 35], and Rigney *et al.* proposed that cycling, as opposed to isothermal high temperature tests, may cause 'retention' of the tetragonal phase in the same manner that thermal cycling of maraging steel causes retention of the high temperature austenitic phase.

The benefit of chemical homogeneity and uniformity of particle size and shape may also have been shown in tests with sol–gel prepared plasma spray powders, where the sol–gel powders, spherically shaped particles of high microhomogeneity and uniform size, were reported to outperform reference commercial powders in isothermal and thermal cycling tests up to 1100 °C [53].

Effect of stabilizer

The properties and high temperature performance of TBCs are mainly determined, however, by the choice and amount of oxide added to stabilize the zirconia. Calcia- and magnesia-stabilized zirconia have mostly been displaced from gas turbine TBC use, for example, because of their phase instability under thermal cycling, lack of corrosion resistance and volatility (for the case of MgO). As is now well known, the thermal cycle life of yttria-stabilized zirconia TBCs for gas turbine engines is highly dependent upon the Y_2O_3 composition. Fully stabilized cubic ZrO_2 (formed at $>18\,mol\%$ $YO_{1.5}$) might be thought the best choice for gas turbine TBCs, since it avoids the monoclinic–tetragonal transformation, but in fact it gives poor thermal cycle performance. Instead, wide experience has shown that maximum thermal cycle life is provided by partially stabilized zirconia within the narrow concentration range of about $6–8\,wt\%$ Y_2O_3 as indicated in Fig. 8.6 by Stecura [60]. The improved thermal cycle life is often explained as resulting from 'microcracking' within the partially stabilized ZrO_2. However, $6–8\,wt\%$ Y_2O_3 corresponds to the composition (about $7–9\,mol\%$ $YO_{1.5}$) of maximum stability of the tetragonal t′ phase in the $Y_2O_3–ZrO_2$ system (Fig. 8.2), and the improved thermal cycle life results from the total properties of the t′ phase, and not just microcracking.

Over the years, other stabilizing oxides have been proposed to give improved thermal performance or corrosion resistance over yttria. Stecura, for example, has developed ytterbium-stabilized zirconia and shown, using 1120 °C thermal cycling, that plasma-sprayed $ZrO_2–12.4\,wt\%$ Y_2O_3 coatings give 80% and 60% longer thermal cycle lives, respectively, than equivalently prepared $ZrO_2–6.1\,wt\%$ Y_2O_3 and $ZrO_2–8.0\,wt\%$ Yb_2O_3 coatings [66]. Ceria-stabilized zirconia has been patented as a hot corrosion-resistant thermal barrier coating [67]. There is also interest in ceria-stabilized zirconia as an ultrahigh temperature thermal barrier coating, with Brandon and Taylor reporting that $ZrO_2–25\,wt\%$ CeO_2 remained wholly tetragonal after 100 h at 1500 °C, and developed only 13% monoclinic phase after 100 h at 1600 °C [68]. Scandia and india have been identified as potential hot corrosion-

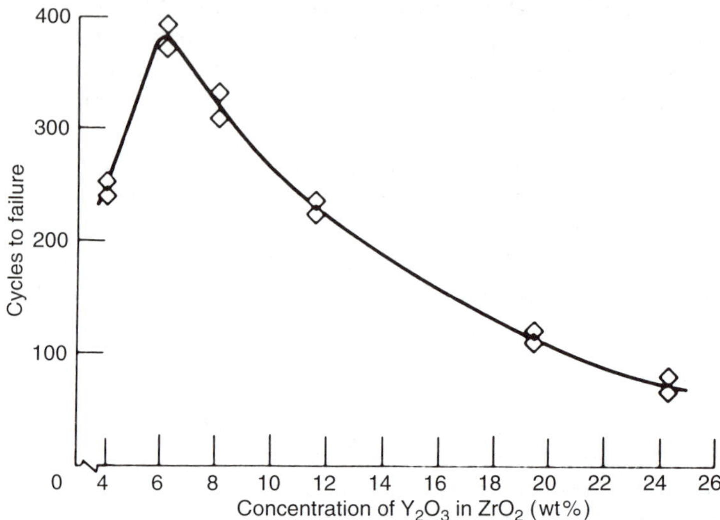

Figure 8.6 Effect of the yttria content on the thermal cycle life of ZrO_2–Y_2O_3 thermal barrier coatings as determined by Stecura [60].

resistant stabilizers from zirconia [69, 70], but the thermomechanical properties and phase stabilities of scandia- and india-stabilized zirconia TBCs have not been established. A compilation and review of the thermodynamic properties of a number of high temperature metal oxide–zirconia systems, useful in searching for other possible stabilized zirconia coatings, has been prepared by Jacobson [71].

Properties of metallic bond coat

Zirconia thermal barriers are permeable to oxygen at high temperature, and must therefore be backed by an oxidation-resistant bond coat. The choice of the bond coat has been found to be as critical as the choice of the zirconia outer coating in determining the success of the TBC. It was earlier thought that the bond coat should be of low strength so as to relieve stress resulting from the difference in the coefficients of thermal expansion for the zirconia ceramic and the superalloy substrate. However, current results suggest that strengthened bond coats with greater creep resistance may be the route to improved thermal cycle life [72, 73]. On the other hand, the bond coat must not form brittle phases, and it must have good resistance to interdiffusion with the substrate alloy.

Bond coats for plasma-sprayed zirconia TBCs are themselves normally produced by plasma spraying using vacuum, low pressure or shrouded plasma spray techniques to minimize formation of metal oxide phases. Low pressure or vacuum plasma spraying also allows increased substrate temperatures of up to 1600–1800 °F (870-980 °C) which promote diffusion of the bond coat into the substrate and so give higher bond strengths [48]. The upper bond coat surface must be of a certain roughness (on the order of 600 µin. (15 µm) average amplitude [37]) for adhesion of the zirconia layer [5]. Plasma spray conditions are set to give, as far as possible, fully dense bond coats.

The bond coats used for aviation engines are usually made of an MCrAlY, an M-based alloy (where M = Ni, Co or NiCo) containing Cr, Al and Y. At high temperature, an oxide film predominantly of alumina forms on the bond coat at the ZrO_2–MCrAlY interface, and acts as the basis of attachment for the ZrO_2 outer layer. The properties (growth rate, composition, integrity, spalling behavior, etc.) of this oxide film are critical, and they are strongly affected by the specific composition of the MCrAlY. Stecura has shown that removing yttrium from the bond coat results in rapid spalling failure of the TBCs, whereas optimizing the yttrium, and to a lesser extent the chromium and aluminum, content of the bond coat can increase TBC high temperature spall life by a factor of 10 or more [74, 75]. An example of one of the better performing bond coat compositions found by Stecura is Ni–35 Cr–6 Al–0.95 Y (wt%). The crucial effect of Y in bond coats is not unexpected. The importance of yttrium in promoting adhesion of the alumina surface oxide, and thereby increasing service life, was recognized during the original development of the MCrAlY materials as high temperature oxidation-resistant coatings for superalloys. A large effort has been made to understand how small additions of 'active' elements such as Y, Hf, Zr, etc., act to improve the alumina surface oxide or its adhesion on an MCrAlY [76–78]. Stecura has recently investigated MCrAlYb (where Yb replaces Y as the active element) bond coats, and found improvement in TBC spall life over the conventional MCrAlY bond coat [79].

There is likely to be a movement away from simple MCrAlY materials as bond coats in the future, especially for specific needs or nonaviation applications. Improved TBC life has been found, for instance, by Wortman *et al.* with NiCoCrAlY bond coats which were modified for higher strength, and given a diffusion aluminide topcoat to enrich the outer surface in aluminum for improved oxidation resistance [72]. For diesel engines, Kvernes has used rapid solidification technology to produce a high silicon Si–Al bond coat having a reduced coefficient of thermal expansion and the capability of being used on light alloys in engines [80]. Also for diesels, graded coatings using mixed CoCrAlY–ZrO_2 or NiCrAlY–ZrO_2 interlayers are being applied to reduce tensile stress at the coating–substrate interface in thick thermal barrier diesel engine coatings [81, 82]. The use of graded coatings is expected to be successful here because temperatures are too low (< 500 °C) for the metal particles to be oxidized. Other proposals for modifying the TBC–bond coat interface in internal combustion engines include the co-plasma spraying of SiC whiskers to improve cracking toughness of the ZrO_2–bond coat interface and ZrO_2 ceramic [83], and the incorporation of various oxides and metals into the ZrO_2 by a slurry bisque method to provide diesel engine coatings having good wear and friction properties, as well as thermal insulating capabilities [84, 85].

8.6.2 Electron beam-physical vapor deposited (EB-PVD) thermal barrier coatings

EB-PVD of metallic coatings

Electron beam-physical vapor deposition is widely used in the gas turbine industry for applying metallic coatings (usually of the MCrAlY type) to turbine blades and vanes for protection against oxidation and hot corrosion. The technique is also capable (see below) of producing zirconia thermal barrier coatings with potentially superior properties for certain applications. A schematic of an EB-PVD production facility for the deposition of MCrAlY coatings is shown in Fig. 8.7 [86]. In a large vacuum chamber, a high power electron beam is focused onto an MCrAlY ingot target which melts the metal locally and produces an MCrAlY vapor. The vapor deposits as a coating on

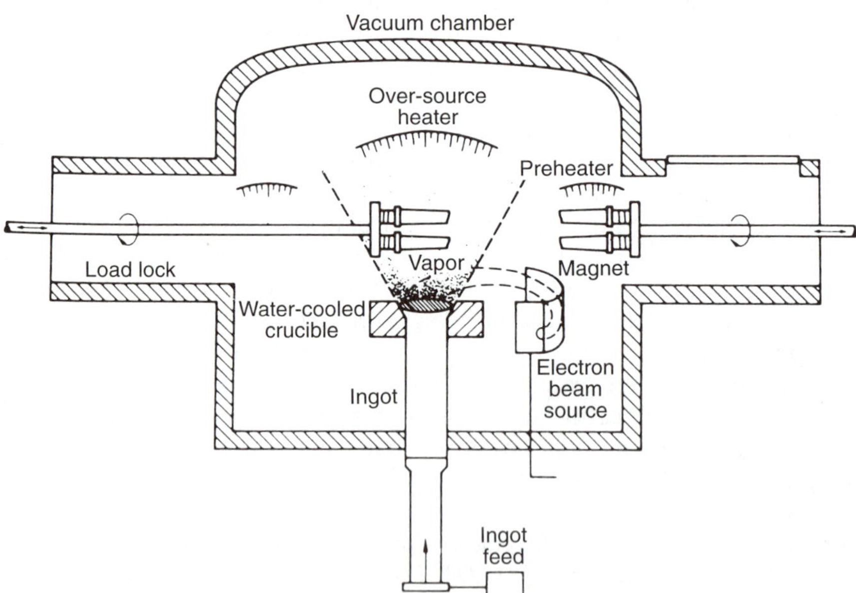

Figure 8.7 Schematic representation of the electron beam-physical vapor deposition (EB-PVD) coating process as given by Goward [86]. Reproduced by permission of G.W. Goward and the Metals and Ceramics Information Center.

the airfoils, which are held in rotatable and retractable fixtures above the vapor source. Preheating and oversource heating are used, as indicated in Fig. 8.7. The process yields an MCrAlY coating with an excellent coating–substrate adhesion, uniform thickness, high density (essentially zero porosity), optimum composition and grain structure and a smooth outer surface for good aerodynamic flow. There is little problem with closing of the surface air-cooling holes of the blade or vane. The latest versions of production airfoil vacuum coatings are large machines (extending through two floors of the production facility), which have sophisticated computer controls and cost several million dollars each.

EB-PVD preparation of zirconia thermal barrier coatings

Zirconia thermal barrier coatings can also be deposited by the EB-PVD process. The general process, and the advantages it is thought to confer, are described by Strangman [48, 87]. The principal modification required for thermal barrier EB-PVD coating is to provide a certain oxygen partial pressure within the vacuum chamber so as to maintain oxygen stoichiometry in the deposited ZrO_2 film. Bond coats for EB-PVD thermal barrier coatings are normally of the MCrAlY type and are themselves deposited by EB-PVD. For best adhesion, the bond coat surface should be smooth, or preferably polished [88], in contrast to plasma-sprayed TBCs, which require a rough bond coat. This implies, according to Strangman, that adhesion of EB-PVD thermal barrier coatings involves a chemical bond between the zirconia and bond coat surfaces, and he points out that high deposition temperatures, or postcoating heat treatments at 1600–2000 °F (870–1090 °C) act (presumably because they strengthen the chemical bonding) to improve bond quality.

Properties of EB-PVD zirconia thermal barrier coatings

The major advantage of EB-PVD thermal barrier coatings lies in their columnar outer structure (Fig. 8.8) which reduces stress buildup within the body of the coating. As described by Strangman, 'strain within the coating is accommodated by free expansion (or contraction) of the columns into the gaps, which results in negligible stress buildup' [48]. The columnar structure of EB-PVD zirconia TBCs has the disadvantage, however, of increasing heat conductivity by a factor of about 2 as compared to plasma-sprayed TBCs [89].

Note that Fig. 8.8 indicates, following the description of Strangman [48], that a thin, dense ZrO_2 layer occurs between the bond coat surface and the upper columnar zirconia structure. This phase grows under oxygen-deficient conditions just at the beginning of ZrO_2 deposition, and its thickness is controlled by how quickly the oxygen bleed is activated after ZrO_2 coating commences. Strangman considers this dense, interfacial ZrO_2 film to be critical to the life of EB-PVD coatings, and proposes that it provides for

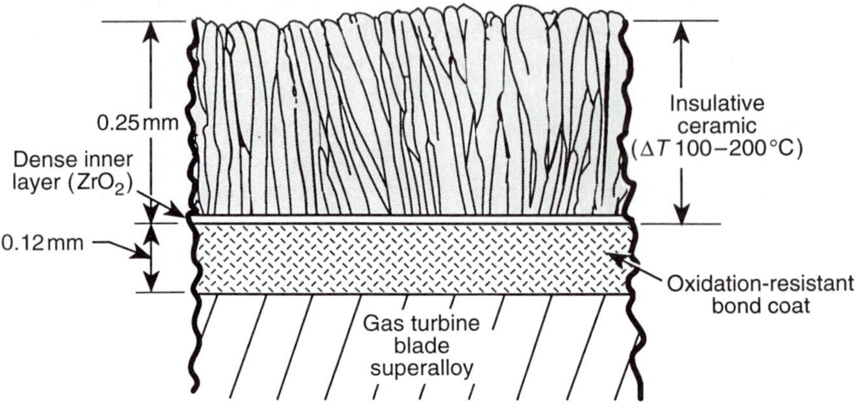

Figure 8.8 Cross section illustrating the strain-tolerant columnar ZrO$_2$ microstructure of EB-PVD zirconia thermal barrier coatings.

chemical bonding between the ZrO$_2$ and bond coat, but if it becomes too thick ($>2\,\mu$m) it may sustain and transmit compressive stresses sufficient to cause cracking within the outer zirconia coating.

Other EB-PVD groups appear less certain of the necessity for a discrete, dense ZrO$_2$ layer at the zirconia–bond coat interface. However, TBC production by EB-PVD is still something of an art, and each EB-PVD production group will tend to develop its own proprietary practices to yield TBCs having superior structures and properties. In this vein, we may note, for example, that Fritscher *et al.* have developed EB-PVD techniques for producing densitometrically graded zirconia layers [90] and chemically graded MCrAlY bond coats [91] which they report to improve TBC life.

The superior spalling performance predicted for EB-PVD thermal barrier coatings has been confirmed in gas turbine flight tests. Toriz *et al.* examined EB-PVD coatings on first-stage vane airfoil surfaces after 5200 h of flight time, and concluded EB-PVD coatings to be superior to plasma-sprayed coatings in having (1) longer thermal cycle lives, (2) smoother surface finishes, (3) better surface finish retention and (4) superior erosion resistance [92]. EB-PVD zirconia was found by Meier *et al.* to have better degradation resistance than either plasma-sprayed zirconia or metallic MCrAlY in 4200 h engine side-by-side flight tests of first-stage blades coated variously with metallic MCrAlY, plasma-sprayed zirconia or EB-PVD zirconia [93]. The thermal life of the second-stage vanes was also determined, in the same tests, to be significantly increased by EB-PVD zirconia coatings. The EB-PVD process thus appears to give superior zirconia thermal barrier coatings for aerodynamic blade and vane surfaces, although plasma-sprayed TBCs are likely to continue to be used for combustors and other nonaerodynamic surfaces.

8.7 CAUSES FOR FAILURE OF ZIRCONIA TBCs

In general, zirconia TBCs are considered to fail predominantly either by oxidation or hot corrosion. Failure by oxidation occurs primarily with TBCs in aviation engines. Failure by hot corrosion is a problem when TBCs are applied in diesel engines or gas turbines burning low quality fuel containing corrosive impurities such as sulfur or, especially, vanadium. The mechanisms of attack for the two modes of failure are substantially different.

8.7.1 Oxidative failure of plasma-sprayed zirconia thermal barrier coatings

The primary cause of plasma-sprayed TBC failure under oxidation conditions is thought to be crack formation in the zirconia near the ZrO_2–bond coat interface, induced by oxidation of the bond coat (see below). Nonetheless, reactions occur within the ZrO_2 layer itself during high temperature cyclic exposure and they too may contribute to the degradation of the TBC.

Degradation of the zirconia layer

Among the reactions that contribute to degradation of the zirconia layer are sintering and grain growth, thermal decomposition of the t′ phase, and the t tetragonal–monoclinic phase transformation. Sintering is influenced by impurity levels (especially of SiO_2), and can improve strength [63], although TBC cycle life may in fact be strongly decreased by silica impurities [64]. Also, excessive amounts of SiO_2 at the grain boundaries can decrease porosity and reduce the TBC creep life. Sintering of the zirconia granular columns was noted as a problem in developmental research on EB-PVD zirconia TBCs by Fritscher and Bunk [94]. Grain growth is detrimental because larger grains lead to spontaneous transformation of the tetragonal t phase, and so exacerbate the tetragonal–monoclinic transformation problem during thermal cycling.

The role of the tetragonal–monoclinic transformation in determining TBC life remains unclear, at least quantitatively. Large concentrations of monoclinic phase in the TBC can undoubtedly be catastrophic, but small to moderate amounts of monoclinic phase may be tolerated. The data of Rigney *et al.* [37] show a general trend for coatings having the shortest life (40 h, 64% monoclinic) and longest life (1000 h, 97% tetragonal), yet some coatings with moderate monoclinic contents survived longer (660–700 h, 17–21% monoclinic) than 100% tetragonal coatings (300–400 h, 100% tetragonal). Moreover, the tendency was for the monoclinic phase to decrease, and the tetragonal phase to increase, with thermal cycling. One coating described by Rigney *et al.*, which survived for 285 h, was originally 47% monoclinic and 53% tetragonal, but became 28% monoclinic and 72% tetragonal by the end of its cycle life. Conversion of monoclinic phase to tetragonal phase with thermal cycling was also seen by Borom and Johnson [95]. However, as Rigney *et al.* and others have emphasized, there are numerous factors in plasma-spraying such as powder characteristics, spraying conditions, zirconia layer porosity, etc., that also strongly affect the stress state and spalling life of the ZrO_2 TBC. In some cases, these factors doubtlessly offset stress induced by the monoclinic–tetragonal transformation. Quantitative determination of the effect of the monoclinic–tetragonal transformation on TBC cycle life will therefore be difficult, except perhaps for comparisons using standardized plasma-spraying regimes.

In high temperature thermal degradation, the nontransformable t' component of a TBC disproportionates over time into the low yttria, transformable t phase and the high yttria cubic phase. The TBC thermal spalling life is then reduced because the resultant t phase undergoes the t $\rightleftarrows$ m transformation during thermal cycling, which produces volume changes and stress within the TBC. However, the phenomenon has not been quantified, and no numbers can currently be put on the correlation between the extent of high temperature phase separation and the reduction in TBC thermal cycle capability.

Borom and Johnson have recently postulated, on the basis of bimaterial dilatometry experiments, that a previously undetected phase transformation occurs at about 1000–1200 °C with ZrO_2–8 wt% Y_2O_3 [95]. The transformation is not believed to be the high temperature t' phase degradation observed by Miller *et al.* [30], but rather to be 'the conversion of a lower-density, metastably retained, high-temperature, tetragonal structure to a higher-density, thermodynamically stable, tetragonal structure.' A change in the tetragonal phase field of the Y_2O_3–ZrO_2 phase diagram is suggested as necessary to account for the newly revealed transformation. However, as noted by Borom and Johnson, further work is required to verify these observations.

Bond coat oxidation in plasma-sprayed TBC failure

Plasma-sprayed zirconia TBCs subjected to thermal cycling under oxidative conditions fail predominantly by spalling [96]. This has been shown for furnace tests, burning rig tests, and engine tests [96]. The spalling is caused by the formation of longitudinal cracks of a critical size that develop, after a certain number of thermal cycles, within the zirconia layer just above the bond coat interface. Specks of the bond coat oxide are generally found on the zirconia fracture surface, which suggests that some cracking occurs within the bond coat oxide [96, 97]. The extended zirconia cracks are believed to be formed by microcrack linkup and slow crack growth, which is driven by a coaxially compressive but radially tensile stress, generated at the zirconia–bond coat interface beause of thermal expansion mismatch and, especially, bond coat oxidation [5]. Interestingly, thermal stress caused by high heat flux across the TBC does not appear to be a major factor, provided that one uses the 'right' TBC composition. In 0.5 s, 3000 °C plasma torch cycle tests by

Miller and Berndt [98], ZrO_2–8 wt% Y_2O_3 withstood 3000 cycles with no sign of distress, whereas ZrO_2–12 wt% Y_2O_3 was severely crazed after 250 cycles and ZrO_2–20 wt% Y_2O_3 spalled after only 1 cycle. Thermal stress tolerance thus depends on the Y_2O_3 concentration, and reducing the Y_2O_3 content from 20 wt% to 8 wt% evidently yields phase structures that provide microcracking, or other mechanisms, that can accommodate the thermal stresses generated even in aggressive TBC cycling.

Considerable evidence indicates oxidation of the bond coat to be the primary generator of the stress that causes formation of the spall-inducing cracks within the zirconia at the bond coat interface. No spalling of the TBC occurs, for instance, when the thermal cycle tests are conducted in an inert argon atmosphere. Nor is there any significant change in weight. In contrast, a progressive weight gain, attributable to bond coat oxidation, is observed for TBC thermal cycle tests carried out in air. Often a critical weight gain of 3–6 mg/cm^2, and therefore some critical thickness of the bond coat oxide, seems to be associated with the point of TBC failure.

Miller has examined the relationship between weight gain and the effective strain at the zirconia–bond coat interface for TBC thermal cycle failure, and proposed an oxidation-based model for predicting thermal barrier coating life [99]. His results indicate that, although cycle frequency (i.e. cyclic heat stress) has an effect, the predominant factor is total time at temperature (i.e. bond coat oxidation). A finite element study supports the theory by indicating that the stresses induced at the rough zirconia–bond coat interface (required for plasma-sprayed TBCs) by thermal mismatch *plus* bond coat oxide growth would promote crack propagation, whereas no propagation should occur under thermal mismatch stress alone [100].

However, McDonald and Hendricks have reported 1040 °C cyclic testing of Y_2O_3–ZrO_2 TBCs where cyclic stress was found to be more important than oxidation in causing TBC spalling [101]. Also, Brindley and Miller have recently shown that the 1150 °C isothermal oxidation behavior of Ni–35 Cr–6 Al–0.95 Y, Ni–18 Cr–12 Al–0.3 Y and Ni–16 Cr–6 Al–0.3 Y (by wt%) does not correlate with their performance as bond coats in 1150 °C cyclic testing of ZrO_2–7 wt% Y_2O_3 TBCs [102]. The Ni–35 Cr–6 Al–0.95 Y alloy was the most oxidized but provided the longest TBC life, whereas the Ni–16 Cr–6 Al–0.3 Y alloy was the least oxidized but gave the shortest TBC life. Brindley and Miller concluded that, although bond coat oxidation had a strong detrimental effect on TBC life, it was not the only bond coat factor affecting TBC life. The effect of bond coat (Co–32 Ni–21 Cr–8 Al–0.5 Y) oxidation on plasma-sprayed ZrO_2–8 wt% Y_2O_3 TBC cycle life has been examined also by Wu *et al.*, who propose that zirconia–bond coat interface cracking initiates at outgrowths of metal oxide composed mostly of Co(Cr, Al)$_2$O$_4$ (and therefore possibly weak) which extend into the ZrO_2 layer [97].

The kinetics of microcrack linkup and critical crack growth at the zirconia–bond coat interface are difficult to study for lack of a real-time, nondestructive analytical technique. One innovative approach has been to monitor the sound produced by the cracking processes via acoustic emission (AE) as explored by Berndt, Herman and Shankar, and reviewed by Berndt [103]. Acoustic emission was coordinated, at least qualitatively, with both TBC adhesion and TBC thermal cycle tests. AE count rates were found to increase, for example, with bond strength in TBC adhesion tests. Acoustic emission also indicated that bond coats confer toughness on TBCs because the AE count data showed high peaks, few events (large cracks occurring) when no bond coat was used, but low peaks, many events (only microcracking) when a bond coat was present. Acoustic emission data are complex and difficult to interpret, but the technique provides information not available by other means.

8.7.2 Oxidative failure of EB-PVD thermal barrier coatings

EB-PVD thermal barrier coatings are sufficiently new that little has been reported on their mechanism of failure under thermal cycle oxidation conditions. The principal results available are by Meier *et al.* [89]. They found that, whereas plasma-sprayed TBCs failed by cracking in the zirconia layer near the zirconia–bond coat interface, EB-PVD coatings failed by cracking at the interface between the thermally grown oxide (TGO) which forms on the bond coat and the bond coat itself. In addition, Meier *et al.* observed that an inelastic deformation of the zirconia layer occurred with plasma-sprayed TBCs during thermal cycling, but that the zirconia layer in EB-PVD TBCs remained elastic (presumably because of its columnar structure), with the life-limiting factor for EB-PVD TBCs appearing to be strain that developed in the thermally growing bond coat oxide.

From these observations, Meier *et al.* developed a life prediction model for EB-PVD thermal barrier coatings which was based (1) on predicting the rate of growth for the thermally grown oxide on the bond coat surface, and (2) on calculating the maximum in-plane TGO tensile mechanical strain that results from the thermal oxide growth. Tests in the investigation showed the model to predict the rate of TGO growth to within about 20%, and the thermal cycle life of EB-PVD coatings to within $\pm 2X$.

8.7.3 Corrosive failure of thermal barrier coatings

Although we are concerned with TBC hot corrosion, yttria-stabilized zirconia (YSZ) in fact appears to be more corrosion resistant than the MCrAlY and aluminide metallic coatings commonly used on gas turbine airfoils. This has been shown in aircraft engine [72] and ship propulsion engine trials [104], in burner rig hot corrosion tests [105], and in general experience with TBCs in marine diesels [8]. Hot corrosion of YSZ thermal barrier coatings is not expected to be a problem in aircraft engines burning aviation quality ($>0.05\,wt\%$ S, $0\,wt\%$ V) fuel.

Calcia- and magnesia-stabilized zirconia, on the other hand, have more tendency to react with combustion gas SO_3 than YSZ, and possibly may be attacked under aviation service conditions. Formation of $CaSO_4$ (and thus depletion of CaO from ZrO_2 matrix) has been noted with $CaO-ZrO_2$ TBCs used on aviation combustors in the past [106]. Reaction of magnesia-stabilized zirconia (MSZ) coatings with combustion gas SO_3 in aircraft engines, if it occurs, is not well known. In any event, the current trend is to move away from the use of calcia- or magnesia-stabilized TBCs in aircraft gas turbine engines.

Hot corrosion of ZrO_2-based TBCs becomes a problem when low quality fuels containing appreciable levels of sulfur, and especially vanadium, are used. Most information available on TBC hot corrosion originates from TBC applications in marine diesel engines, or from U.S. DOE-sponsored research initiated during the fuel crises of the 1970s to examine the compatibility of thermal barrier coatings with low quality, or 'alternate source' coal- or shale-derived fuel. Kvernes conducted much of the early marine diesel work, which he has covered in his publications and reviews [8, 50]. The DOE-sponsored reasearch has been mostly reported in the proceedings of the 1979, 1982, 1987, 1990 and 1992 conferences on coatings for advanced heat engines, organized by J. Fairbanks of DOE and (for the earlier conferences) J. Stringer of the Electricity Producers Research Institute (EPRI). Much of the early work aimed at the use of TBCs in electric utility gas turbine engines was reviewed by Miller [107]. More TBC corrosion performance data are now becoming available, as engine manufacturers, TBC vendors and others again begin to explore TBCs for corrosive environments.

Hot corrosion of stabilized ZrO_2 TBCs is mainly driven by reaction between the fuel contaminant oxides (Na_2O, SO_3, V_2O_5 etc.) – produced by oxidation of Na,V,S fuel impurities during combustion – and the oxides employed to stabilize the zirconia (Y_2O_3, MgO). Formation of $MgSO_4$ by reaction with SO_3, for example, removes the MgO stabilizer from the ZrO_2 matrix, and so causes destabilization and failure of the TBC. The corrosion reactions appear to be mostly determined by Lewis acid–base oxide interactions [108]. The literature of the hot corrosion reactions of TBCs has been briefly reviewed by Jones [109], who has also described some aspects of the development of hot corrosion-resistant TBCs [110].

Hot corrosion studies of TBCs have been made by numerous groups around the world, and generally good agreement is seen. Data and observations taken mostly from the work of Bratton *et al.* [31, 111] are given here as an illustrative example. In 800 °C (metal temperature) burner rig tests using GT No.2 fuel doped with various levels of Na, V, P, Ca, Fe, Mg and S, Bratton *et al.* found MgO-stabilized ZrO_2 to be attacked and destabilized by

$$ZrO_2(MgO) + SO_3(g) \rightarrow ZrO_2 \text{ (monoclinic)} + MgSO_4 \qquad (8.1)$$

For fuels containing no Mg, destabilization of MgO-stabilized ZrO_2 occurs also by reactions of the type:

$$ZrO_2(MgO) + V_2O_5(l) \rightarrow ZrO_2 \text{ (monoclinic)} + Mg_3V_2O_8 \qquad (8.2)$$

When both V_2O_5 and SO_3 are present, they compete for reaction with the MgO stabilizer, much as is the case when MgO inhibitors are used with V- and S-containing fuels [112], and destabilization may occur by either $MgSO_4$ or $Mg_3V_2O_8$ formation, depending upon the specific SO_3 and V_2O_5 chemical activities in the corrodent melt.

For Y_2O_3-stabilized ZrO_2 TBCs, Bratton *et al.* concluded the destabilization reaction (when both Mg and V are present in the fuel and form $Mg_3V_2O_8$) to be

$$ZrO_2(Y_2O_3) + Mg_3V_2O_8 \rightarrow ZrO_2 \text{ (monoclinic)} + 2YVO_4 + 3MgO \qquad (8.3)$$

In the absence of Mg in the fuel, the destabilization reaction as determined by Kvernes [38], Hamilton and Nagelberg [47], and others is simply

$$ZrO_2(Y_2O_3) + V_2O_5 \rightarrow ZrO_2 \text{ (monoclinic)} + 2YVO_4 \qquad (8.4)$$

Details of the specific course of reaction of V_2O_5 with YSZ, as well as the diffusion behavior of V and Y within the YSZ matrix, are given by Hertl [113]. The crystallographic aspects of the reaction of V_2O_5 with Y_2O_3-stabilized ZrO_2 have also been examined by transmission electron microscopy by Susnitzky *et al.* [114].

Note that reactions 8.3 and 8.4 indicate Y_2O_3 to be a stronger base than MgO; that is, the stronger base, Y_2O_3, displaces the weaker base, MgO, from reaction with acidic V_2O_5 by

$$Mg_3V_2O_8 + Y_2O_3 \rightarrow 2YVO_4 + 3MgO \qquad (8.5)$$

Reaction of Y_2O_3 with SO_3 in the presence of Na_2SO_4 to give destabilization of YSZ is possible, as shown by Pettit and Barkalow [115], via the reaction:

$$ZrO_2(Y_2O_3) + 3SO_3 \, (+ Na_2SO_4) \rightarrow Y_2(SO_4)_3 \text{ (in } Na_2SO_4 \text{ soln)} \qquad (8.6)$$

A relatively high SO_3 partial pressure is required, however, and in general, yttria-stabilized zirconia TBCs will tend to resist sulfur degradation but be susceptible to vanadium degradation. On the other hand, magnesia-stabilized zirconia is more susceptible to sulfur degradation, but probably superior to YSZ in resistance to attack by V_2O_5, as indicated by reactions 8.3 to 8.5.

No reaction of ZrO_2 itself with V_2O_5 or SO_3-Na_2SO_4 is normally detected in the hot corrosion of TBCs. However, Bratton *et al.* [111] found that reaction of ZrO_2 may occur when phosphorus is present as a fuel contaminant. (Petroleum fuels tend to contain little phosphorus, but phosphorus may be found in alternative fuels such as derived by coal liquefaction.) Destabilization of YSZ by P_2O_5 occurs by the reaction:

$$ZrO_2(Y_2O_3) + P_2O_5 \rightarrow ZrO_2 \text{ (monoclinic)} + 2YPO_4 \qquad (8.7)$$

When sodium is present, Bratton *et al.* report that there can also be reaction of the ZrO_2 by

$$8ZrO_2 + 4Na + O_2 + 6P_2O_5 \rightarrow 4NaZr_2(PO_4)_3 \qquad (8.8)$$

Jones [110] has demonstrated that ZrO_2 reacts readily with P_2O_5 by the reaction:

$$ZrO_2 + P_2O_5 \rightarrow ZrP_2O_7 \qquad (8.9)$$

Note that while both P_2O_5 and V_2O_5 are classed as acidic oxides, P_2O_5 reacts strongly with ZrO_2, whereas essentially no reaction is seen between V_2O_5 and ZrO_2. Reaction 8.9 of acidic P_2O_5 with 'acidic' ZrO_2 appears to be an exception to the hypothesis that TBC hot corrosion is mainly controlled by Lewis acid–base oxide reactions where basic and acidic oxides interact, but little reaction occurs between oxides of nearly equivalent acidic strength [108]. However, the acid–base reaction influence may be overridden in this case by other factors such as lattice energy; instances of this happening were reported by Dent-Glasser and Duffy in their studies of acid–base reactions between oxides and oxysalts [116].

A recent development in TBC hot corrosion has been concern with silicates derived from desert sand as a threat to aircraft engines using TBCs at very high ($>1200\,^\circ$C) temperatures. Stott *et al.* [117] have studied the attack of various desert sand melts and silicate glasses at 1400–1600$\,^\circ$C on specimens of both APS and EB-PVD ZrO_2–8 wt% Y_2O_3. Two mechanisms of attack were identified, depending primarily on the amount of CaO in the melt. Although both ZrO_2 and Y_2O_3 were extracted from the YSZ by low CaO melts, proportionately more Y_2O_3 was extracted, and the YSZ became partially monoclinic because of Y_2O_3 depletion. In the high CaO melts, less Y_2O_3 was extracted, and there was diffusion of CaO into the ZrO_2 matrix; the net effect was that the YSZ tended to remain in the tetragonal structure. But substantial damage to the YSZ ceramic resulted in both cases.

With the exception of high temperature silicate attack, virtually all TBC hot corrosion to date has been diagnosed as resulting from reaction between the corrodent oxides (V_2O_5, SO_3–Na_2SO_4) and the zirconia-stabilizing oxide. Damage by other mechanisms such as corrosion of the bond coat by salts penetrating the ZrO_2 layer, or freeze–thaw cracking by molten deposits in the ZrO_2 pores, has been seldom documented, apart from a few exceptions [107, 118, 119]. In fact, Nagaraj *et al.* [105] specifically note that there was no penetration of salt into the TBC, nor corrosion of the underlying bond coat, in burner rig tests of plasma-sprayed and EB-PVD yttria-stabilized zirconia TBCs at 1300$\,^\circ$F (700$\,^\circ$C) (2 wt% S, 10 ppm sea salt) and 1700$\,^\circ$F (925$\,^\circ$C) (0.4 wt% S, 5 ppm sea salt). This was true even for EB-PVD coatings which had been feared to be possibly prone to salt infiltration down the intercolumnar porosity. Similarly, in aircraft engine tests of YSZ thermal barrier coatings, no infiltration of salt through the TBC (although Na_2SO_4 was found on the TBC surface) nor bond coat corrosion was detected despite engine gas pressures of approximately 10 atm (1 MPa) and corrosion conditions aggressive enough to significantly hot corrode aluminide coatings on adjacent second-stage HP vanes [72]. On the other hand, Strangman and Schienle report that they have observed salt deposit penetration into the TBC zirconia in certain tests [120].

One would intuitively expect molten deposits to penetrate the porous zirconia layer of TBCs, especially if the deposits were heavy and perhaps of a certain composition (V_2O_5-rich vanadate–sulfate melts are particularly wetting, for example). If no penetration is observed, this implies that the deposits are not heavy, or for some reason not sufficiently molten and wetting. There could also be a problem in detection; moderate hot corrosion of the bond coat could be manifested, for instance, only as an increased bond coat 'oxidation' rate, with no unique corrosion phase being discernible. Also, if bond coat corrosion is found at a ZrO_2 spall site, it is difficult to ascertain whether the corrosion occurred before, or after, ZrO_2 spallation.

8.8 ALLEVIATION OF TBC HOT CORROSION

The approaches taken to alleviate TBC hot corrosion are relatively few, and have centered mostly on the development of corrosion-resistant stabilizers such as CeO_2 [67], and possibly Sc_2O_3 [69] and In_2O_3 [70]. Chu and Rohr [121] have proposed adding SiO_2 and Al_2O_3 to YSZ so that the $ZrO_2(Y_2O_3)$ crystallites are enveloped and protected by zirconium and aluminum silicates. This solution may be compromised, however, by the now known deleterious effect of silica on the thermal cycle life of Y_2O_3–ZrO_2 TBCs [64]. A recent idea is to use additions of a yttrium acetonylacetonate complex to the fuel to inhibit hot corrosion by sulfur- and vanadium-containing fuels [122]. This treatment has the potential to protect yttria-stabilized zirconia better than the normal MgO treatment of vanadium-containing fuels, which tends to be ineffective because the 'inert' salt, $Mg_3V_2O_8$, formed in MgO inhibition can react with the Y_2O_3 stabilizer in YSZ as shown in reaction 8.3.

Densification of the upper ZrO_2 surface by seal coats or laser glazing also has frequently been advocated as a means for improving TBC hot corrosion resistance. Seal coats generally suffer from having unavoidable pinholes. In burner rig tests using S- and V-contaminated fuel, Nagaraj and Wortman [123] found, for example, that whereas Al_2O_3, HfO_2 or Pt sputtered seal coats did not suffer corrosion themselves, there was corrosion of the zirconia beneath the seal coat in each case, indicating penetration of the molten deposit through seal coat pinholes. Laser-glazed TBC surfaces normally develop shrinkage cracks upon cooling that spoil the imperviousness of the surface. However, Petitbon *et al.* [124] have shown that spraying alumina powder into the laser beam during surface melting avoids shrinkage cracks, and produces a surface that is resistant to hot corrosion attack itself, and which reduces molten deposit penetration into the TBC body. Although initial tests were only at 400–500 °C and therefore not strongly corrosive, penetration of S and V into the laser-glazed TBC was much less than for an as-sprayed TBC. Informative papers on the formation and thermal stability of the tetragonal t′ phase in laser-sealed ZrO_2–8.5 wt% Y_2O_3 TBCs [125], and on pulsed laser sealing of Y_2O_3–ZrO_2 TBCs [126], have also been published recently by West *et al.* Surface densification by electrical biasing during the last stages of EB-PVD has been reported as another means for reducing molten deposit wicking into the TBC [120]. As regards improvement in corrosion life, one should note, however, that Hertl [113] found single-crystal YSZ (i.e. analogous to dense, laser-glazed YSZ) to be as readily corroded by V_2O_5 as polycrystalline YSZ.

8.9 FUTURE USES AND TRENDS FOR ZIRCONIA TBCs

The primary areas of usage for TBCs in the future will probably continue to be in engines, particularly gas turbines and diesels. Since the requirements, and indeed prospects, for TBCs in these two engine types are different, they will be discussed separately.

8.9.1 Thermal barrier coatings for aviation and industrial gas turbines

Zirconia-based thermal barrier coatings are currently a vital part of aviation gas turbine technology. They are widely used on static components such as combustion chambers, fuel vaporizers, transition ducts, afterburner flameholders, vane platforms and airfoils. In these applications, their benefits in extending component service life, in increasing engine durability and in reducing operating costs have been well established. No modern aero gas turbine can be competitive without thermal barrier coatings.

Future research and development with aviation TBCs will undoubtedly involve a continued optimization of processing and material compositions for both the bond coat and ZrO_2 ceramic. Thermal barrier coatings are highly processing and composition intensive; that is, their properties depend critically upon processing and composition in ways not easy to foresee. The success of TBCs in aviation engines, for example, is primarily due to the large increase in thermal cycle life that occurs at the unique composition of 6–8 wt% Y_2O_3 in YSZ coatings, but this composition could not be predicted *a priori*, and was found only through developmental research. New coating techniques, or modifications of old methods, may be important. Miller and Brindley [73] have recently reported a new plasma-spraying procedure, whereby depositing the initial ZrO_2 by low pressure plasma-spraying (which allows high substrate preheat and plasma gun power) and then the remainder of the ZrO_2 layer by air plasma spraying, it is possible to successfully deposit plasma-sprayed TBCs on smooth substrates. According to Miller and Brindley, this method improves thermal cycle life by eliminating the rough ceramic–bond coat interface, while retaining the high thermal insulation properties and low production costs of plasma-sprayed TBCs. And, they believe, the method might even allow the omission of the bond coat on highly oxidation-resistant substrates.

The ultimate goal of engine manufacturers is to use TBCs, at their full thermal advantage, on the rotating blades of gas turbines. It is estimated that this could provide savings of as high as 10 million gallons per year of fuel for a fleet of 250 aircraft [93]. Producing TBCs that can be relied upon to operate for at least some fixed number of hours without catastrophic spalling or other failure is the problem. A NASA-sponsored study on thermal barrier coating life prediction was completed recently by Meier *et al.* [89]. Their report suggests that EB-PVD TBCs have several advantages for life prediction: (1) their structures are uniform and reproducible (i.e. they have a smooth ceramic–bond coat interface, and the ZrO_2 is composed of highly oriented, and uniformly sized and spaced columnar grains); (2) under thermal cycling the ZrO_2 layer remains apparently unaffected and elastic, and its adhesion to the bond coat is not critically degraded; and (3) the *single* life-limiting mechanism appears to be bond coat oxide growth which, because it occurs on a smooth surface under largely controlled conditions, should have a predictable growth rate. Models developed by Meier *et al.* were in fact successful in predicting oxide growth to within 20%, and TBC life to within $\pm 2X$. So the task of developing a TBC with a reliable engineering lifetime for blade airfoils remains daunting, but there are indications that it may not be impossible.

Another area of future development in gas turbine TBCs is likely to be hot corrosion-resistant thermal barrier coatings, which would allow TBC advantages of durability and fuel economy in industrial gas turbines burning low quality fuel, or in military tank or ship gas turbines which may be forced to use low quality or contaminated fuel in wartime action. There will also be an emphasis on finding thermal barrier coatings with better high temperature stability than the present Y_2O_3–ZrO_2 TBCs, which are limited to temperatures below 1200 °C because of the thermal instability of the YSZ tetragonal t′ phase.

8.9.2 Thermal barrier coatings for diesel engines

There are large incentives for applying TBCs in diesel engines in prospect of improving fuel efficiency. In the United States, for example, approximately 3 000 000 barrels of diesel fuel per day were used for diesel engine powered transportation in 1987 [127]. A 5% improvement in fuel economy would yield more than $1.5 billion per year in fuel cost savings. To reduce highway fuel consumption, the U.S. Department of Energy established an advanced heavy-duty diesel engine program which has as one goal the development of a demonstration truck transportation diesel engine, the LHR-25 (where

LHR stands for low heat rejection), having a specific fuel consumption of 0.25 lb/BHP h (42 kg/GJ) [127]. This engine is intended to use not only thermal insulation, but also advanced technologies in lubrication, combustion and thermomechanical systems to achieve the 0.25 lb/BHP h (42 kg/GJ) goal, which is about 25% better fuel economy than given by current truck diesels.

Substantial controversy has arisen, however, as to how much, or even whether, TBCs actually improve fuel economy in diesels. In early U.S. Army-sponsored research, an Army five-ton truck engine was converted to adiabatic operation by application of zirconia coatings and other modifications and, when driven from Detroit to Washington D.C., exhibited an increase in fuel economy from about 6 mpg to 9 mpg (2.5 km/l to 3.8 km/l) [128]. The engine was run without radiator, water pump or fan, and used only oil cooling. Power from the exhaust gas was recovered by a turbocompound system geared to the crankshaft. However, subsequent U.S. laboratory engine studies indicated fuel economy improvements of only 2–4% for TBCs in diesels [129], and it was reported that a later analysis attributed the near 50% increase in fuel economy for the Army diesel engine to the reduction of parasitic cooling losses, and to other engine and drivetrain modifications, with only 1% being from thermal insulation [129]. (Note however that in a 1991 publication, the developers of the Army adiabatic diesel state the engine to have shown an approximately 5% improvement in fuel economy on the dynamometer and, more importantly in their view, a conclusive 16–38% superiority in fuel economy over conventional water-cooled engines in vehicle road tests [85].) No benefit in fuel economy at all was seen in another set of TBC developmental experiments [81].

Worse than this, the use of TBCs was found actually to increase fuel consumption in diesel engines in European tests. According to Woschni *et al.* [130, 131], this occurs because the heat transfer coefficient h increases rapidly as the cylinder gas temperature rises; this controls the heat flux between the engine gas and combustion chamber walls according to the equation:

$$\dot{q} = h(T - T_{\mathrm{w}}) \tag{8.10}$$

where $\dot{q}$ is the heat flux, h is the heat transfer coefficient, T is the engine gas temperature and T_{w} is the wall temperature. At high temperature and under heavy load, the heat transfer coefficient becomes so large that it overrides the reduction in the $(T\text{-}T_{\mathrm{w}})$ factor, and heat flux to the combustion chamber walls is in fact *greater* than for an uninsulated engine. Laboratory engine tests of brake specific fuel consumption (BSFC) versus engine load by Woschni *et al.* indicate the BSFC of an insulated engine to be slightly improved at low engine load, but to be degraded by 6–7% at the highest engine loads [130]. Woschni *et al.* believe that the heat transfer coefficient increases with temperature mainly because, as the combustion chamber temperature rises, the combustion flame quenches closer to the combustion chamber wall. This brings high temperature sources nearer to the wall, and effectively reduces the thickness of the thermal boundary layer at the wall surface, so that the thermal gradient (and thereby the heat flux) at the combustion chamber wall is actually higher for insulated engines than for uninsulated engines. Therefore, Woschni *et al.* conclude that it is 'practically impossible to obtain a gain in fuel consumption by insulation measures.'

As might be imagined, the results and theories by Woschni *et al.* have had a tremendous impact in the diesel TBC community. The Woschni effect was the topic of a panel discussion at the 1988 ASME conference in New Orleans [132], and has been addressed in a review by Fairbanks [133]. The crucial point seems to be that both combustion and heat transfer occur at, or very near, the combustion chamber walls. Therefore, the combustion processes may affect heat transfer – the viewpoint espoused by Woschni *et al.* – or the heat transfer factors (i.e. the high wall and combustion chamber temperatures) may affect combustion. With regard to flame impingement reducing the thermal boundary layer so as to increase heat transfer, Morel points out that (1) the flame adjoins any given surface for only a very short fraction of the time of the total diesel cycle, and (2) if the increase in heat transfer results only from thinning of the thermal boundary layer, this implies that 'the thermal boundary layer would have to be thinner by a factor of 625 at all surfaces and at all crank angles' [132]. Woschni responds to these issues by maintaining that, although the increase in the heat transfer coefficient may be of short duration, it is so large that it nonetheless affects heat transfer over the total engine cycle. A possible factor here may be the 'rate of recovery' of the thermal boundary layer after it has been thinned by momentary impingement of the flame front.

The alternative possibility, i.e. that the hot walls and high temperatures in the combustion chamber of an insulated engine may adversely influence the combustion process, has been investigated by Alkidas [134]. In laboratory engine tests, Alkidas was able either to duplicate the results of Woschni *et al.*, wherein the BSFC is improved at low load but worsened under heavy load, or to obtain sustained improvement in the BSFC over the entire load range simply by increasing the air consumption. Alkidas suggests therefore that the cause of the deterioration in fuel economy in insulated engines is adverse combustion brought about by insufficient air–fuel mixing. Moreover, Alkidas showed that high heat losses (greater than from the uninsulated engine) began to occur at about the same load point that the BSFC deteriorated in the engine suffering adverse combustion. Alkidas does not explain, however, nor is it obvious, why the heat rejection rate should increase as combustion deteriorates. This question needs to be resolved if deterioration of combustion is to be accepted as the total cause of the Woschni effect.

Findings reported recently by Kawamura are also indicative of combustion degeneration caused by high engine temperatures [135]. In this case, monitoring of the combustion process in an engine fitted with a quartz window revealed that fuel injection was being impaired so that the jet of fuel did not mix properly with the combustion air. This resulted in adverse combustion and soot formation. The injection problem was presumed to be caused by increased viscosity of the air, brought about by the high temperatures and pressures within the combustion chamber. Use of a swirl type precombustion chamber was said to remedy the fuel–air mixing problem and to allow up to 5% improved thermal efficiency; it also reduced HC and NO_x emissions from the insulated engine.

It cannot yet be concluded that the Woschni effect has been resolved, and further research is unquestionably merited, both for the obvious technological importance and for scientific curiosity. It may be appropriate to note here that there is no evidence of the Woschni effect with TBCs in gas turbines. The application of properly engineered zirconia TBCs to combustor cans, vanes, air-cooled blades, etc., invariably has been found to reduce heat transfer to the substrate, therefore lowering the temperature of the substrate metal (for all other conditions remaining the same). The reason for this is that the combustion and critical heat transfer effects are separated in gas turbines, with combustion occurring in combustor cans (which are designed so that flame does not impinge upon the TBC-coated combustor walls), and the critical heat transfer processes taking place at the vane and blade surfaces.

In the midst of the turmoil on TBC fuel economy effects, there have been accounts of long-term field service tests of thin (>0.5 mm or 0.020 in.) TBCs in towboat diesels [136], natural gas engines in pipeline pumping stations [137] and bus diesel engines [138], in which improved fuel economies of 10% or more were reported. Several other instances of reported fuel economy improvements of 5–15% in field service or in engine tests are cited in the review papers by Hoag [129] and Fairbanks [133]. Such tests have been criticized as not being well controlled (in one case, the engine was overhauled at the same time the TBCs were applied) or as being improperly interpreted, but taken in sum they represent accumulated experience that should not necessarily be totally discounted. The towboat engine tests demonstrated, for instance, that thin TBCs could successfully survive 14 000 h of diesel service with little damage [136]. It must also be acknowledged that small variations in TBC parameters can have effects that are difficult to anticipate. For example, in the engine tests by Assanis *et al.*, TBCs of 0.5 mm (0.020 in.) thickness gave 2–3% efficiency gains at low rpm, whereas 1.0 mm TBCs gave only equal or worse efficiency than the baseline metal engine across the entire rpm range [139]. One of the most recent reports on TBC fuel effects indicates that a 4.6% increase in fuel economy was achieved using a thick (3.5 mm or 0.140 in.), sealed-surface TBC under turbocompound engine test conditions [140]. An accurate analysis of TBC effects in diesel combustion is still hindered, as discussed above and as pointed out by Hoag [129], by lack of an adequate understanding of in-cylinder heat transfer mechanisms, including the role of the thermal boundary layer.

Beyond fuel economy, there are many other benefits that are cited as being conferred by TBCs in diesel engines. They include superior corrosion protection against low quality fuels, better heat management and avoidance of hot spotting, easier starting and smoother running, reduced cetane requirements and increased power [7, 138]. Because of these cited benefits, an industry of some size now exists which applies after-market TBCs to land and marine diesel engines. One major U.S. diesel engine company also presently features thermal barrier coatings on the cylinder heads, valves and piston domes of its high performance series of 2500 hp (1.9 MW) off-road engines [141].

Another critical, and equally controversial, property of TBCs is their effect on emissions from diesel engines. Since TBCs raise the combustion chamber temperature, an initial assumption would be that TBCs should reduce soot, unburned hydrocarbon and CO emissions, but raise NO_x emissions since the formation of NO_x increases rapidly with temperature. However, results from laboratory and engine tests have been variable. Assanis *et al.* [139] found slightly increased smoke levels with 0.5 mm TBCs (although both initial and TBC-affected smoke levels were very low), and they cite other laboratory engine research papers where TBCs gave increased soot and smoke formation. On the other hand, significant decreases in both soot and smoke, as well as BSFC improvements, were observed by Alkidas when the insulated engine configuration was adjusted to give proper air–fuel mixing [134]. Road tests of TBCs in certain bus diesel engines have also been reported as positive, and TBCs are scheduled to be installed in the engines of some community bus systems to reduce soot and smoke pollution [138]. The critical question of NO_x formation in TBC-fitted engines is equally or more uncertain, with several in-service and laboratory tests showing NO_x reduction [135, 137–139], whereas various other studies revealed increases in NO_x emission [133, 142]. Winkler *et al.* hypothesize that TBCs may reduce NO_x emissions by altering the combustion process so as to in fact lower the peak temperature (and therefore NO_x formation) that occurs in the diesel cycle [138].

The future of thermal barrier coatings in diesel engines – particularly as to whether they will be routinely used in the millions of highway truck and bus diesel engines – is difficult to assess. The wide discrepancies found in research results to date on the effect of TBCs on fuel economy and engine emissions are discouraging. On the other hand, diesel combustion in engines fitted with thermal barrier coatings is a complex process (Kvernes and Lugscheider mention a count of some 150 variables affecting deposition of the ZrO_2 coating alone [143]), and it is not unreasonable to expect that continued optimization of the TBC, and of the diesel engine for TBC usage, would yield real and consistent TBC benefits, including fuel efficiency and possibly reduction in particulate and gaseous emissions. Much depends, of course, upon the world fuel situation; if fuel shortages develop once again, the appeal of diesel TBCs will undoubtedly increase.

REFERENCES

1. Fawley, R.W. (1972) Superalloy progress, in *The Superalloys* (eds C.T. Sims and W.C. Hagel), Wiley, New York, pp. 3–29.
2. Kool, G.A. (1994) Current and future materials in advanced gas turbine engines. *ASME Publication 94-GT-475*, ASME, New York.
3. Cole, G.S. and Cremisio, R.S. (1972) Solidification and structure control in superalloys, in *The Superalloys* (eds C.T. Sims and W.C. Hagel), Wiley, New York, pp. 479–508.
4. Plunkett, J.D. (1964) NASA Contributions to the technology of inorganic coatings. *Technology Survey NASA SP-5014*, NASA, Washington D.C., p. 145.
5. Miller, R.A. (1987) *Surf. Coat. Technol.*, **30**(1), 1–11.
6. Miller, R.A. (1990) Assessment of fundamental materials needs for thick thermal barrier coatings (TTBCs) for truck diesel engines, in *Proceedings of the 1990 Coatings for Advanced Heat Engines Workshop* (ed. J. Fairbanks), Castine ME, U.S. Dept. of Energy, Washington D.C. pp. II7–II34.
7. Nagaraj, B.A., Wortman, D.V. and Lindblad, N. (1992) Private communication, General Electric Aircraft Engines.
8. Kvernes, I. (1987) Ceramic coatings as thermal barriers in diesel and gas turbine components, in *High Tech Ceramics – Part C: Materials Science Monographs 38C* (ed P. Vincenzini), Elsevier, Amsterdam, pp. 2519–36.
9. Plunkett, J.D. (1964) NASA contributions to the technology of inorganic coatings. *Technology Survey NASA SP-5014*, NASA, Washington D.C., p. 123.
10. Richerson, D.W. (1982) *Modern Ceramic Engineering: Properties, Processing, and Use in Design*, Marcel Dekker, New York, pp. 38–45 and 139–142.
11. Vogan, J.W. and Stetson, A.R. (1979) Thick ceramic coating development for industrial gas turbines, in *Proceedings of the First Conference on Advanced Materials for Alternative Fuel Capable Directly Fired Heat Engines* (eds J.W. Fairbanks and J. Stringer), Castine ME, U.S. Dept. of Energy, Washington D.C., pp. 542–81.
12. Bratton, R.J., Lau, S.K., Lee, S.Y. and Andersson, C.A. (1979) Ceramic coating evaluations and developments, in *Proceedings of the First Conference on Advanced Materials for Alternative Fuel Capable Directly Fired Heat Engines* (eds J.W. Fairbanks and J. Stringer), Castine ME, U.S. Dept. of Energy, Washington D.C., pp. 582–605.
13. Grot, A.S. and Martyn, J.K. (1981) *Ceram. Bull.*, **60**(8), 807–11.
14. Bratton, R.J., Lau, S.K., Andersson, C.A. and Lee, S.Y. (1982) Studies of thermal barrier coatings for heat engines, in *Proceedings of the Second Conference on Advanced Materials for Alternative-Fuel-Capable Heat Engines* (eds J.W. Fairbanks and J. Stringer), Monterey CA, Electric Power Research Institute, Palo Alto, CA, Ch. 6, pp. 82–101.
15. Hirschfeld, D.A., Liu, D.M. and Brown, J.J. (1992) CMZP – a new high temperature thermal barrier material, in *Proceedings of the 4th International Symposium on Ceramic Materials and Components for Engines* (eds R. Carlsson, T. Johansson and L. Kahlman), Gothenburg, Sweden, Elsevier Applied Science, London, pp. 372–80.
16. Anonymous (1993) *Amer. Ceram. Soc. Bull.*, **72**(10), 22.
17. Hellman, J.R. and Stubican, V.S. (1983) *J. Amer. Ceram. Soc.*, **66**(4), 260–4.
18. Yoshimura, M. (1988) *Ceram. Bull.*, **67**(12), 1950–5.
19. Sheu, T.-S., Tien, T.-Y. and Chen, I.-W. (1992) *J. Amer. Ceram. Soc.*, **75**(5), 1108–16.
20. Sasaki, K., Bohac, P. and Gauckler, L.J. (1993) *J. Amer. Ceram. Soc.*, **76**(3), 689–98.
21. Nettleship, I. and Stevens, R. (1987) *Int. J. High Tech. Ceram.*, **3**, 1–32.
22. Scott, H.G. (1975) *J. Mater. Sci.*, **10**, 1527–35.

23. Jue, J.F., Chen, J. and Virkar, A.V. (1991) *J. Amer. Ceram. Soc.*, **74**(8), 1811–20.
24. Garvie, R.C. (1978) *J. Phys. Chem.*, **82**(2), 218–24.
25. Heuer, A.H., Claussen, N., Kriven, W.M. and Ruhle, M. (1982) *J. Amer. Ceram. Soc.*, **65**(12), 642–50.
26. Heuer, A.H., Chaim, R. and Lanteri, V. (1988) Review: phase transformations and microstructural characterization of alloys in the Y_2O_3–ZrO_2 system, in *Advances in Ceramics, Vol. 24: Science and Technology of Zirconia III* (eds S. Somiya, N. Yamamoto and H. Yanagida), The American Ceramic Society, Westerville OH, pp. 3–20.
27. Ruh, R., Garrett, H.J., Domagala, R.F. and Patel, V.A. (1977) *J. Amer. Ceram. Soc.*, **60**(9/10), 399–403.
28. Igawa, N., Ishii, Y., Nagasaki, Y. *et al.* (1993) *J. Amer. Ceram. Soc.*, **76**(10), 2673–6.
29. Bhattacharjee, S., Syamaprasad, U., Galgali, R.K. and Mohanty, B.C. (1992) *Mater. Lett.*, **15**, 281–4.
30. Miller, R.A., Garlick, R.G. and Smialek, J.L. (1983) *Ceram. Bull.*, **62**(12), 1355–8.
31. Hernandez, M.T., Jurado, J.R., Duran, P. and Fierro, J.L.G. (1991) *J. Amer. Ceram. Soc.*, **74**(6), 1254–8.
32. Virkar, A.V. and Matsumoto, R.L.K. (1986) *J. Amer. Ceram. Soc.*, **69**(10), C224–6.
33. Noma, T., Yoshimura, M., Somiya, S. *et al.* (1988) Stability of diffusionlessly transformed tetragonal phases in rapidly quenched ZrO_2–Y_2O_3, in *Advances in Ceramics, Vol. 24: Science and Technology of Zirconia III* (eds S. Somiya, N. Yamamoto and H. Yanagida), The American Ceramic Society, Westerville OH, pp. 377–84.
34. VanValzah, J.R. and Eaton, H.E. (1991) *Surf. Coat. Technol.*, **46**, 289–300.
35. Brandon, J.R. and Taylor, R. (1991) *Surf. Coat. Technol.*, **46**, 75–90.
36. Shankar, N.R., Herman, H., Singhal, S.P. and Berndt, C.C. (1984) *Thin Solid Films*, **119**, 159–71.
37. Rigney, D.V., Mantkowski, T.E. and Froning, M.J. (1987) Influence of raw materials on the performance characteristics of ceramic coatings, in *Proceedings of the 1987 Coatings for Advanced Heat Engines Workshop* (ed. J.W. Fairbanks), Castine ME, Dept. of Energy, Washington D.C., pp. V45–60.
38. Kvernes, I., Solberg, J.K. and Lillerud, K.E. (1979) Ceramic coatings on diesel engine components, in *Proceedings of the First Conference on Advanced Materials for Alternative Fuel Capable Directly Fired Heat Engines* (eds J.W. Fairbanks and J. Stringer), Castine ME, Dept. of Energy, Washington D.C., pp. 233–57.
39. Schmid, H.K. (1987) *J. Amer. Ceram. Soc.*, **70**(5), 367–76.
40. Miller, R.A., Smialek, J.L. and Garlick, R.G. (1981) Phase stability in plasma-sprayed, partially stabilized zirconia–yttria, in *Advances in Ceramics, Vol. 3, Science and Technology of Zirconia* (eds A.H. Heuer and L.W. Hobbs), The American Ceramic Society, Westerville OH, pp. 241–53.
41. Taylor, R., Brandon, J.R. and Morrell, P. (1992) *Surf. Coat. Technol.*, **50**, 141–9.
42. Froning, M.J. and Jayaraman, N. (1986) Quantitative phase analysis by X-ray diffraction of ZrO_2–8% Y_2O_3 system, in *High Temperature Coatings* (eds M. Khobaib and R.C. Krutenat), The Metallurgical Society, Warrendale PA, pp. 179–91.
43. Srinivasan, R., De Angelis, R.J., Ice, G. and Davis, B.H. (1991) *J. Mater. Res.*, **6**(6), 1287–92.
44. Howard, C.J. and Hill, R.J. (1991) *J. Mater. Sci.*, **26**, 127–34.
45. Scardi, P., Lutterotti, L. and Galvanetto, E. (1993) *Surf. Coat. Technol.*, **61**, 52–9.
46. Srinivasan, R., Harris, M.B., Simpson, S.F. *et al.* (1988) *J. Mater. Res.*, **3**(4), 787–97.
47. Hamilton, J.C. and Nagelberg, A.S. (1984) *J. Amer. Ceram. Soc.*, **67**(10), 686–90.
48. Strangman, T.E. (1985) *Thin Solid Films*, **127**, 93–105.

49. Bennett, A. (1986) *Mater. Sci. Technol.*, **2**, 257–61.
50. Burgel, R. and Kvernes, I. (1986) Thermal barrier coatings, in *Proceedings of the Conference on High Temperature Alloys for Gas Turbines and Other Applications* (eds W. Betz *et al.*), Liège, D. Reidel, Dordrecht, pp. 327–56.
51. Goward, G.W. (1987) Seventeen years of thermal barrier coatings, in *Proceedings of the 1987 Coatings for Advanced Heat Engines Workshop* (ed. J.W. Fairbanks), Castine ME, Dept. of Energy, Washington D.C., pp. III1–9.
52. Rhys-Jones, T.N. and Toriz, F.C. (1989) *High. Temp. Technol.*, **7**(2), 73–81.
53. Scott, K.T. and Restall, J.E. (1988) Some aspects of the development and performance of ceramic thermal barrier coatings for gas turbines, in *Thermal Spray Technology, New Ideas and Processes*. Proceedings of the National Thermal Spray Conference (ed D.L. Houck), Cincinnati OH, ASM International, Metals Park OH, pp. 255–61.
54. Movchan, B.A., Malashenko, I.S., Yakovchuk, K.Y. *et al.* (1994) *Surf. Coat. Technol.*, **67**, 55–63.
55. Patten, J.W., Prater, J.T., Hays, D.D., Moss, R.W. and Fairbanks, J.W. (1980) *Thin Solid Films*, **73**, 463–70.
56. Hocking, M.G., Vasantasree, V. and Sidky, P.S. (1989) *Metallic & Ceramic Coatings: Production, High Temperature Properties & Applications*, Wiley, New York. pp. 252–67.
57. Hecht, R.J. (1987) Plasma spray processing and equipment requirements for thermal barrier coatings in advanced heat engines, in *Proceedings of the 1987 Coatings for Advanced Heat Engines Workshop* (ed. J.W. Fairbanks), Castine ME, Dept. of Energy, Washington D.C., pp. V1–7.
58. Taylor, T.A., Appleby, D.L., Weatherill, A.E. and Griffiths, J. (1990) *Surf. Coat. Technol.*, **43/44**, 470–80.
59. Joshi, S.V. and Srivastava, M.P. (1993) *Surf. Coat. Technol.*, **56**, 215–24.
60. Stecura, S. (1985) Optimization of the $NiCrAl–Y/ZrO_2–Y_2O_3$ thermal barrier system. *NASA Tech. Memo. 86905*, NASA, Cleveland OH.
61. McPherson, R. (1981) *Thin Solid Films*, **83**, 297–310.
62. McPherson, R. (1989) *Surf. Coat. Technol.*, **39/40**, 173–81.
63. Eaton, H.E. and Novak, R.C. (1987) *Surf. Coat. Technol.*, **32**, 227–36.
64. Miller, R.A., Brindley, W.J., Goedjen, J.G. *et al.* (1994) The effect of silica on the cycle life of a zirconia–yttria thermal barrier coatings, in *Proceedings of the 7th National Thermal Spray Conference*, 20–24 June 1994, Boston MA, pp. 49–54.
65. Crabos, F., Monge-Cadet, P. and Piraggi, B. (1995) Influence of powder microstructure on the thermal cycling behavior of $ZrO_2–8\%$ Y_2O_3 plasma coating, in *Elevated Temperature Coatings: Science and Technology I* (eds N.B. Dahotre, J.M. Hampikian and J.J. Stiglich), The Metallurgical Society, Warrendale, PA. pp. 63–72.
66. Stecura, S. (1987) *Thin Solid Films*, **150**, 15–40.
67. Siemers, P.A. and McKee, D.W. (1982) U.S. Patent 4,328,285.
68. Brandon, J.R. and Taylor, R. (1991) *Surf. Coat. Technol.*, **46**, 91–101.
69. Jones, R.L. (1989) *Surf. Coat. Technol.*, **39/40**, 89–96.
70. Jones, R.L. and Mess, D. (1992) *J. Amer. Ceram. Soc.*, **75**(7), 1818–21.
71. Jacobson, N.S. (1991) Thermodynamic properties of some metal oxide–zirconia systems. *NASA Tech. Memo. 102351*, NASA, Cleveland OH.
72. Wortman, D.J., Nagaraj, B.A. and Duderstadt, E.C. (1989) *Mater. Sci. Engng*, **A121**, 433–40.
73. Miller, R.A. and Brindley, W.J. (1992) Plasma sprayed thermal barrier coatings on smooth surfaces, in *Proceedings of International Thermal Spray Conference and Exposition*, Orlando FL, ASM International, Metals Park OH, pp. 493–98.

74. Stecura, S. (1980) *Thin Solid Films*, **73**, 481–9.
75. Stecura, S. (1989) *Thin Solid Films*, **182**, 121–39.
76. Huntz, A.M. (1987) *Mater. Sci. Engng*, **87**, 251–60.
77. Smeggil, J.G. (1987) *Mater. Sci. Engng*, **87**, 261–5.
78. Stott, F.H. and Wood, G.C. (1987) *Mater. Sci. Engng*, **87**, 267–74.
79. Stecura, S. (1986) *Thin Solid Films*, **136**, 241–56.
80. Kvernes, I. and Noerholm, O. (1987) Coatings for diesel engines and associated problems, in *Procedings of the 1987 Coatings for Advanced Heat Engines Workshop* (ed. J.W. Fairbanks), Castine ME, Dept. of Energy, Washington D.C., pp. III73–81.
81. Yonushonis, T.M., Hg, H.K. and Novak, R.C. (1990) Thick thermal barrier coatings for diesel engines, in *Proceedings of the 1990 Coatings for Advanced Heat Engines Workshop* (ed. J. Fairbanks), Castine ME, Dept. of Energy, Washington D.C., pp. II45–51.
82. Beardsley, M.B. (1990) Application of thick thermal barrier coatings to diesel engines, in *Proceedings of the 1990 Coatings for Advanced Heat Engines Workshop* (ed. J. Fairbanks), Castine ME, Dept. of Energy, Washington D.C., pp. II53–6.
83. Lutz, J.C. and Harris, D.H. (1988) Development of thermal barrier coatings for the internal combustion engine, in *Thermal Spray Technology, New Ideas and Processes*. Proceedings of the National Thermal Spray Conference (ed. D.L. Houck), Cincinnati OH, ASM International, Metals Park OH, pp. 437–42.
84. Kamo, R., Woods, M. and Sutor, P. (1987) Development of tribological system and advanced high-temperature in-cylinder components for advanced high-temperature diesel engines, in *Proceedings of the 1987 Coatings for Advanced Heat Engines Workshop* (ed. J.W. Fairbanks), Castine ME, Dept. of Energy, Washington D.C., pp. IV73–92.
85. Kamo, R. and Bryzik, W. (1992) Tribological and thermal ceramic coatings for advanced adiabatic engine, in *Proceedings of the 4th International Symposium on Ceramic Materials and Components for Engines* (eds R. Carlsson, T. Johansson and L. Kahlman), Gothenburg, Sweden, Elsevier Applied Science, London, pp. 1260–75.
86. Goward, G.W. (1974) Coatings and coating processing for gas turbine airfoils operating in a marine environment, in *Proceedings of the 1974 Gas Turbine Materials in the Marine Environment Conference* (eds J.W. Fairbanks and I. Machlin), Castine ME, Metals and Ceramics Information Center, Battelle, Columbus OH, pp. 277–96.
87. Strangman, T.E. (1982) U.S. Patent 4,321,331.
88. Ulion, N.E. and Ruckle, D.L. (1982) U.S. Patent 4,321,310.
89. Meier, S.M., Nissley, D.M. and Sheffler, K.D. (1991) Thermal barrier coating life prediction model development. *NASA Contractor Report 189111*, NASA Lewis Research Center, Cleveland OH.
90. Fritscher, K. and Schulz, U. (1993) Burner-rig performance of density-graded EB-PVD processed thermal barrier coatings, in *Ceramic Coatings* (ed. K. Kokini), ASME New York, MD-Vol. 44, pp. 1–8.
91. Shulz, U., Fritscher, K., Peters, M. and Kaysser, W.A. (1994) Processing and behavior of chemically graded EB-PVD MCrAIY bond coats, in *Proceedings of the 3rd International Symposium on Structural and Functional Gradient Materials* (eds B. Ilschner and N. Cherradi), 10–12 Oct. 1994, Lausanne, Switzerland, pp. 441–6.
92. Toriz, F.C., Thakker, A.B. and Gupta, S.K. (1989) *Surf. Coat. Technol.*, **39/40**, 161–72.
93. Meier, S.M., Gupta, D.K. and Sheffler, K.D. (1991) *J. Mineral*, **43**(3), 50–3.
94. Fritscher, K and Bunk, W. (1990) Density-graded TBCs processed by EB-PVD,

in *Proceedings of the 1st International Symposium on Functionally Gradient Material*, 8–9 Oct. 1990, Sendai, Japan.
95. Borom, M.P. and Johnson, C.A. (1992) *Surf. Coat. Technol.*, **54/55**, 45–52.
96. Miller, R.A. and Lowell, C.E. (1982) *Thin Solid Films*, **95**, 265–73.
97. Wu, B.C., Chang, E., Chang, S.F. and Chao, C.H. (1989) *Thin Solid Films*, **172**, 185–96.
98. Miller, R.A, and Berndt, C.C. (1984) *Thin Solid Films*, **119**, 195–202.
99. Miller, R.A. (1984) *J. Amer. Ceram. Soc.*, **67**(8), 517–21.
100. Chang, G.C., Phuchareon, W. and Miller, R.A. (1987) *Surf. Coat. Technol.*, **30**, 13–28.
101. McDonald, G. and Hendricks, R.C. (1980) *Thin Solid Films*, **73**, 491–6.
102. Brindley, W.J. and Miller, R.A. (1990) *Surf. Coat. Technol.*, **43/44**, 446–57.
103. Berndt, C.C. (1985) *Trans. ASME, J. Engng Gas Turbines and Power*, **107**, 142–6.
104. Grossklaus, W.D., Katz, G.B. and Wortman, D.J. (1986) Performance comparison of advanced airfoil coatings in marine service, in *Proceedings of the Symposium on High Temperature Coatings* (eds M. Khobaib and R.C. Krutenat), Orlando FL, The Metallurgical Society, Warrendale PA, pp. 67–83.
105. Nagaraj, B.A., Maricocchi, A.F., Wortman, D.J. *et al.* (1992) Hot corrosion resistance of thermal barrier coatings. *Tech. Paper 92-GT-44*, ASME, New York.
106. Goward, G.W. (1987) Private communication.
107. Miller, R.A. (1986) Ceramic thermal barrier coatings for electric utility gas turbines. *NASA Tech. Memo. 87288*, NASA Lewis Research Center, Cleveland OH, pp. 1–10.
108. Jones, R.L., Williams, C.E. and Jones, S.R. (1986) *J. Electrochem. Soc.*, **133**(1), 227–30.
109. Jones, R.L. (1988) *High Temp. Technol.*, **6**(4), 187–93.
110. Jones, R.L. (1991) *Mater. at High Temp.*, **9**(4), 228–36.
111. Singhal, S.C. and Bratton, R.J. (1980) *Trans. ASME, J. Engng Power*, **102**(10), 770–5.
112. Rhys-Jones, T.N., Nicholls, J.R. and Hancock, P. (1983) *Corr. Sci.*, **23**, 139–49.
113. Hertl, W. (1988) *J. Appl. Phys.*, **63**(11), 5514–20.
114. Susnitzky, D.W., Hertl, W. and Carter, C.B. (1988) *J. Amer. Ceram. Soc.*, **71**(11), 992–1004.
115. Barkalow, R.H. and Pettit, F.S. (1979) Mechanisms of hot corrosion attack of ceramic coatings materials, in *Proceedings of the First Conference on Advanced Materials for Alternative Fuel Capable Directly Fired Heat Engines* (eds J.W. Fairbanks and J. Stringer), Castine ME, Dept. of Energy, Washington D.C., pp. 704–14.
116. Dent-Glasser, L.S. and Duffy, J.A. (1987) *J. Chem. Soc. Dalton Trans.*, **1987**, 2323–8.
117. Stott, F.H., de Wet, D.J. and Taylor, R. (1994) *MRS Bull.*, Oct. 1994, pp. 46–9.
118. McKee, D.W., Luthra, K.L., Siemers, P. and Palko, J.E. (1979) Resistance of thermal barrier ceramic coatings to hot salt corrosion, in *Proceedings of the First Conference on Advanced Materials for Alternative Fuel Capable Directly Fired Heat Engines* (eds J.W. Fairbanks and J. Stringer), Castine ME, Dept. of Energy, Washington D.C., pp. 258–69.
119. Hodge, P.E., Miller, R.A. and Gedwill, M.A. (1980) *Thin Solid Films*, **73**, 447–53.
120. Strangman, T.E. and Schienle, J.L. (1990) *Trans. ASME, J. Engng Gas Turbines and Power*, **112**, 531–5.
121. Chu, W.-F. and Rohr, F.J. (1988) *Adv. Ceram. Mater.*, **3**(3), 222–4.
122. Bornstein, N., Roth, H. and Pike, R. (1993) Vanadium corrosion studies. *UTRC Report R93-918120-2*, United Technologies Research Center, E. Hartford CT.
123. Nagaraj, B.A. and Wortman, D.J. (1990) *Trans. ASME, J. Engng Gas Turbines and Power*, **112**, 536–42.
124. Petitbon, A., Boquet, L. and Delsart, D. (1991) *Surf. Coat. Technol.*, **49**, 57–61.

125. Mohammed Jasim, K., Rawlings, R.D. and West, D.R.F. (1992) *Mater. Sci. Technol.*, **8**, 83–91.
126. Mohammed Jasim, K., Rawlings, R.D. and West, D.R.F. (1992) *J. Mater. Sci.*, **27**, 3903–10.
127. Alpaugh, R.T. (1987) Overview of DOE's transportation conservation heat engine program, in *Proceedings of the 1987 Coatings for Advanced Heat Engines Workshop* (ed. J.W. Fairbanks), Castine ME, Dept. of Energy, Washington D.C., pp. I1–8.
128. Kamo, R. and Bryzik, W. (1984) Cummins/TACOM advanced adiabatic engine. *SAE Tech. Paper 840428*, Society of Automotive Engineers, Warrendale PA.
129. Hoag, K.L. (1987) A perspective on low heat rejection diesel engine development, in *Proceedings of the 1987 Coatings for Advanced Heat Engines Workshop* (ed. J.W. Fairbanks), Castine ME, Dept. of Energy, Washington D.C., pp. I19–15.
130. Woschni, G., Spindler, W. and Kolesa, K. (1987) Heat insulation of combustion chamber walls – a measure to decrease the fuel consumption of IC engines? *SAE Tech. Paper 870339*, Society of Automotive Engineers, Warrendale PA.
131. Woschni, G. and Spindler, W. (1988) *Trans. ASME, J. Engng Gas Turbines and Power*, **110**, 482–8.
132. Round table discussion of Woschni and Spindler paper (1988) *Trans. ASME, J. Engng Gas Turbines and Power*, **110**, 488–502.
133. Fairbanks, J.W. (1987) The enigma of the adiabatic or low heat rejection diesel engine concepts, in *Proceedings of the 1990 Coatings for Advanced Heat Engines Workshop* (ed. J. Fairbanks), Castine ME, Dept. of Energy, Washington D.C., pp. I49–60.
134. Alkidas, A.C. (1989) Performance and emissions achievements with an uncooled heavy-duty, single cylinder diesel engine. *SAE Tech. Paper 890144*, Society of Automotive Engineers, Warrendale PA.
135. Kawamura, H. (1991) The study of a heat insulated engine constructed by ceramic engine parts, in *Proceedings of the 4th International Symposium on Ceramic Materials and Components for Engines* (eds R. Carlsson, T. Johansson and L. Kahlman), Gothenburg, Sweden, Elsevier Applied Science, London, pp. 8–31.
136. Levy, A.V. (1988) The performance of ceramic coatings on diesel engine combustion zone components, in *Thermal Spray Technology, New Ideas and Processes*. Proceedings of National Thermal Spray Conference (ed. D.L. Houck), Cincinnati OH, ASM International, Metals Park OH, pp. 263–72.
137. Holloman, L. and Levy, A.V. (1990) The use of ceramic coatings on combustion zone components to enhance the performance and durability of natural gas combustion engines, in *Proceedings of the 1990 Coatings for Advanced Heat Engines Workshop* (ed. J. Fairbanks), Castine ME, Dept. of Energy, Washington D.C., pp. I81–96.
138. Winkler, M.F., Parker, D.W. and Bonar, J.A. (1992) Thermal barrier coatings for diesel engines: ten years of experience. *SAE Tech. Paper 922438*, Society of Automotive Engineers, Warrendale PA.
139. Assanis, D., Wiese, K., Schwarz, E. and Bryzik, W. (1990) Investigation of the effects of thin ceramic coatings on diesel engine performance and exhaust emissions, in *Proceedings of the 1990 Coatings for Advanced Heat Engines Workshop* (ed. J. Fairbanks), Castine ME, Dept. of Energy, Washington D.C., pp. I25–33.
140. Beardsley, M.B. and Larson, H.J. (1992) Thick thermal barrier coatings for diesel components. NASA CR-190759, Caterpillar Corp., Peoria IL, prepared for NASA, Cleveland OH.
141. Wilson, R. (1993) *Diesel Progress, Engines & Drives*, Mar. 1993, pp. 50–3.

142. Reichenbach, D.H. (1990) Ceramics in diesel engines – outlook for the future, in *Proceedings of the 1990 Coatings for Advanced Heat Engines Workshop* (ed. J. Fairbanks), Castine ME, Dept. of Energy, Washington D.C., pp. I17–24.
143. Kvernes, I. and Lugscheider, E. (1990) High quality TBC and wear resistant coatings – problems and requirements, in *Proceedings of the 1990 Coatings for Advanced Heat Engines Workshop* (ed. J. Fairbanks), Castine ME, Dept. of Energy, Washington D.C., pp. II109–35.

9

Pack cementation diffusion coatings

Robert Bianco and Robert A. Rapp

9.1 PRINCIPLES AND HISTORY

9.1.1 Introduction

Pack cementation is an *in situ* chemical vapor deposition (CVD) batch process that has been used to produce corrosion- and wear-resistant coatings on inexpensive or otherwise inadequate substrates for over 75 years [1]. The traditional pack consists of four components: the substrate or parts to be coated; the masteralloy (i.e. a powder of the element or elements to be deposited on the surface of the parts, such as Cr and/or Al, Cr and/or Si); a halide salt activator or energizer (e.g. NaCl, NaF, NH_4Cl, etc.); and a relatively inert filler powder (e.g. Al_2O_3, SiO_2, or SiC). The masteralloy powder, the halide salt activator, and the inert filler are mixed thoroughly, and the parts to be coated are buried in this mixture within a heat-resistant retort. Commercially, inversely stacked Inconel boxes are used as the retort. Inversely stacking the boxes creates pathways for the inert carrier gas. Depending on the size of the parts to be coated, the retort size ranges from small boxes approximately $35\,cm \times 35\,cm \times 35\,cm$ to larger boxes approximately $90\,cm \times 90\,cm \times 50\,cm$. As an inlet for the carrier gas, an Inconel tube is attached to the outer box of the retort. The assembled retort is then sealed with fusible silicate. Experimentally, a cylinder of commercially pure alumina (Al_2O_3) with one end closed is used as the retort. The alumina retort size ranges from small cylinders approximately $5\,cm$ outer diameter by $10\,cm$

Metallurgical and Ceramic Protective Coatings. Edited by Kurt H. Stern.
Published in 1996 by Chapman & Hall, London. ISBN 0 412 54440 7

long to larger cylinders approximately 10 cm outer diameter by 20 cm long. The alumina retort is closed by an alumina lid and sealed using an alumina-base cement. The sealed retort is induction or resistance heated to a high temperature (e.g. 800–1100 °C) while protected internally by an inert or reducing atmosphere. Depending on the size of the retort, heating rates typically vary over the range 4–12 h. At the elevated temperature, the masteralloy reacts with the halide salt activator to produce volatile metal halides that diffuse through the gas phase of the porous pack, such that they deposit and diffuse into the substrate. The process temperature is so chosen that both adequate halide vapor pressures are generated and solid-state diffusion occurs. Because generally the vapor pressures of the volatile halides for different elements differ greatly in magnitude (for a given halogen activity), the deposition of only one element is usually attempted, i.e. only aluminizing, or chromizing, or siliconizing, etc. However, recent process advances permit codeposition of two or more elements. The chemistry and kinetics of pack cementation are discussed in section 9.3.

9.1.2 Aluminizing

Aluminizing of heat-resistant alloys was invented by Van Aller in 1911 [2]. The original process consisted of placing pieces of Fe or Cu in a powder mixture of Al, NH_4Cl and graphite which was then heated to 450 °C for 2 h followed by a postdeposition interdiffusion treatment at 700–800 °C. The resulting coatings comprised either an Al-rich surface composition or an intermetallic compound, depending on the substrate, the halide salt activator, and the chemical potential of the Al source. Aluminizing was initially used to coat Fe wire or ribbon for heating elements and to coat Cu tubes for use in steam power plants. The formation of a protective Al_2O_3 scale resulting from elevated temperature exposures in air contributed to the enhanced oxidation resistance. However, until its use for gas turbine vane airfoils in 1957 [3], the development and application of aluminizing was not well documented.

For the coating of Ni- and Co-base superalloys, the microstructures of aluminized surfaces have been categorized as low or high activity processes based on the relative chemical activity of the Al source and the mechanism of the coating growth [4, 5]. For Ni-base alloys, high activity aluminized coatings are produced from a powder pack mixture containing a pure Al source (unit thermodynamic activity) heated at 700–900 °C. Because of the large Al chemical potential gradient, the inward-grown coating consists of an Ni_2Al_3 matrix with a dispersion of Cr- and Mo-rich precipitates, refractory-rich MC carbide particles and a number of other substrate elements dissolved in solid solution (e.g. Ti, Co, Cr, Mo, Nb, W, Hf). The growth rate is controlled by the inward diffusion of Al through Ni_2Al_3. A subsequent interdiffusion treatment at 1000–1100 °C is used to transform the Ni_2Al_3

coating into a hyperstoichiometric (Al-rich) β-NiAl phase. This transformation involves both the inward diffusion of Al from the Al-rich Ni_2Al_3 phase and the outward diffusion of Ni from the Ni-base substrate. In addition, an intermediate region or interdiffusion zone results from this transformation, as is discussed later.

The low activity aluminizing process for superalloys is produced from a pack powder mixture containing a binary aluminum alloy powder source of reduced chemical activity, e.g. Ni–Al, Fe–Al, or Cr–Al. The vapor pressures of the aluminum halide species generated by the low activity pack are significantly lower, and thus higher temperatures (1000–1150 °C) are required to produce aluminized coatings comparable to those from a high activity process. In fact, the higher temperature supports the simultaneous growth of the coating, since solid-state diffusion is the dominant rate-controlling step. The low activity process provides the necessary chemical potential gradient for vapor transport and coating growth, but since the chemical activity of the source is reduced, the resultant coating is an outward-grown, hypostoichiometric (Ni-rich) β-NiAl phase. The dominant diffusing element in the Ni-rich β-NiAl phase is Ni [6]. Other alloying elements also diffuse outward and remain in solid solution or precipitate as second-phase particles in the β-NiAl matrix. Aluminide coatings on Co-base alloys are produced by an analogous pair of processes.

Two observations should be noted about the low activity process. Because Ni has a higher mobility in the Ni-rich NiAl coating, Kirkendall voids are formed in the metal beneath the coating–substrate interface by condensation of the countercurrent flux of vacancies created to balance the unequal fluxes of Ni outward and Al inward. Kirkendall voids are forms of microscopic porosity in a material resulting from the supersaturation of vacancies, i.e. vacancy concentration exceeds equilibrium vacancy concentration. Kirkendall voids will form unless a mechanism for their annihilation exists. For a pure Ni substrate, Kirkendall voids are frequently observed, whereas no voids are observed in coated Ni-base superalloy substrates because the annihilation of vacancies occurs at the interfaces in the multiphase interdiffusion zone (Fig. 9.1a). Depending on the superalloy composition, the interdiffusion zone is comprised of long blocky MC-type or $M_{23}C_6$-type (M = Ti, Nb, Mo, W, Hf, Ta) carbides near the original substrate surface and platelet or acicular σ-phase particles rich in Cr, Mo and Co (Fig. 9.1b). The interdiffusion zone results from the outward diffusion of Ni and from both the diffusion of C from the bulk alloy and the degeneration of carbides below the coating. Carbon reacts with the refractory metal alloying elements in the zone depleted of Ni, precipitating MC-type carbides and, lacking further C, σ-phase is precipitated.

As a result of dominant NiAl growth at the external surface for low activity aluminide coatings, pack powders (i.e. masteralloy and inert pack particles) are entrapped by the coating. Pack entrapment is undesirable as it can

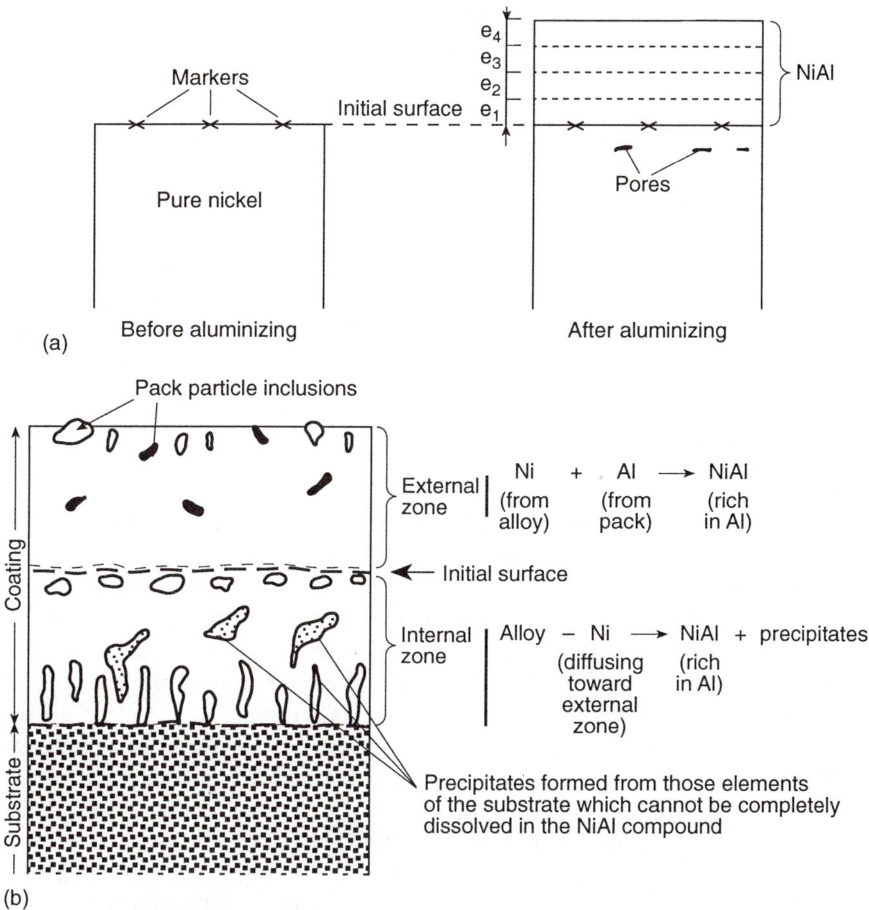

Figure 9.1 Schematic illustration of the low activity process to produce an NiAl coating by outward diffusion on (a) pure Ni or (b) Ni-base superalloy [5].

seriously degrade the mechanical and chemical integrity of the coating. However, by physically isolating the substrates from the pack powder mixture, a process procedure called *above-the-pack* or *out-of-pack*, entrapment of pack particles can be eliminated to produce cleaner coatings [7].

Conversely, aluminide coatings on Fe-base alloys using both processes are essentially inward-grown, except for the low activity process on low and high alloy steels (e.g. 2.25 Cr–1 Mo steels, Incoloy 800 or austenitic stainless steels). Aluminide coatings consist of either an Al-enriched ferrite surface layer or external phases of FeAl or $FeAl_2$ [8].

The coating elements Al, Cr and Si are ferrite stabilizers for steels. Therefore, ferritic stainless steels such as 409 SS remain ferritic at the coating temperature when Al, Cr + Al, or Cr + Si coatings are diffused into the ferritic alloy. Plain carbon steels and low alloy steels are austenitic at the coating temperature (1000–1050 °C), and the incorporation of Al, Cr + Al or Cr + Si may result in the transformation of the austenite to a single-phase ferrite layer at the surface. However, high alloy austenitic alloys can incorporate only limited concentrations of the coating elements into solid solution, and higher concentrations lead to a two-phase microstructure at the surface. Upon adding Al or Cr + Al to austenitic alloys such as 304 SS and 316 SS, the inward diffusion of Al results in the precipitation of the NiAl compound within a ferrite matrix at the surface [9, 10]. The resulting two-phase microstructure is mechanically tough and forms an Al_2O_3 scale at steady state upon exposure.

9.1.3 Chromizing

The chromizing process was invented by Kelley [11]. It consisted of a powder mixture containing pure Cr, NH_4Cl and an Al_2O_3 filler powder which was heated to 950 °C for 16 h. Chromized coatings were first applied to low carbon and low alloy steels, where they proved to be highly resistant to aqueous and high temperature corrosion. Complications arise when chromized coatings are applied to C-containing alloys, such as carbon steels and Ni- or Co-base superalloys [12–16]. Because of the high affinity of Cr for C and the relatively high diffusivity of carbon in the substrates, an outward-grown $Cr_{23}C_6$ carbide is formed at the external surface by the chromizing process. Not only does the extraction of C to form the surface carbide weaken the substrate, but this external carbide acts as a diffusion barrier to rapid Cr enrichment. In fact, the coating growth rates are strongly dependent on the C content of the substrate [13]. Therefore, chromizing is carried out above 1050 °C so that dissolution of the carbide at the carbide–alloy interface supplies Cr to the substrate [16]. One remedy to improve deposition rates and surface compositions of the steels is to codeposit a second element which has a much lower affinity for C, e.g. Al or Si, thereby transforming the surface into a phase with low C solubility, e.g. α-Fe, and pushing C into the bulk substrate [17]. Another solution is to codeposit a second element having a greater affinity for C, e.g. V; this serves to tie up free C, which otherwise diffuses to the surface [15].

Deposition of Cr into Fe-base alloys via $CrCl_2(v)$ results in a displacement reaction to release $FeCl_2(v)$ [15]. Therefore, this deposition mechanism results in an overall negligible weight or dimensional change for the substrate.

In recent years, Cr additions to bulk Ni-base alloys or to aluminized alloys have further improved their resistance to hot corrosion and high temperature oxidation [3]. The microstructure of chromized coatings on Ni-base alloys is much simpler than that of an aluminide coating. The Ni–Cr binary phase diagram [18] shows a high solubility for Cr ($\approx$ 50 at%) in the Ni-rich matrix and no complex intermetallic phases. Therefore, high activity chromized coatings are comprised of a Cr-enriched Ni-base substrate with an external α-Cr layer [19, 20].

9.1.4 Siliconizing

Silicide coatings have been applied to Ni- and Fe-base alloys by a number of methods, including chemical vapor deposition (CVD) [21, 22], fusion slurry techniques [23], and pack cementation [24]. Siliconizing of pure Fe and plain carbon steels was first produced via halide-activated pack cementation [24]. The pack consisted of a powder mixture containing an Fe-base masteralloy with 6–20 wt% Si, an NH_4Cl salt activator and an Al_2O_3 filler; it was heated to 1500 °C for 5–30 h. Silicide coatings were found to have enhanced resistance to both low temperature (type II) and high temperature (type I) hot corrosion by fused Na_2SO_4, resulting from the formation of an external silica scale.

According to the Ni–Si binary phase diagram [25], the addition of Si to Ni-base substrates results in the formation of several low melting point eutectics and complex intermetallic compounds. The typical microstructure of silicide coatings on Ni-base substrates is comprised of a multiphase, duplex layer containing an outer NiSi phase and an inner Ni_3Si phase. Conversely, the typical microstructure of silicide coatings on Fe-base substrates is comprised of an outer, single-phase ferrite layer enriched with 6–11 wt% Si. These silicide coatings are very brittle and susceptible to stresses induced by thermal cycles [26]. Therefore, the use of silicide coatings on Ni- and Fe-base alloys is not very practical. To utilize the beneficial effects of Si in lower concentrations, aluminide and chromized coatings have been modified to deposit Si into the surface of these coatings, either simultaneously or during a subsequent treatment [27–30].

9.1.5 Deposition of a reactive element

Small fractions of a percent of certain reactive elements (e.g. Y, Zr, Hf, Ce, La, Th) have been found to improve the adherence of protective oxide scales, such as Al_2O_3 and Cr_2O_3, on heat-resistant alloys [31–35]. Overlay coatings (plasma spray, electron beam evaporation, etc.) applied to Ni-base alloys for use in gas turbines have always included a reactive element, but only a few investigators have deposited reactive elements (REs) by other methods.

Jedlinski *et al.* [36] incorporated Y and Ce into the surface of NiAl coatings on Ni-base alloys by an ion implantation technique. Fuhui *et al.* produced a Y-modified aluminide coating on IN 738 using a standard high activity aluminizing treatment followed by a fusion slurry technique to deposit Y [37]. Three investigations have reported pack cementation methods:

(a) Tu *et al.* produced a Y-modified aluminide coating by a two-step process: high activity aluminizing followed by yttrizing [38]. The modified coatings consisted of an external aluminide layer with Y enrichment, as detected by wavelength dispersive spectroscopy (WDS), near the surface and along grain boundaries. The process to deposit Y consisted of a pack powder mixture containing pure Y, NH_4Cl and Y_2O_3 which was heated at 1050 °C for 4 h. The alumina filler was not inert in this case because the high thermodynamic stability of the aluminum chlorides permitted some reaction with the yttrium chloride species.

(b) Rapp *et al.* replaced some of the alumina filler with yttria in a pack mixture to codeposit Cr and Al plus Y on pure Fe and low alloy steel substrates [17]. Energy dispersive spectroscopy (EDS) analysis detected about 0.4 wt% Y in the coating surface [39].

(c) LePrince *et al.* used a high activity process with a powder mixture of pure source elements to codeposit Al and Hf onto Ni-base fibers [40]. The resulting coating consisted of an NiAl matrix with Ni_5Hf precipitates.

In all cases, the RE-doped aluminide coatings exhibited improved oxide scale adherence during cyclic oxidation testing.

9.1.6 Codeposition

In general, for resistance to oxidation, a ternary alloy using the interaction of two oxidation-resistant elements is more effective than a simple binary alloy [41]. For example, when an alloy containing Fe–20 Cr–5 Al (compositions are expressed in wt% unless otherwise indicated) is exposed to a high temperature oxidizing environment, initially a continuous adherent Cr_2O_3 scale is formed on the surface of the alloy. This scale prevents the rapid oxidation of the Fe in the alloy, thereby eliminating the dissolution and inward diffusion of oxygen atoms that would otherwise cause the external oxidation of Al. Escaping internal oxidation, the Al atoms diffuse to the Cr_2O_3–alloy interface and form an even slower-growing, more protective, stable Al_2O_3 scale at steady state. Actually, the transient Cr_2O_3 and steady-state Al_2O_3 scales are quite thin, usually on the order of 1–5 μm or ~0.5 mg/cm² oxygen weight gain. Therefore, a pack cementation process which would deposit two elements simultaneously, producing a diffusion coating with a surface containing approximately 20–30 wt% Cr with 4–6 wt% Al or 2–4 wt% Si should be very effective in protecting low alloy steel from corrosive environments at high temperatures. For Ni-base alloy substrates, a surface containing approximately 20–30 wt% Al, 10–15 wt% Cr and 0.1–0.2 wt% RE is required. These coatings should also provide protection from the accelerated oxidation suffered when a fused salt deposit contacts the alloy surface, i.e. hot corrosion, as occurs for certain conditions in gas turbines and in power plants burning fossil fuels. Furthermore, alumina and silica scales are known to be more resistant than chromia in gaseous environments containing mixed oxidants, e.g. carbon plus sulfur.

Single elements are commonly deposited into metallic substrates by the pack cementation method, but a single-step process using a single pack mixture which would simultaneously codeposit multiple elements into metallic substrates would be more effective because the resulting coatings would exhibit superior environmental resistance. Galmiche [42] developed a single-step, high activity process to deposit Cr and Al simultaneously from a pack powder mixture containing both pure Cr and Al sources with an NH_4Cl activator salt. The resulting coating consisted of a single-phase β-NiAl layer with only limited Cr enrichment (≈ 5 wt%). In fact, although the powder mixture obtained both pure Cr and Al sources, the process probably proceeds as a two-step process. Aluminizing occurs first because of the higher thermodynamic stability of the volatile aluminum chlorides. However, once Al is depleted from this dilute pack mixture, chromizing of the aluminide coating can occur. In general, the prospect of true simultaneous deposition of Al and Cr is not yet practiced commercially, although it has been developed experimentally [14, 17, 43]. However, dual vapor transport of Al and Si is thermodynamically feasible [44] and simultaneous deposition has been practiced commercially (e.g. Si-Mod) on Ni-base alloys for over 10 years. Nickel-base superalloys have been aluminized/siliconized using a powder mixture containing a source of both pure Al and Si, an NH_4F activator salt, and an Al_2O_3 filler treated at 1050–1065 °C for 4 h [45]. The resulting coating consisted of a β-NiAl layer with substantial Si enrichment (3–6 wt%) as Si-rich precipitates.

9.2 RANGE OF APPLICABILITY

Halide-activated pack cementation (HAPC) is a relatively inexpensive and commercially feasible diffusion coating process. HAPC is very versatile, capable of coating large objects such as long pipes, large plates or the complex shapes of turbine hardware, although small internal passages represent a problem. Because the coatings are grown at the substrate surface with a continuous gradient in composition, the epitaxially bonded coatings provide excellent adherence and an associated resistance to thermal fatigue. However, the coating compositions produced with HAPC do not possess the flexibility or control offered by overlay or PVD methods. Traditional chemical vapor deposition (CVD) is also a process used commercially, especially in the semiconductor industry. Diffusion coatings, such as HAPC and CVD, are deposited at high temperatures to support simultaneous solid–vapor reaction (deposition) and interdiffusion with the substrate. However, CVD processes require special furnaces and gas manifold systems to uniformly hold and coat substrates.

9.3 TECHNICAL ASPECTS

In a cementation pack at the process temperature, the masteralloy and the halide salt activator react to form volatile metal halide species of significant partial pressures according to the following equation:

$$Me(alloy) + AX_x(s \text{ or } 1) = MeX_x(v) + A(1 \text{ or } v) \qquad (9.1)$$

where Me is Cr, Al or Si; A is Na, NH_4, etc.; and X is F, Cl or Br [46]. In fact, multiple vapor species are formed for each element (e.g. Al, AlX, AlX_2, AlX_3; Cr, CrX_2, CrX_3, CrX_4; and Si, SiX_2, SiX_3, SiX_4). A partial pressure gradient for each vapor species, which supports vapor transport from the pack to the substrate surface, results from a higher thermodynamic activity of the element in the powder mixture and a lower activity at the substrate surface. At the surface, deposition of the desired coating element(s) occurs via the dissociation or disproportionation of the halide molecules, or by a displacement reaction with the substrate [46]. Finally, the coating elements interdiffuse with the metallic substrate, producing some specific surface composition and microstructure, or perhaps another phase (compound).

The thermodynamics and kinetics of pack cementation have been studied rather extensively [14, 46–53]. The earlier work concentrated on the aluminization of Ni and Ni-base alloys, whereas Nciri and Vandenbulcke concentrated on Fe-base alloy substrates [51]. According to Fig. 9.2, the driving force for gaseous diffusion through the porous powder mixture is the negative gradient in partial pressure of the individual halide vapor species [46]. The kinetics of vapor diffusion through a pack depleted at the surface were modeled using a Fick's first law expression. These calculations rationalize whether a particular halide vapor species will diffuse toward or away from the substrate. However, solid-state diffusion is usually the dominant rate-controlling step for the kinetics of the process.

Single elements are commonly deposited onto metallic substrates by the HAPC method, but a single-step process which would simultaneously codeposit multiple elements (e.g. Cr–Al or Cr–Si) onto metallic substrates would be more effective because the resulting coatings could exhibit superior resistance to oxidation and corrosion. According to the explanation above, one might suppose that a mixture of several pure elemental powders should produce simultaneous deposition. However, this prospect may fail in practice. As seen in Fig. 9.3, comparable partial pressures for two (or more) elements, as required for dual gaseous diffusion, essentially never occur in cementation packs because of large differences in the standard Gibbs energies of formation for their respective halide species [44, 53, 54]. However, binary Cr-rich alloys (i.e. Cr–Al and Cr–Si alloys) exhibit highly negative deviations from ideal thermodynamic behavior, so that such masteralloys can be used to reduce the activities of the Al or Si components by several orders of magnitude, Fig. 9.4a and b [55, 56]. Hence, by using such binary alloy powders, codeposition

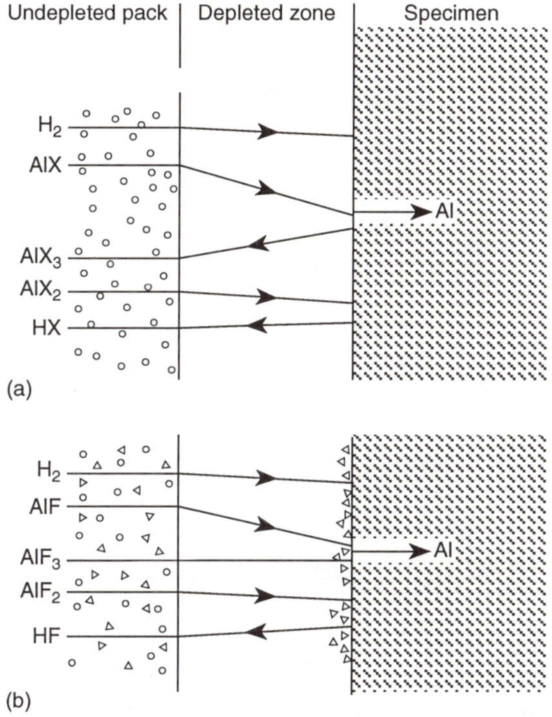

Figure 9.2 (a) Activator circulation and (b) activator condensation mechanisms of Al deposition proposed by Levine and Caves [46], where circles are Al(l) and triangles are $AlF_3(c)$ [49].

of Al and Cr into Ni- or Fe-base alloys [9, 17, 39, 57], or Cr and Si into Fe-base alloys [58] can be achieved if a halide salt activator of appropriate stability is also provided. As shown in Fig. 9.5, the dilute Cr–Al or Cr–Si masteralloy powders, with reduced thermodynamic activity for Al and Si, respectively, generate lower vapor pressures for the otherwise favored halide species, i.e. AlX_x or SiX_y [39, 43, 53, 58]. In this way, the fluxes for the components to be codeposited are brought to comparable magnitudes so that simultaneous deposition results in the desired surface composition.

In addition, a third (or fourth) minor element might also be incorporated into the coating [39]. Two methods have been used to produce this further doping. A small amount of an oxide source of the desired element can be added to the pack mixture, replacing some of the inert filler. In this case, the dopant oxide must be converted in the high halide activity of the pack, producing additional metal halide species according to equation 9.2:

$$xMO_y(s) + yMeX_x(v) = xMX_y(v) + yMeO_x(s) \tag{9.2}$$

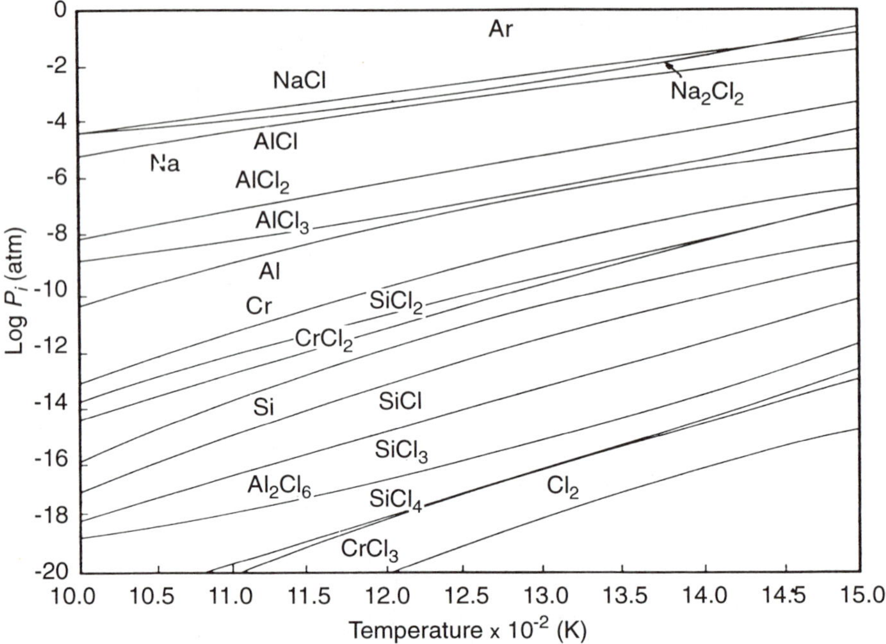

Figure 9.3 Log (partial pressure) of gaseous halides as a function of temperature in NaCl-activated packs containing pure Al, Cr and Si [53].

where M is Zr, Y or Hf; Me is Al or Si. Calculations of pack equilibria using the computer program ITSOL [59] and experiments using an atmospheric pressure sampling mass spectrometer have substantiated these conversion reactions for ZrO_2 and Y_2O_3 additions to Al_2O_3 [60]. By a second method, the additional element could be introduced as the halide salt activator (e.g. $ZrCl_4$, YCl_3 or $HfCl_4$). For example, $ZrCl_4$ can react with the Cr–Al binary masteralloy to produce both $AlCl_x$ and $CrCl_x$ species as well as a significant vapor pressure of a $ZrCl_x$ species according to equation 9.3:

$$\underline{Me}(s) + ZrCl_4(v) = MeCl_x(v) + ZrCl_{4-x}(v) \tag{9.3}$$

If the substrate does not contain the RE, the necessary driving force exists for some dissolution of the RE.

9.4 POSSIBILITIES FOR IMPROVEMENT

The engineering of pack chemistries to deposit simultaneously two or three elements has been achieved experimentally in the past several years [9, 17, 39, 58]. The subject was reviewed recently [61]. A variety of alloys have been coated with Cr + Al (low carbon and low alloy steels; 410, 304 and 316

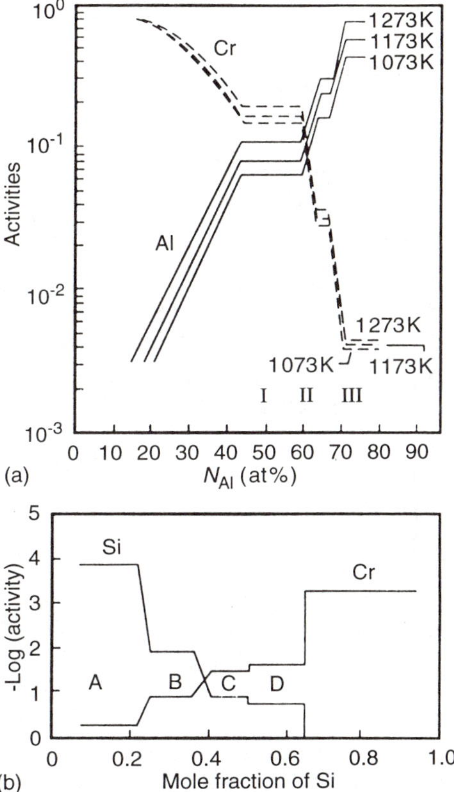

Figure 9.4 Activity data as a function of solute content for (a) Al and Cr in a Cr–Al alloy at 1073 K, 1173 K and 1273 K [56]; and (b) Si and Cr in a Cr–Si alloy at 1373 K [55, 58].

stainless steels; Ni-base superalloys) and Cr + Si (low carbon and low alloy steels; 304 and 409 stainless steels; Incoloy 800). Because of their unique metallurgical characteristics, each alloy requires a specific pack chemistry to obtain the optimum coating composition. The ideal coating contains sufficient concentrations of the two elements for superior steady-state resistance to oxidation and corrosion without excessive contents which could lead to brittleness, deleterious phase changes, and oxidant penetration in service.

9.4.1 Chromizing/aluminizing

Low alloy steels

Low alloy steels (e.g. Fe-2.25 Cr–1.0 Mo–0.15 C) are commonly used in utility boilers, petrochemical plants and coal gasification systems because of

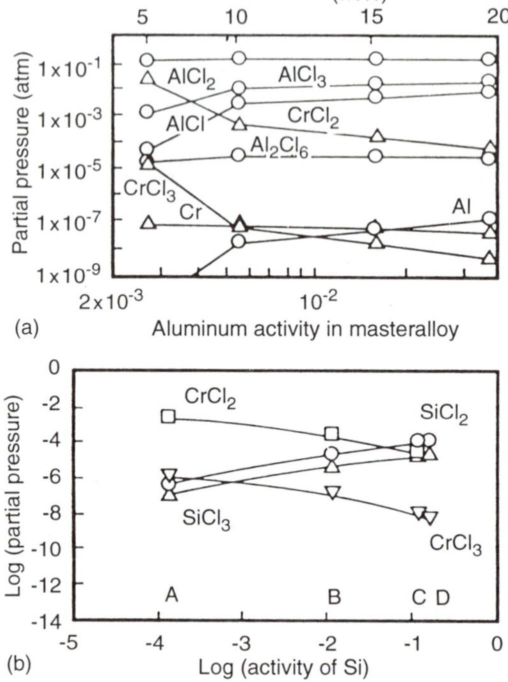

Figure 9.5 Equilibrium partial pressures of metallic halides in (a) NH_4Cl-activated packs containing Cr–Al masteralloys and Al_2O_3 filler at 1150 °C [61]; and (b) NaCl-activated packs containing Cr–Si masteralloys and SiO_2 filler at 1050 °C [55, 58].

their excellent creep strength in service. Exposed to a variety of corrosive environments, they require surface modification to improve their corrosion resistance. A single-step pack cementation process has been developed to produce the Kanthal composition (Fe–20 Cr–4.5 Al) on the surface of Fe–2.25 Cr–1.0 Mo–0.15 C alloys [62]. These alloys were successfully coated in a pack containing 8 wt% of a Cr–10 Al masteralloy, 2 wt% NH_4Cl salt activator and the balance Al_2O_3 filler heated at 1150 °C for 6 h. A ferrite layer is formed initially on the substrate surface, allowing for subsequent diffusion of Cr and Al into the ferrite matrix. Because ferrite has a very low C solubility, the rejection of C into the austenite phase at the ferrite–austenite interface minimizes decarburization and prevents the formation of a $Cr_{23}C_6$ carbide at the surface. The resulting Cr–Al enriched surface provides excellent isothermal and cyclic oxidation resistance by an Al_2O_3 scale, even at temperatures as low as 700 °C [62].

Stainless steels

Chromized/aluminized coatings were achieved on several austenitic steels, e.g. 304 SS, 310 SS, 316 SS and Incoloy 800 [9]. Sufficient amounts of Cr and Al to produce Al_2O_3-forming scales were incorporated into each alloy. The coating morphologies consisted of β-NiAl precipitates in a Cr-rich ferrite matrix, thus producing a mechanically tough microstructure. Compared to the initial work by Bangaru and Krutenat [10], the use of a Cr–Al masteralloy reduces the tendency to form an outward-grown, external β-NiAl layer that is brittle and entraps pack particles.

Nickel-base superalloys

Nickel-base superalloy turbine blades are commonly aluminized in a cementation pack for oxidation resistance. However, such aluminide coatings lack adequate resistance to hot corrosion caused by deposits of fused alkali sulfates. A Cr-modified aluminide coating would promise substantial improvements in the hot corrosion resistance of the coating [63, 64]. In addition, a small amount of a reactive element (Zr, Hf or Y) is known to improve the adherence of the protective Al_2O_3 scales [65, 66].

Two pack arrangements have been used in the laboratory to produce Cr-enriched aluminide coatings on several commercial superalloys [67]. By the simpler *powder contacting* arrangement, the substrate is buried in direct physical contact with the powder mixture. By this method, an outward-grown, β-NiAl plus α-Cr coating layer is formed, but some inert filler (Al_2O_3), reactive element oxide source (ZrO_2) and masteralloy particles are entrapped within the outer, two-phase layer of the coated substrate [39]. Pack entrapment can be eliminated by an *above-pack* arrangement, whereby the test coupon is held isolated from contact with the pack by a porous alumina enclosure. However, substrate isolation reduces both the growth rate of the coating and the amount of the Cr enrichment [67].

A typical microstructure of an above-pack coating (Fig. 9.6) consists of an Al_2O_3-free, Cr-alloyed β-NiAl layer containing some reactive element (RE) dopant, also deposited from the gas phase [39]. In this outward-grown β-NiAl coating, a dispersion of α-Cr particles is precipitated near the original substrate surface and also grown inward from the gas phase at the coating surface. An interdiffusion zone between the outer layer and the original $Ni + Ni_3Al$ substrate is comprised of β-NiAl and Ni_3Al, $M_{23}C_6$, MC-type carbides and σ-phase [4, 68]. The average bulk surface compositions for coated René 80 alloy substrates resulting from this above-pack arrangement were approximately $Ni–(33–42 \text{ at}\%)Al –(6–8 \text{ at}\%)Cr –(0.05–0.3 \text{ at}\%)RE$, i.e. RE as Y or Zr, respectively. These coatings were produced from a pack mixture containing either an NH_4Cl salt activator plus an RE oxide source or else an RE-base chloride salt activator with a suitable Cr–Al masteralloy

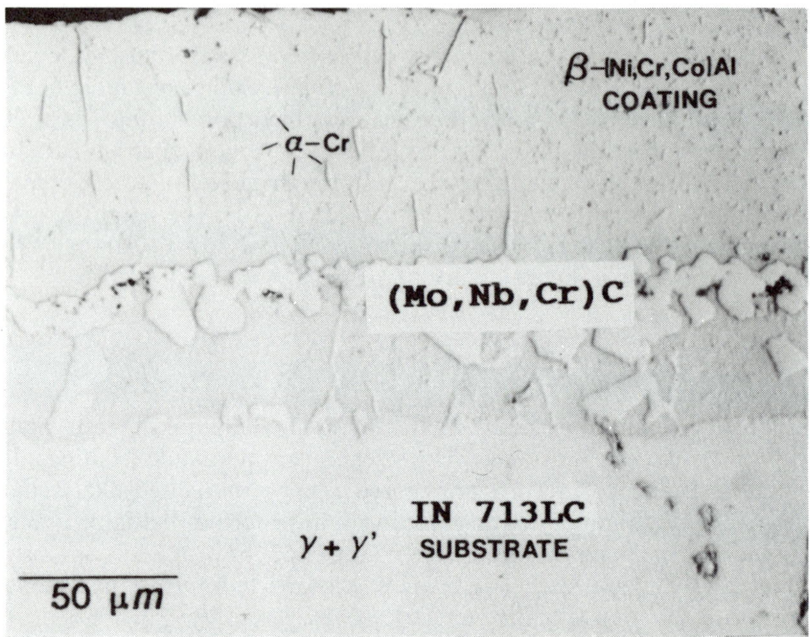

Figure 9.6 Cross-sectional optical micrograph of an IN713LC alloy treated at 1150 °C for 24 h in a (2:1) YCl$_3$CrCl$_2$-activated pack containing 25 wt% of Cr–7.5 wt% Al masteralloy and 73 wt% of an Al$_2$O$_3$ filler; Above-pack method.

at 1150 °C for 24 h. In addition, oxidation-resistant RE(Zr, Hf, Y)-doped aluminide coatings were developed. The microstructures of these coatings were analogous to those described above, except that second-phase α-Cr particles were absent from the coating surface and the average surface compositions were about Ni–(40–45 at%)Al–(2–2.5 at%) Cr–0.05–0.30 at%)RE, i.e. RE as Y, Zr or Hf.

Both hot corrosion and cyclic oxidation tests were performed on the RE-doped, Cr-modified aluminide coatings, whereas only cyclic oxidation tests were performed on the RE-doped aluminide coatings. Figure 9.7 plots the weight changes for isothermal hot corrosion tests conducted for various exposure times using a 5 mg/cm^2 film of Na$_2$SO$_4$ applied before each exposure on Cr–RE modified aluminide diffusion coatings produced from the powder contacting arrangement. A commercial low activity aluminide diffusion coating (GE Codep C), lower-Cr ZrCl$_4$ activated coatings and lower Cr, above-pack coatings did not adequately protect the Ni-base superalloys during hot corrosion simulation. However, the higher Cr, aluminide diffusion coatings provided a substantial improvement in coating lifetime [39, 64, 69].

Figure 9.8 is a plot of the weight changes for cyclic oxidation tests at 1100 °C in static air of both Cr–RE (Zr or Y) modified and RE (Zr, Hf or

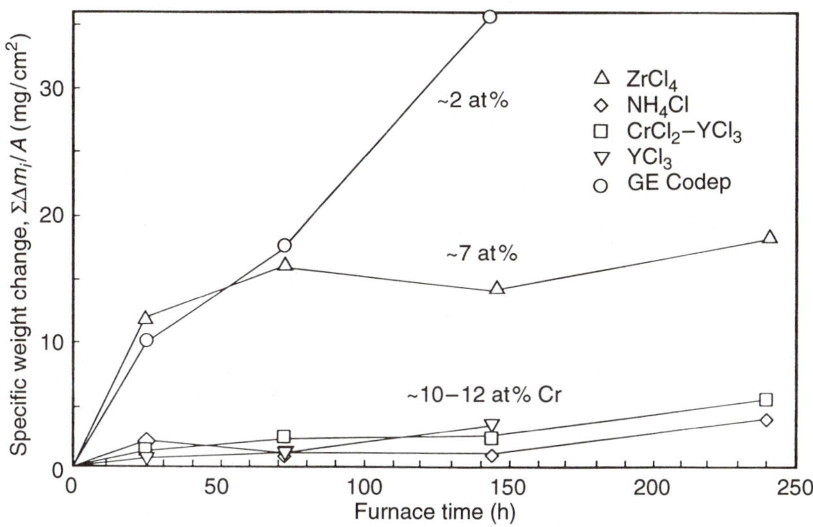

Figure 9.7 Isothermal hot corrosion of a René 80 alloy substrate coated by a powder-contacting method in 0.1% SO$_2$/O$_2$ at 900 °C with periodic Na$_2$SO$_4$ recoating [69].

Y) doped aluminide diffusion coatings prepared by the above-pack arrangement. Several pack chemistries of each type were identified which produced acceptable cyclic oxidation behavior resulting from the formation of adherent α-Al$_2$O$_3$ protective scales for 500 hour-long cycles. Coating failures occurred for RE-free and RE-lean aluminides and for carbide coatings (e.g. NH$_4$Cl plus ZrO$_2$ in Fig. 9.8b). Severe Al consumption resulted from the formation of less adherent α-Al$_2$O$_3$ scales; eventually less protective NiAl$_2$O$_4$ and NiO scales are formed [61].

Intermetallic compounds

Intermetallic compounds (e.g. Fe$_3$Al, Ti$_3$Al and TiAl) have been candidate structural materials for fossil fuel and high temperature aerospace applications because of their high strength-to-weight ratios. For example, pure Fe$_3$Al has poor room temperature properties, resulting from hydrogen embrittlement. Additions of Cr have been shown to improve its room temperature ductility. However, the resistance to high temperature sulfidation drops considerably because the formation of a protective alumina scale is impeded by Fe–Cr–Al sulfides [70]. A single-step, pack aluminizing treatment using a powder mixture containing 20 wt% of Fe–42 Al masteralloy and 2 wt% NaF salt activator treated at 925 °C has produced a Cr-enriched, FeAl coating layer using an above-pack arrangement [71]. The resulting FeAl layer forms a

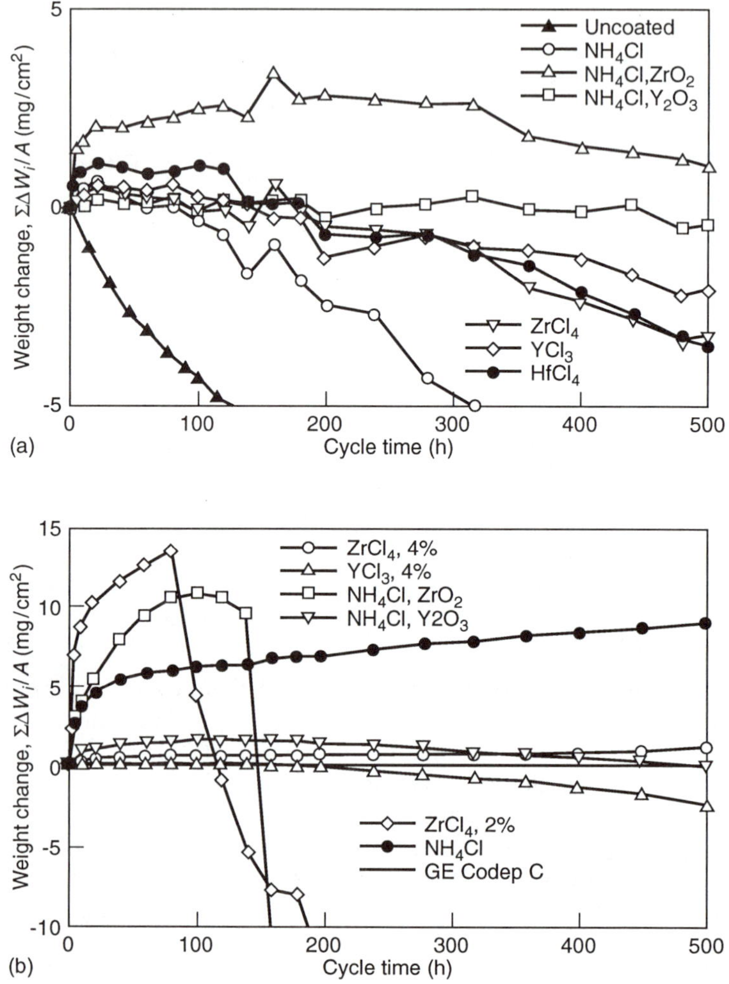

Figure 9.8 Weight changes for 1100 °C cyclic oxidation in static air of coated IN713LC alloy substrates versus cycle time for the above-pack method [69]: (a) RE doped and (b) Re–Cr modified aluminide coatings.

protective Al_2O_3 scale during oxidation exposure and provides excellent oxidation and sulfidation resistance to the Cr-alloyed Fe_3Al substrate.

However, the FeAl layer is also a brittle phase. Trace additions of boron are known to segregate to grain boundaries and thereby improve the room temperature fracture toughness of bulk FeAl [72]. Therefore, the FeAl coating layer was doped with small additions of boron (1–3 at%) by codeposition of Al and B in an aluminizing pack mixture containing 0.5–1.0 wt% FeB

[71]. The B additions reduced the microhardness of the FeAl layer from a Vickers hardness number (VHN) of approximately 500 to 350. As for bulk B-doped FeAl [73], the presence of B in the FeAl coating increased scale spallation during cyclic oxidation in static air at 900 °C [71].

Conversely, $TiAl_3$-base coatings have also been successfully produced on TiAl [74], Ti_3Al [75, 76], and $Ti_3Al + Nb$ [77] by a conventional high activity pack aluminizing process. The $TiAl_3$ phase is the only phase in the Ti–Al system which produces a protective α-Al_2O_3 scale during oxidation exposure without TiO_2 formation [74]. $TiAl_3$-base coatings produced excellent cyclic oxidation behavior compared to uncoated substrate materials up to 1000 °C in static air. Small, but permeable, cracks in these coatings, resulting from volume expansion during coating growth and coefficient of thermal expansion mismatch during thermal cycles, contributed to the degradation of thicker coatings during oxidation [77].

9.4.2 Chromizing/siliconizing

Ferritic stainless steels

A ferritic 409 SS alloy was chromized/siliconized using a Cr–10 Si masteralloy powder, an NaF salt activator and an SiO_2 filler heated at 1050 °C for 16 h [55, 58]. Generally, an inward-grown coating is formed with an average surface composition of Fe–23 Cr–2.55 Si. In cross-sectional micrographs, a thin, dark horizontal interface identifies the original substrate surface, showing some minor contribution by outward diffusion to coating growth. In the absence of any phase change during the coating process, no substrate–coating interface is introduced, and the dominant inward diffusion excludes both porosity and pack entrapment. Although no corrosion testing has yet been performed for the coating, this composition should provide excellent resistance to corrosion in aqueous and in high temperature, gaseous environments.

Low-alloy steels

An Fe–0.5 Cr–0.5 Mo–0.1 C low alloy steel was chromized/siliconized using a Cr–10 Si masteralloy, an NaF–3 mol% NaCl mixed (dual) salt activator and an SiO_2 filler heated at 1050 °C for 16 h. The dual activator promotes the simultaneous deposition of both Cr and Si. Fluoride activator salts are essentially siliconizing, whereas chloride activator salts are chromizing. The proper proportion of fluoride to chloride activator salt produced an inward-grown ferrite coating layer with an optimum average surface composition high in Cr (≈ 25 wt%) with approximately 3 wt% Si, while eliminating the formation of the deleterious σ-phase (>35 wt% Cr). Similar results were

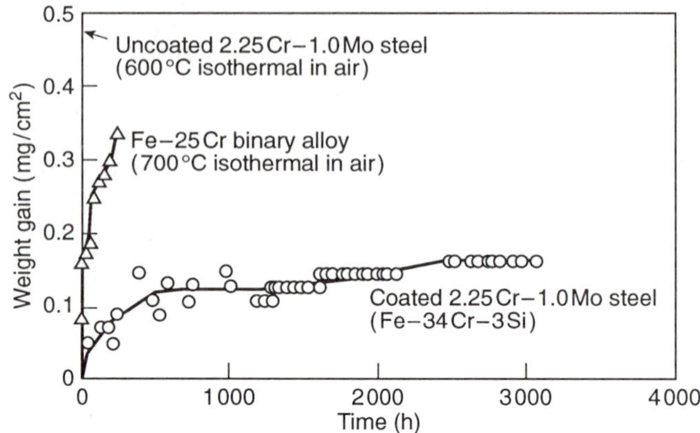

Figure 9.9 A plot of weight gain versus time for the cyclic oxidation in static air at 700 °C of coated samples with surface compositions of Fe–34 Cr–3 Si (wt%) [55, 58].

obtained for an AISI 1018 and an Fe–2.25 Cr–1.0 Mo–0.15 C low alloy steel [55, 58].

As shown in Fig. 9.9, this Cr–Si coated steel has been tested in cyclic oxidation in static air at 700 °C and has exhibited extremely slow oxidation kinetics with no detectable spalling of the oxide scale. The remarkably slow weight gain ($<0.2\,\text{mg/cm}^2$ for over 4.4 months at 700 °C) for the coated Fe–2.25 Cr–1.0 Mo–0.15 C alloy has also been found for a similarly coated AISI 1018 steel. The very slow kinetics result from a slow-growing silica layer beneath a thin, outer chromia scale. Also shown for comparison in Fig. 9.9 are the isothermal oxidation kinetics for an uncoated Fe–2.25 Cr–1.0 Mo–0.15 C [78] and an Fe–25 Cr binary alloy [79]. The Fe–25 Cr–3 Si coating has also shown negligible corrosion in an aerated 3.5 wt% NaCl aqueous solution at room temperature [80]. However, the Fe–20 Cr–3 Si coating on AISI 1018 steel substrates was less resistant than 304 SS in a 0.5 M H_2SO_4 solution at room temperature. The Fe–32 Cr–3 Si coating on AISI 1018 substrates was isothermally oxidized in air at 700 °C. A parabolic rate constant, $K_p = 4.6 \times 10^{-15}\,\text{gm}^2/\text{cm}^4$ s, measured for this coating was an order of magnitude smaller than that for other commercial chromia-forming alloys, as presented in Fig. 9.10 [81–84]. Similar coatings on both AISI 1018 and Fe–2.25 Cr–1.0 Mo–0.15 C low alloy steel substrates were also tested in an erosion–corrosion simulator at the Lawrence Berkeley Laboratory in Berkeley CA. The specimens were eroded by angular SiO_2 particles of average size 150 μm, at a particle velocity of 20 m/s, at impact angles of 30° and 90°, and at 25, 650 and 850 °C for 5 h exposures. According to weight loss and cross-sectional thickness losses, the Cr–Si coatings performed better than other coated or uncoated steels [80].

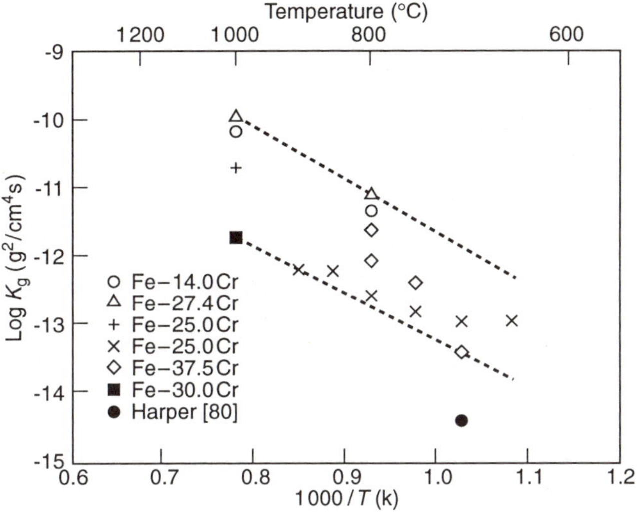

Figure 9.10 Arrhenius plot comparing the measured gravimetric rate constant of a coated AISI 1018 steel coupon oxidized in air at 700 °C [80] with reported rate constants in the literature for the growth of Cr_2O_3 on binary Fe–Cr alloy [81–84].

Austenitic steels

As mentioned above, austenitic alloys can be coated by diffusing the coating elements into the solid solution or by transforming the matrix into a ferrite + austenite two-phase field. For example, about 3 wt% Si was diffused into the surface of the austenitic alloy 800 without a significant change in microstructure. In contrast, an austenitic 304 SS alloy was chromized/siliconized uing a Cr–10 Si masteralloy, an NaF salt activator, and an Al_2O_3 filler treated at 1150 °C for 16 h. This coating process promotes the transformation of the austenitic surface into a ferrite (Fe–25 Cr–7 Ni–2 Si) plus austenite (Fe–20 Cr–11 Ni–2 Si) microstructure [55, 58]. Figure 9.11 presents the oxidation kinetics for these two coated alloys compared to the uncoated alloys upon cyclic oxidation at 1035 °C in static air. The coated alloy 800 does suffer some scale spallation upon temperature cycling (approximately 1% of the surface area), but this behavior is an important improvement over the uncoated alloy. The coated 304 SS exhibited excellent high temperature oxidation resistance with no evidence of spalling, whereas the uncoated alloy did not form an effective adherent protective scale.

A key element in the reduction of oxidation kinetics is the presence of ferrite, with its much higher diffusion coefficients, at the surface. The high diffusivity for ferrite promotes the early formation and retention of the thermodynamically more stable oxide scales (i.e. chromia rather than iron oxide at an early stage of oxidation, and silica rather than chromia at steady state).

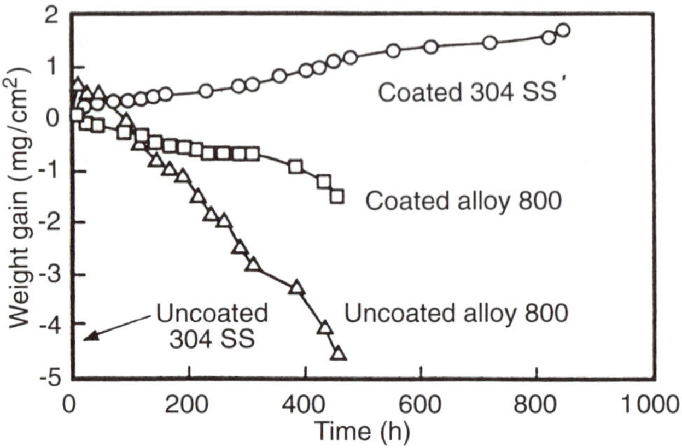

Figure 9.11 A plot of weight gain versus time for the cyclic oxidation in static air at 1035 °C of coated and uncoated 304 SS and alloy 800 [55, 58].

Niobium-base alloys

Because of their high solidus temperatures and reasonable creep strengths, Nb alloys have been considered as candidate materials for gas turbine applications. However, Nb and its alloys have inherently poor oxidation resistance because the oxide (Nb_2O_5) formed is not compact or resistant to permeation by molecular oxygen; therefore, a protective coating is necessary [85]. A new, two-step coating process has been developed to provide oxidation resistance for high temperature (1100–1500 °C) service [86]. First, a diffusion barrier of Mo or Mo–W alloy is applied to the Nb substrates using a conventional sputtering or CVD technique. Then the substrate is silicided to form an $MoSi_2$ or $(Mo, W)Si_2$ layer by an NaF-activated pack cementation method. Molybdenum disilicide has excellent high temperature oxidation resistance, providing a thin protective SiO_2 oxygen barrier.

Because of the large mismatch in coefficient of thermal expansion between $MoSi_2$ and SiO_2, the siliciding treatment has been modified to codeposit Si and Ge, and thereby produce an $Mo(Si, Ge)_2$ or $(Mo, W)(Si, Ge)_2$ diffusion layer. Because the stabilities for the volatile fluorides of Si and Ge are reasonably similar, codeposition of Si and about 3–8 at% Ge to form $(Mo, W)(Si, Ge)_2$ has been achieved using mixtures of pure Si and Ge powders with an NaF salt activator and treated at 1150 °C for 16 h. Small germania additions to silica greatly increase its CTE, and thereby decrease the CTE mismatch between the scale and the coating; this improves its cyclic oxidation behavior [87]. In a recent study, six out of eight coupons of pure Nb coated with $(Mo, W)(Si, Ge)_2$ survived cyclic oxidation in static air for 200 hour-long periods at 1370 °C with weight gains of only 1.2–1.6 mg/cm². Premature failures occurred on two coupons due to excessive growth of Nb_2O_5 initiated at cracks in the coating, but the Nb_2O_5 formation was not catastrophic. One of the cyclic oxidation tests of a coupon coated with $(Mo, W)(Si, Ge)_2$ was extended to 500 h without failure [86]. These results indicate that this multicomponent silicide coating offers significant promise to protect Nb-base alloys in oxidizing environments at very high temperatures.

9.5 SUMMARY

Halide-activated pack cementation (HAPC) has been used in industry for nearly 75 years to apply protective diffusion coatings to otherwise inadequate substrates. HAPC is a relatively inexpensive and highly versatile diffusion coating process for enriching the surface with Al, Cr, Si or B, and recently, combinations of these elements have also been achieved.

The codeposition of two or more elements in a halide-activated cementation pack is inherently difficult because of large differences in the thermodynamic stabilities of their volatile halides. However, with the aid of computer-assisted analysis of the pack equilibria, combinations of suitable binary alloys and halide activator salts can effectively achieve codeposition. The codeposition of Cr + Al or Cr + Si by pack cementation has yielded diffusion coatings with excellent resistance to high temperature oxidation and corrosion for a wide range of alloy substrates.

ACKNOWLEDGMENTS

The authors acknowledge the support from the Oak Ridge National Laboratory (R.R. Judkins, contract FWP-FEAA028) and from the Naval Air Development Center/Office of Naval Research (M. Thomas, T. Kircher, J. Sedricks, grant N00014-90-J-1765). The authors also acknowledge important contributions to this research by E.L. Courtright (Battelle Pacific Northwest Laboratory), by J.L. Smialek and N.S. Jacobson (NASA Lewis Research Center) and by A.J. Mueller, E.R. Naylor, F.D. Geib, S.C. Kung, V.A. Ravi, D.M. Miller and M.A. Harper during their studies at The Ohio State University.

REFERENCES

1. Allison, G. and Hawkins, M.K. (1914) *General Electric Review*, **17**, 947–51.
2. Van Aller, T. (1911) Treatment of metals. U.S. Patent 1,155,974.
3. Goward, G.W. and Cannon, L.W. (1988) *Trans. ASME*, **110**(1), 150–4.
4. Goward, G.W. and Boone, D.H. (1971) *Oxid. Metals*, **3**(5), 475–95.
5. Pichoir, R. (1978) Influence of the mode of formation on the oxidation and corrosion behavior of NiAl-type protective coatings, in *Materials and Coatings to Resist High-Temperature Corrosion* (eds D.R. Holmes and A. Rahmel), Applied Science Publishers, London, pp. 271–91.
6. Shankar, S. and Seigle, L.L. (1978) *Metall. Trans. A*, **9**(10), 1467–76.

  7. Parzuchowski, R.S. (1977) *Thin Solid Films*, **45**, 349–55.
  8. Akuezue, H.C. and Stringer, J. (1989) Multiphase diffusion and growth during aluminizing (Fe–Cr–Al system), in *Diffusion Analysis and Applications* (eds A.D. Romig, Jr. and M.A. Dayananda), The Metallurgical Society of AIME, pp. 337–60.
  9. Miller, D.M., Kung, S.C., Scarberry, S.D. and Rapp, R.A. (1988) *Oxid. of Metals*, **29**(3/4), 239–54.
 10. Bangaru, N.V. and Krutenat, R.C. (1984) *J. Vac. Sci. Technol.*, **2**(4), 806–15.
 11. Kelley, F.D. (1923) *Trans. Amer. Electrochem. Soc.*, **43**, 351–70.
 12. Hoar, T.P. and Croom, E.A.G. (1951) *J. Iron Steel Inst.*, Oct. 1951, pp. 101–7.
 13. Menzies, I.A. and Mortimer, D. (1965) *Corr. Sci.*, **5**, 539–58.
 14. Ravi, V.A. (1988) Simultaneous chromizing–aluminizing of nickel and nickel-base alloys by halide-activated pack cementation. Ph.D. dissertation, The Ohio State University.
 15. Meier, G.H., Cheng, C., Perkins, R.A. and Bakker, W. (1989) *Surf. Coat. Technol.*, **39/40**, 53–64.
 16. Miller, D.M. (1990) Simultaneous chromizing–aluminizing of an Fe–2.25 Cr–1 Mo steel and cobalt-base alloys by halide-activated pack cementation. Ph.D. dissertation, The Ohio State University.
 17. Rapp, R.A., Wang, D. and Weisert, T. (1987) Simultaneous chromizing–aluminizing of iron and iron-base alloys by pack cementation, in *High Temperature Coatings* (eds M. Khobaib and R. Krutenat), The Metallurgical Society of AIME, pp. 131–41.
 18. Nash, P. (1986) The Ni–Cr System, in *Binary Alloy Phase Diagrams* (ed. T.B. Massalski) The American Society for Metals, p. 389.
 19. Sun, W.P., Lin, T.H. and Hon, M.H. (1987) *Metall. Trans. B*, **18**(9), 617–19.
 20. Mazille, H.M. (1980) *Thin Solid Films*, **65**, 67–74.
 21. Felix, P.C. and Erdös, E. (1972) *Werkstoffe und Korrosion*, **23**(8), 627–36.
 22. Itzhak, D., Tuller, F.R. and Schieber, M. (1980) *Thin Solid Films*, **73**, 379–84.
 23. Smialek, J.L. (1974) Fused silicon-rich coatings for superalloys. *NASA Technical Memorandum TM X-3001.*
 24. Fitzer, E. (1954) *Archiv Eisenhüttenwesen*, **25**, 455.
 25. Singleton, M.F., Murray, J.L. and Nash, P. (1986) The Ni–Al system, in *Binary Alloy Phase Diagrams* (ed. T.B. Massalski), The American Society for Metals, p. 141.
 26. Grünling, H.W. and Bauer, R. (1982) *Thin Solid Films*, **95**, 3–20.
 27. Höch, P. and Bren, J. (1969) *Metalloberfläche*, **23**(1), 13–16.
 28. Pennisi, F.J. and Gupta, D.K. (1981) *Thin Solid Films*, **84**(1), 49–58.
 29. Schneider, K., Bauer, R. and Grünling, H.W. (1978) *Thin Solid Films*, **54**(3), 359–67.
 30. Young, S.G. and Deadmore, D.L. (1980) *Thin Solid Films*, **73**, 373–8.
 31. Golightly, F.A. *et al.* (1976) *Oxid. Metals*, **10**(3), 163–87.
 32. Kumar, A., Nasrallah, M. and Douglass, D.L. (1974) *Oxid. Metals*, **8**(4), 227–63.
 33. Allam, I.M., Whittle, D.P. and Stringer, J. (1978) *Oxid. Metals*, **12**(1), 35–66.
 34. Tien, J.K. and Pettit, F.S. (1972) *Metall. Trans.*, **3**(6), 1587–99.
 35. Whittle, D.P. and Stringer, J. (1980) *Phil. Trans. R. Soc. Lond. A*, **295**, 309–29.
 36. Jedlinski, J., Godlewski, K. and Mrowec, S. (1989) *Mater. Sci. Engng*, **A121**, 539–43.
 37. Fuhui, W., Hanyi, L., Linxiang, B. and Weitao, W. (1989) *Mater. Sci. Engng*, **A121**, 387–9.
 38. Tu, D.C., Lin, C.C., Liao, S.J. and Chou, J.C. (1986) *J. Vac. Sci. Technol.*, **4**(6), 2601–8.
 39. Bianco, R. and Rapp, R.A. (1990) Simultaneous chromizing–aluminizing of nickel-base superalloys with reactive element additions, in *High Temperature Chemistry V* (eds W.B. Johnson and R.A. Rapp), The Electrochemical Society, pp. 211–22.
 40. LePrince, G., Alperine, S., Vandenbulcke, L. and Walder, A. (1989) *Mater. Sci. Engng*, **A121**, 419–25.

41. Wagner, C. (1965) *Corr. Sci.*, **5**(11), 751–64.
42. Galmiche, P. (1968) *Metals and Materials*, **2**(7), 241–8.
43. Ravi, V.A., Choquet, P.A. and Rapp, R.A. (1988) Chromizing–aluminizing coating of Ni- and Fe-base alloys by the pack cementation technique, in *Oxidation of High-Temperature Intermetallics* (eds T. Grobstein and J. Doychak), The Metallurgical Society of AIME, pp. 127–45.
44. Walsh, P.N. (1973) Chemical aspects of pack cementation, in *Proceedings of the 4th International Conference on Chemical Vapor Deposition* (eds G.F. Wakefield and J.M. Blocher) The Electrochemical Society, pp. 147–68.
45. Fischer, T. (1992) Chromizing Arizona. Private communication, 11 Nov. 1992.
46. Levine, S.R. and Caves, R.M. (1974) *J. Electrochem. Soc.*, **121**(8), 1051–64.
47. Gupta, B.K., Sarkhel, A.K. and Seigle, L.L. (1976) *Thin Solid Films*, **39**, 313–20.
48. Gupta, B.K. and Seigle, L.L. (1980) *Thin Solid Films*, **73**, 365–71.
49. Pennisi, F.J., Kandasamy, N. and Seigle, L.L. (1981) *Thin Solid Films*, **84**(1), 17–27.
50. Seigle, L.L. (1984) Thermodynamics and kinetics of pack cementation processes, in *Surface Engineering* (eds R. Kossowsky and S.C. Singhal), Applied Science Publishers, London, pp. 345–69.
51. Nciri, B. and Vandenbulcke, L. (1983) *J. Less Common Metals*, **95**, 55–72.
52. Kung, S.C. and Rapp, R.A. (1988) *J. Electrochem. Soc.*, **135**(3), 731–41.
53. Kung, S.C. and Rapp, R.A., (1989) *Oxid. Metals*, **32**(1/2), 89–109.
54. Restall, J.E. and Hayman, C. (1987) Process for diffusion coating metals. U.S. Patent 4,687,684.
55. Harper, M.A. and Rapp, R.A. (1991) Codeposition of chromium and silicon in diffusion coatings for iron-base alloys using pack cementation, in *Surface Modification Technologies IV* (eds T.S. Sudarshan, D.G. Bhat and M. Jeandin), The Metallurgical Society of AIME, pp. 415–28.
56. Johnson, W., Komarek, K. and Miller, E. (1968) *Trans. Metall. Soc. AIME*, **242**, 1685–8.
57. Marijnessen G.H. (1983) Codeposition of chromium and aluminum during a pack process, in *High-Temperature Protective Coatings* (ed. S.C. Singhal), The Metallurgical Society of AIME, pp. 27–35.
58. Harper, M.A. and Rapp, R.A. (1991) Chromized/siliconized pack cementation diffusion coatings for heat-resistant alloys, in *Proceedings of the First International Conference on Heat Resistant Materials* (eds K. Natesan and D.J. Tillack), The American Society for Metals, pp. 379–86.
59. Flynn, H., Morris, A.E. and Carter, D. (1986) An interactive gas-phase removal version of SOLGASMIX, in *Proceedings of the 25th CIM Conference of Metallurgists*, TMS-CIM.
60. Bianco, R., Rapp, R.A. and Jacobson, N.S. (1992) *Oxid. Metals*, **38**(1/2), 33–43.
61. Bianco, R., Harper, M.A. and Rapp, R.A. (1991) *J. Metals*, **43**(11), 68–73.
62. Geib, F.D. and Rapp, R.A. (1993) *Oxid. Metals*, **40**, 213–28.
63. Godlewska, E. and Godlewski, K. (1984) *Oxid. Metals*, **22**(3/4), 117–31.
64. McCarron, R.L., Lindblad, N.R. and Chatterji, D. (1976) *Corrosion*, **32**(12), 476–81.
65. Barrett, C.A. (1988) *Oxid. Metals*, **30**(5/6), 361–90.
66. Jedlinski, J. and Mrowec, S. (1987) *Mater. Sci. Engng*, **87**, 281–7.
67. Bianco, R. and Rapp, R.A. (1993) *J. Electrochem. Soc.*, **140**(4), 1181–90.
68. Gale, W.F. and King, J.E. (1992) *Metall. Trans. A*, **23**(9), 2657–65.
69. Bianco, R., Rapp, R.A. and Smialek, J.L. (1993) *J. Electrochem. Soc.*, **140**(4), 1191–1203.
70. DeVan, J.H. (1988) Oxidation behavior of Fe_3Al and derivative alloys, in *Oxidation of High-Temperature Intermetallics* (eds T. Grobstein and J. Doychak), The Metallurgical Society of AIME, pp. 107–15.

71. Geib, F.D. and Rapp, R.A. (1992) Diffusion coatings for iron aluminide Fe_3Al via halide-activated pack cementation, in *Processing and Manufacturing of Advanced Materials for High-Temperature Applications* (eds T.S. Srivatsan and V.A. Ravi), The Metallurgical Society of AIME, pp. 347–60.
72. Liu, C.T., Lee, E.H. and McKamey, C.G. (1990) *Scripta Metall.*, **23**(6), 875–80.
73. Smialek, J.L., Doychak, J. and Gaydosh, D.J. (1990) *Oxid. Metals*, **34**(3/4), 259–75.
74. Subrahmanyan, J. (1988) *J. Mater. Sci.*, **23**(6), 1906–10.
75. Mabuchi, H., Asai, T. and Nakayama, Y. (1989) *Scripta Metall.*, **23**(4), 685–9.
76. Kung, S.C. (1990) *Oxid. Metals*, **34**(3/4), 217–28.
77. Smialek, J.L., Gedwill, M.A. and Brindley, P.K. (1990) *Scripta Metall.*, **24**, 1291–96.
78. Christl, W., Rahmel, A. and Schutze, M. (1989) *Oxid. Metals*, **31**(1/2), 1–34.
79. Natesan, K. and Park, J.H. (1989) Role of alloying additions in the oxidation–sulfidation of Fe-base alloys, in *Corrosion & Particle Erosion at High Temperatures* (eds V. Srinivasan and K. Vedula), The Metallurgical Society of AIME, pp. 49–64.
80. Harper, M.A. (1992) Codeposition of chromium and silicon onto iron-base alloys via pack cementation. Ph.D. dissertation, The Ohio State University.
81. Felton, E.J. (1961) *J. Electrochem. Soc.*, **108**(6), 490–5.
82. Francis, J.M. and Whitlow, W.H. (1965) *Corr. Sci.*, **5**(10), 701–10.
83. Tedmon, C.S., Jr. (1967) *J. Electrochem. Soc.*, **113**(8), 788–95.
84. Wood, G.C. and Whittle, D.P. (1967) *Corr. Sci.*, **7**(11), 763–82.
85. Perkins, R.A. and Meier, G.H. (1990) *J. Metals*, **42**(8), 17–21.
86. Mueller, A.J., Wang, G., Rapp, R.A. and Courtright, E.L. (1992) *J. Electrochem. Soc.*, **139**(5), 1266–75.
87. Schlicting, J. and Neumann, S. (1982) *J. Non-Cryst. Solids*, **48**(1), 185–94.

10

Thermal spray coatings

H. Herman and S. Sampath

10.1 INTRODUCTION

Thermal spray processing is a well-established means of forming coatings of thicknesses greater than about 50 µm, so-called thick coatings. A wide range of materials can be thermally sprayed for a variety of applications, ranging from gas turbine technology (heat engines) to the electronics industry. Thermal spray coatings have been produced for at least 40 years, but the last decade has seen a virtual revolution in the capability of the technology to produce truly high performance coatings of a great range of materials on many different substrates [1]. This enhancement of the technology has been achieved largely through the introduction of new spray techniques, the enhancement of spray process controls, the employment of state-of-the-art methods of feedstock materials production and the use of modern techniques of quality assurance.

Although the use of advanced thermal spray coating has largely occurred within the aircraft industry, newer, extended applications of the technique have demonstrated its versatility. Applications include protection from wear, high temperatures, chemical attack, and the more mundane uses of environmental corrosion protection in infrastructure maintenance engineering. Of greatest importance, and making thermal spray processing uniquely important to an ever increasing engineering community, are (i) improved spray footprint definition versus wide spray beam; (ii) high throughput versus competitive techniques; (iii) significantly improved process control; (iv) lower cost-per-mass

Metallurgical and Ceramic Protective Coatings. Edited by Kurt H. Stern.
Published in 1996 by Chapman & Hall, London. ISBN 0 412 54440 7

of applied material, together with overall competitive economics. This chapter gives an overview of thermal spray coatings, as well as the fundamentals and application areas for the technology. Finally, future directions of torch design and applications will be discussed.

10.2 THERMAL SPRAY PROCESSES

The principle behind thermal spray is to melt material feedstock (wire or powder) to accelerate the melt to impact on a substrate where rapid solidification and deposit buildup occur. Thus, a heat source and a means of accelerating the material are required. This is pictured schematically in Fig. 10.1 [2]. Here, the feedstock material, in the form of a powder, is melted and propelled in the effluent of a flame. The high temperature for melting is achieved chemically (through combustion) or electrically (an arc); these melting methods also accelerate the molten particles to the target substrate, where the material solidifies, forming a deposit. The deposit is built up by successive impingement of these individual flattened particles or *splats*.

The various thermal spray processes are distinguished on the basis of the feedstock characteristics (wire/powder) and the heat source employed for melting. The following sections describe the various thermal spray processes in use today.

10.2.1 Combustion flame spraying

Combustion flame spraying employs compressed air or oxygen, mixed with one of a variety of fuels (e.g. acetylene, propylene, propane, hydrogen), both to melt and to propel the molten particles. Generally (with an exception to be noted shortly), the process yields low performance coatings, and it is not employed where high density, well-bonded coatings are required. The reasons for these deficiencies have to do with the relatively low flame velocity of

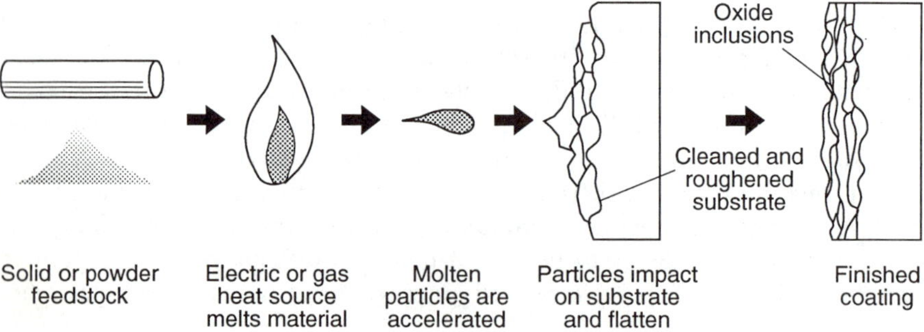

Figure 10.1 A schematic overview of the thermal spray process.

about 50 m/s and the low temperature achieved within the combustion flame, well below 3000 °C. Combustion flame spraying uses either powder, wire or rod as the feedstock material and has found widespread use around the world for its relative simplicity and cost-effectiveness.

10.2.2 High velocity oxyfuel spraying (HVOF)

A novel variation on combustion spraying, HVOF has had a dramatic influence on the field of thermal spray. This technique is based on special torch designs, in which a compressed flame undergoes free expansion upon exiting the torch nozzle, thereby experiencing dramatic gas acceleration, to perhaps over Mach 4. A schematic of a typical HVOF torch (similar to Stellite's Jet-Kote) is shown in Fig. 10.2 [3]. By properly injecting the feedstock powder from the rear of the torch, and concentrically with the flame, the particles are also subjected to velocities so high that they will achieve supersonic values. Therefore, upon impact onto the substrate, the particles spread out very thinly, and bond well to the substrate and to all other splats in its vicinity, yielding a well-adhered, dense coating, comparable, if not superior, to plasma spray coatings. But the powder particles are limited in the temperature they can achieve, due to the relatively low temperature combustion flame, so it is rarely possible to process refractory ceramics using this technique. Of particular importance is the fact that HVOF hardfacings (e.g. WC–Co, Stellite) have hardnesses and wear resistances superior to such materials plasma sprayed in air.

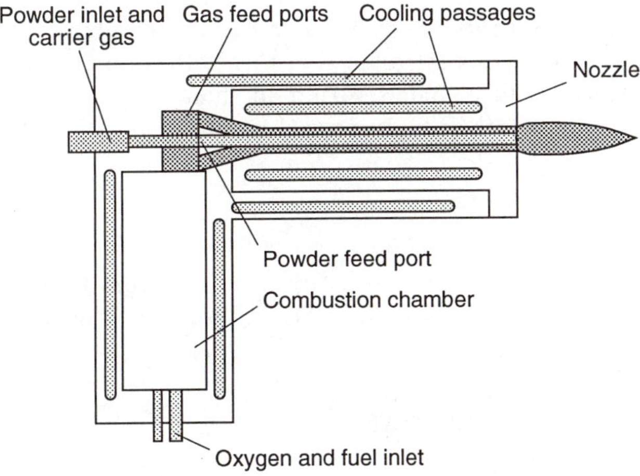

Figure 10.2 An illustration of a Jet-Kote HVOF torch [4].

10.2.3 Two-wire electric arc spraying

Two-wire electric arc spraying involves two current-carrying electrically conductive wires fed into a common arc point at which melting occurs, Fig. 10.3 [4]. The molten material is continuously atomized by compressed air, forming a spray of molten metal with a very high material throughput, as high as 40 kg/h. This rate, which is the highest in standard industrial thermal spray guns, yields highly cost-effective processing, allowing the competitive spraying of corrosion-resistant zinc and aluminum for the marine industry and for infrastructure applications. Modern developments have led to electric arc guns which can operate in inert atmospheres, using argon or nitrogen atomization, and spraying reactive metals such as zirconium and titanium for corrosion protection in the chemical industry.

10.2.4 Plasma spraying

Plasma spraying is relatively straightforward in concept but rather complex in function. The gun, pictured schematically in Fig. 10.4 [5], operates on direct current, which sustains a stable nontransferred eleric arc between a thoriated tungsten cathode and an annular water-cooled copper anode. A plasma gas (generally argon or other inert gas, complemented by a few percent of an enthalpy-enhancing gas, such as hydrogen) is introduced at the back of the gun interior; it swirls in a vortex and leaves at the front exit of the anode nozzle. The electric arc from the cathode to the anode completes the circuit, generally on the outer face of the anode, forming an existing plasma flame which rotates due to the vortex momentum of the plasma gas. The temperature of the plasma just outside the nozzle exit is effectively in excess of 15000 K for a typical DC torch operating at 40 kW. The plasma temperature drops off rapidly from the exit of the anode, and therefore the powder to be processed is introduced at this hottest part of the flame. The powder particles, approximately 40 μm in diameter, are accelerated and melted in the flame on their high speed (100–300 m/s) path to the substrate,

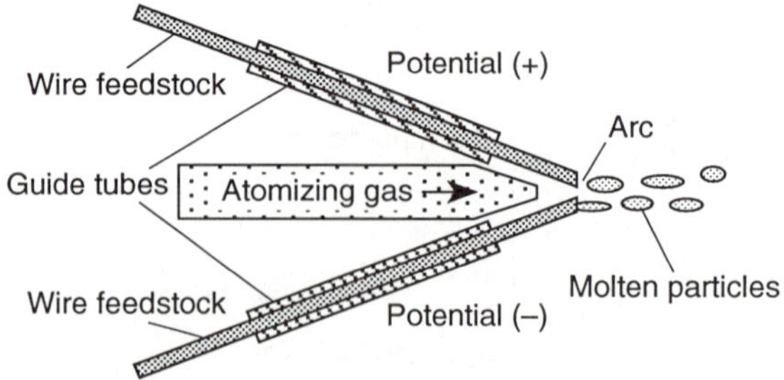

Figure 10.3 A schematic of a two-wire arc spray gun [4].

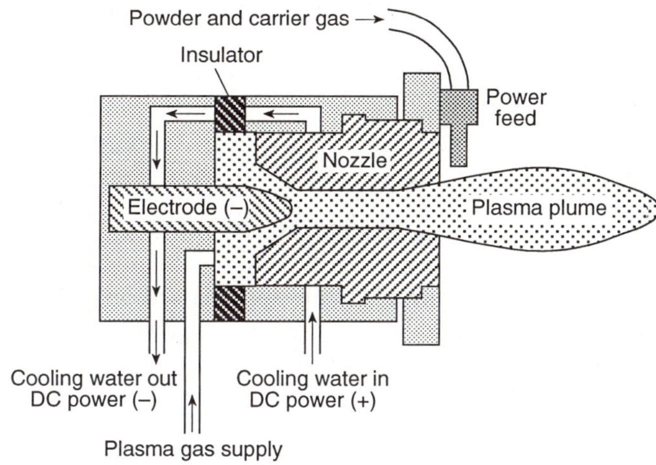

Figure 10.4 A schematic of a plasma spray process [4].

where they impact and undergo rapid solidification (10^6 K/s). Plasma spray is used to form deposits of greater than $50\,\mu m$ of a wide range of industrial materials, including nickel and ferrous alloys as well as refractory ceramics, such as aluminum oxide and zirconia-based materials.

For high performance applications, in order to approach theoretical bulk density and extremely high adhesion strength, plasma spray is carried out using a shrouded flame or in a reduced pressure, inert gas chamber, vacuum plasma spray (VPS). Vacuum plasma, or low pressure plasma, spraying operates at pressures of approximately 50–200 mbar (5–20 kPa). A shrouded flame might use argon or nitrogen to exclude oxygen from the vicinity of the flame and the workpiece. Shrouding has no known effect on particle velocity, whereas gas and particle velocity are increased within the reduced pressure chamber. This, as for HVOF, yields a higher density deposit. The benefits of reduced pressure and shrouded plasma spraying will increase substantially in the future.

It is useful to list the key features of plasma spraying:

- Deposits metals, ceramics or any combinations of these materials
- Forms microstructures with fine, equiaxed grains and without columnar boundaries
- Produces deposits that do not change in composition with thickness (length of deposition time)
- Can change from depositing a metal to a continuously varying mixture of metals and ceramics (i.e. functionally graded materials)
- High deposition rates ($>4\,kg/h$)

- Fabricates freestanding forms of virtually any material or any materials combinations
- Processes materials in virtually any environment, e.g. air, reduced pressure inert gas, high pressure (see the next section)

10.2.5 Vacuum plasma spray (VPS)

Low pressure plasma spraying was developed by Muehlberger in the early 1970s and gained widespread commercial use in the mid-1980s, to a large extent displacing electron beam-physical vapor deposition (EB-PVD) for the production of high quality metallic (MCrAlY) coatings [6]. The compositional flexibility afforded by VPS and the high coating rates achieved through liquid droplet transfer versus the limitations of evaporation in EB-PVD caused a major shift to VPS during the 1980s. As with all thermal spray processes, VPS is limited to line of sight. With the VPS process, individual parts are fixtured on a manipulator with a load-locked transfer chamber. The load-lock is pumped down and the parts are preheated to about 900–1000 °C before being transferred to the coating chamber. Before initiation of the plasma spray, the part is usually treated through reverse transferred arc sputtering to remove any traces of oxide that may have formed during preheat. The part is then plasma sprayed in a nontransferred mode. The coating distribution is determined by computer-controlled gun and part motion. Typical parameters for turbine blade coating would have a gun–substrate distance of $\sim$4–7 cm at a chamber pressure of 30–60 torr (4–8 kPa) and a gun powder of 80 kW. Powder feed rates vary from $\sim$3–5 kg/h depending on the application.

Of further importance is the ability of VPS to process oxygen-sensitive material, such as reactive metals and intermetallic compounds [7, 8]. For example, considerable work has been carried out on the VPS processing of nickel aluminides and molybdenum disilicide, which have potential uses in the aerospace industry. It was demonstrated that the VPS process was capable of producing dense, freestanding forms which showed impressive mechanical properties [7, 8]. The deposits were ultrafine grained and illustrated the capability of VPS in manufacturing of rapidly solidified intermetallics. Some studies have been published on the VPS processing of composites based on Ni_3Al and $MoSi_2$. High density deposits are obtained and some promising toughness increases are found [9]. There is a clear and important potential for VPS in the processing of intermetallics as both protective coatings and as freestanding forms.

10.2.6 Recent advances in thermal spray processes

Traditional plasma spray guns are gas vortex stabilized and operate in the 40–80 kW power range, but by using water stabilization it is possible to operate at considerably higher power levels, around 160 kW and beyond, allowing a mechanical throughput of some 30 times that of gas-stabilized torches [10]. With this enhanced material-handling capability, it becomes practical to consider spray forming of structural shapes and high production, thick thermal barrier coatings, such as those required in abradable seal applications.

At the other end of the thermal spray spectrum is the limitation in obtaining thin films. A major shortcoming of traditional thermal spray technology has been the thinness to which a deposit could be formed. Fine-sized feedstock particles would be required for the production of thin film deposits ($<1\mu m$). To force fine particles into such a flame would require increasing the carrier gas pressure, leading to increased flame turbulence, and thus to a disturbed particle trajectory. These problems, as well as others, could be largely solved by using axial injection of the materials feedstock. New axial-feed plasma gun designs can avoid many of the limitations imposed by nonaxial plasma devices. The operation of one such axial gun is based on electromagnetic coalescence (EMC), allowing four off-centered coalescing 1 kW pilot plasmas to transfer energy to a downwind central anode [11]. The feedstock material, in the form of fine powder or sol–gel fluid, for example, is injected axially using essentially any carrier gas, such as oxygen, hydrocarbon, nitride- or diamond-forming reactive gas (enabling, for example, thermal plasma CVD reactions). The nozzle is designed to preclude internal contact of the feed material with the inert diameter of the anode.

A synthesis of the traditional two-wire electric arc method and the plasma arc is a single-wire plasma spray gun, which uses a single metallic conductor wire to deposit coatings displaying hardnesses and wear resistances of high performance plasma (powder) deposits [12]. Although mainly recognized by and evaluated for the automotive industry, there is no question that wear coatings of high temperature materials could be built up using this and related processes. For example, it would be interesting to evaluate this technique with cored wires (e.g. $NiCr–Cr_3C_2$).

10.3 FEED MATERIALS

A large literature exists on the wide diversity of materials used as thermal spray feedstock [13, 14]. Many different types of ferrous alloys are thermally sprayed, including stainless steel and cast iron. Light metal alloys are thermally sprayed, e.g. titanium and aluminum alloys. Copper alloys, including a wide range of bronzes, are sprayed, as are reactive metals, including niobium and zirconium. Very important both industrially and scientifically, intermetallic alloy powders have recently been plasma sprayed in environmental chambers, for example, on niobium such as oxidation-resistant protective coatings feed as freestanding forms. Wide ranges of cermets are available, that is, blends of metals and ceramics, which when plasma sprayed produce very high strength deposits. Novel ceramics, glasses and various high performance materials have begun to be available. These materials will be the basis for new plasma applications in the future. Table 10.1 is a feedstock classification according to chemistry [15].

Metal alloys and ceramics differ in the manner in which they are fed into thermal spray guns. Metals are employed as powders and wires; ceramics are used as powders and sintered rods (sintered rods fit special proprietary guns), usually 1/4 in. in diameter. Combustion flame spray uses powder or wire of metals and ceramic powders. The HVOF process uses fine powder. In plasma spraying, metal and ceramic powders are used. As indicated above, in plasma spray, very careful control of feedstock powder chemistry and particle size distribution is required.

There have been debates concerning the part played by powder characteristics

Table 10.1 Classification of feedstock according to chemistry and methods of producing thermal spray powders [15]

Classification	Material constituents		
Metallic alloys	MCrAlY (M = Ni, Co, Fe)		
	Co–Cr–W–Co based stellites		
	Ni/Co based self-fluxing alloys		
	Ni/Co–Cr–Mo–Si based Laves phase alloys		
	Ni–Cr–Fe based Inconels		
	Ni–Cr–Mo based Hastelloys		
	Cu based alloys (Cu–Zn, Cu–Al, Cu–Ni)		
Metallic composites		Cladding material	Core material
		Al	Ni
		Al–Mo	NiCr
		Co–Al	NiCrFe
		Ni–Al	FeCr
		Ni	Al
		$Co–Al–Y_2O_3$	Fe–Ni
Intermetallics	Ni–Al		
	Ni–Ti		
	M–Cr–Al–Y types (M = Ni, Co, etc.)		
	Triballoys		
	Superalloys		
Cermets	Ni–Co–Fe based alloys + Al_2O_3, ZrO_2, etc.		
	Ni–Co–Fe based materials + WC–Co, TiC, Cr_3C_2, etc.		
	Al–Ni based materials + graphite (abradables)		
Refractory metals	Carbides of Ti, Zr and Hf (group IVA elements)		
	Nitrides of V, Nb and Ta (group VA elements)		
	Borides of Cr, Mo and W (group VIA elements)		
Nonmetallic hard materials	Oxides of Al_2O_3, Cr_2O_3, TiO_2 and ZrO_2		
	Nonoxides of B_4C, SiC and Si_3N_4		

on deposit properties. Spherical, monosized and chemically homogeneous powder particles are generally preferable to particles having faceted shapes, wide size distributions and a nonuniform distribution of components. Of course, it would be impractical and costly to employ an ideal feedstock in most applications. However, in recent years, it has become possible to approach the ideal, allowing the economical deposition of high performance coatings for an ever increasing range of demanding applications.

10.4 PARTICLE–PLASMA INTERACTION AND PROCESS CONTROL

Over the last decade a number of research laboratories have been examining particle-plasma interactions, employing sophisticated noninvasive diagnostic tools [16, 17]. These include schlieren for turbulence and gas dynamics analysis, laser-strobe vision for particle trajectory and injection studies, enthalpy probe for plasma temperature measurements, laser-Doppler velocimetry for determination of particle velocity, and optical pyrometry for particle temperature analysis. These measurements, coupled with appropriate modeling efforts, have provided important information on the operation of the torch device, mechanics of plasma stabilization, plasma parameters and particle history. This has led to considerable improvements in process technology, ranging from plasma gas selection to particle feed parameters and, thus, has led to improved deposit quality and overall efficiency. Furthermore, the implementation of these tools has led to a considerable reduction in time and effort required to develop processing parameters for new thermal spray powders and applications. This has had a significant impact on the industrial sector as a whole.

We will here discuss only the plasma spray process because most of the other thermal spray processes are not very process sensitive. HVOF is likely to be process sensitive; however, little at this stage is known about its dynamics. This state is rapidly changing due to the needs of industry for HVOF [18]. In the plasma spray process, the interaction between a solid particle and the plasma flame is of vital importance because it should result in proper melting of the feedstock powder. The particle, on being introduced, must collect sufficient kinetic energy in order to accelerate to some velocity approaching that of the flame, and to absorb sufficient enthalpy from the flame to melt. Ideally, the plasma effluent experiences no turbulence and the particle, on entering the flame, causes no perturbation to the flame. But in fact, the flame is generally nonlaminar, and the carrier gas, which drags along the particle, does disturb the flame, distorting it significantly. In reality, thermal plasmas are very different from the ideal, making model building so difficult. But it is useful to consider the ideal nontransferred DC thermal plasma torch because it will give us an agenda for discussion of more complex processes.

An ideal plasma gun should have the ability to create the following conditions for plasma spraying [14]:

1. Particle velocity should be sufficient to achieve a highly dense deposit without leading to an 'explosion' of the particle on impact.
2. The particles should all have the same uniform velocity upon impact.
3. The particles should all be uniformly heated.
4. The particles should be fully molten or plastic without significant vaporization and without undesired chemical reactions.
5. The plasma should be stable without significant variations in power.

10.4.1 Plasma spray process control and modeling

As indicated above, the microstructure of a plasma spray coating will depend on the starting material and its particle size distribution, as well as processing parameters. Processing parameters include plasma power, plasma gas composition, pressures and flow rates, powder injection details and carrier flow and torch–substrate distance. These parameters are sometimes interconnected in complex ways, leading to unavoidable cross-terms in the process of parameterization. They can be independent, making statistical analysis more straightforward. This subject has been covered rather extensively in the papers contained within the annual proceedings of the National Thermal Spray Conference of ASM International [19]. Process control is becoming more common in plasma spraying of high performance coatings. A clear goal is to achieve on-line, feedback control of the operation. This requires a much more detailed understanding of the process.

Accurate and comprehensive *in situ* measurements of the plasma spray process are difficult and sometimes impossible, leading to the need for process modeling. The problem is complex. In the typical DC plasma spray system, powder particles are injected at approximately right angles into a high velocity gas flame. The powder carrier gas increases turbulence within the flame and also disturbs its temperature distribution. The ideal particle becomes entrained in the flame, rapidly increases in temperature, melts (without excessive vaporization) and impacts on the substrate, where rapid solidification occurs.

In order to formulate models for the above particle–plasma interactions, simplifying assumptions are made. These involve plasma–particle momentum and plasma–particle heat transfer, heat conduction within the particle, particle vaporization and, perhaps, spallation. Particularly important considerations are particle trajectory and temperature. It is clear that particle injection strategies will significantly influence the above interactions. Pfender, in a series of papers, has reviewed these issues [20].

Closely connected with modeling is the need for diagnostics of the particle within the plasma flame. Particle velocities are measured using laser methodologies. And effective particle temperatures can be estimated from atomic spectroscopy measurements [21]. More recently, nanopulsed laser strobes have been introduced to image particle trajectories from injection to impact on the substrate [22]. This represents real-time process control, the ultimate goal of these modeling and diagnostic studies.

10.5 DEPOSIT FORMATION DYNAMICS

Thermal spray deposits are comprised of cohesively bonded splats which result from high rate impact and rapid solidification of a high flux (millions of particles per cm^2/s) of flame-melted particles with sizes of $10-100\,\mu m$. A schematic representation is shown in Fig. 10.5, where splat-like structures are entwined in complex arrays. The physical properties and behavior of

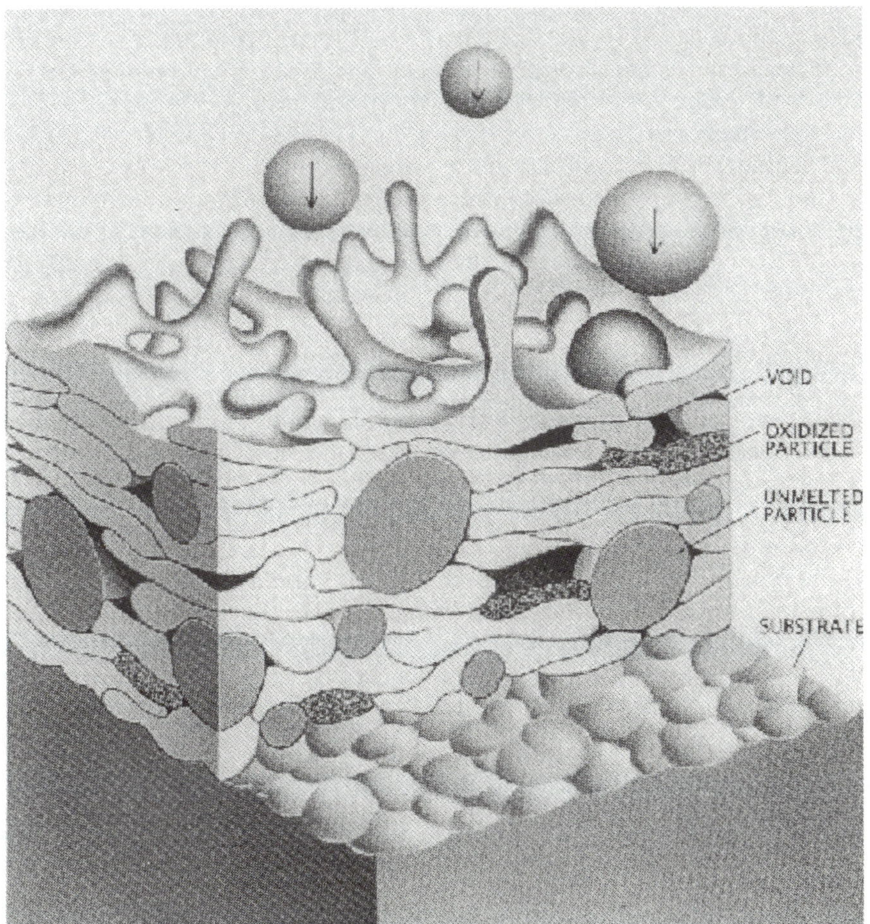

Figure 10.5 A schematic overview of the thermal spray process.

such a deposit will be expected to depend on the cohesive strengths among the splats, the size and morphology of the porosity, the occurrence of cracks and defects and, finally, on the ultrafine-grained microstructure within the splats themselves.

The microstructures of thermal spray deposits are ultimately based on the solidification of many individual molten droplets. A splat results when a droplet of molten material, tens of micrometers in diameter and melted in the flame, strikes a surface, flattens out then solidifies. The collection of these splats forms the deposit. There are numerous considerations relative to the dynamics of deposit evolution during thermal spraying. The mechanistic or physical aspects of splat formation deal with the spreading of the molten droplet, interactions with the substrate, etc. These characteristics are affected by the temperature of the splat, the splat viscosity, its surface tension and other considerations. Splat morphology will depend on a variety of things, the most important of which are particle velocity, temperature, diameter and substrate surface profile. Further considerations involve the physical properties of the splat, which deals with cooling rate, solidification criteria, nucleation and growth of crystals, phase formation, etc. These aspects of splat formation and solidification are complex and interrelated.

Houben has described in considerable detail certain physical aspects of splat formation through heat transfer and mechanical models [23]. He has identified the various types of splat morphologies, described broadly as 'pancake type' and 'flower type'. He has further shown that splat morphology is affected by the velocity of the impinging droplet. Increased velocity leads to enhanced flattening and spreading of the droplet. Montavon *et al.* further explored the influence of spray parameters on splat formation by utilizing shape factors for the splats [24]. Figure 10.6 shows three shape factors defined as follows: *equivalent diameter* (the diameter of a circle with the same area as the selected feature); *circularity* (the elongation of the selected feature; unity is a perfect circle); and *sphericity* (the importance of peripheral material projections at the impact point; unity is for absence of such projections). Using optical microscopy and image analysis Montavon *et al.* determined the influence of spray parameters and impact angle on the splat characteristics for vacuum plasma sprayed, Ni-based alloy [24]. It was further shown that the splat shape factors, especially as influenced by spray angle, have a strong effect on deposit characteristics, such as porosity, deposit efficiency and microhardness.

The cooling and solidification rate for thermal spray deposits are dependent on the solidification of individual splats and substrate conditions. It has been suggested that cooling rates achieved in air plasma spraying are on the order of 10^7 K/s [25, 26]. Moreau *et al.* have used two-color pyrometric techniques to directly measure temperature profiles and cooling curves for solidifying molybdenum and niobium particles during plasma spray deposition [27, 28]. They reported cooling rates on the order of 10^8 K/s for solidification on

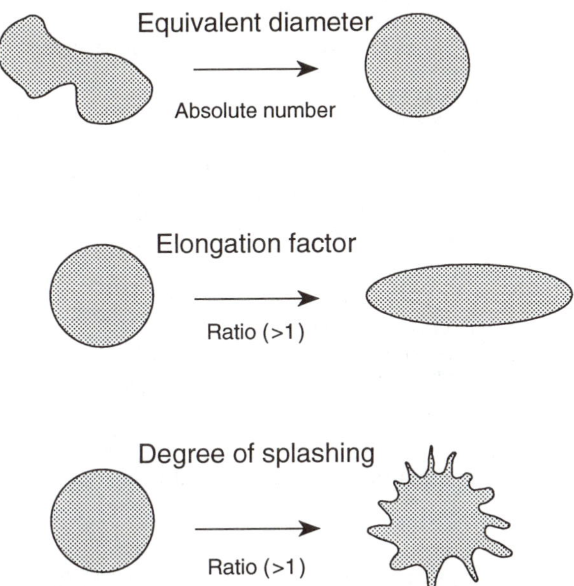

Figure 10.6 Shape factors associated with formation of a thermal spray splat [24].

conducting substrates. This clearly indicates the splats undergo ultrarapid quenching thermal spraying. Table 10.2 shows the solidification parameters of direct relevance to structure which can be calculated from these cooling rates values [29].

The high solidification rates and the G_l/R (degree of constitutional supercooling) values suggest a plane front growth as the likely mode of solidification. Figure 10.7 shows a model for the formation and solidification of a single splat [29]. Each column within a single splat is produced by plane front solidification, arising from multiple nucleation sites within the splat–substrate and splat–splat interfaces. Figure 10.8 is a cross-sectional

Table 10.2 Solidification variables derived from cooling rates [29]

Material	Average cooling rate (K/s)	Heat transfer coefficient (W/m^2)	Solidification rate, R (cm/s)[a]	G_l/R (K s/cm^2)
Nickel	7×10^7	1.4×10^6	46	2.2×10^4
Aluminum	1.5×10^8	3.5×10^6	15	3.7×10^4
Molybdenum	2×10^8	1.8×10^6	150	1.2×10^4

[a] Assuming no isothermal delay.

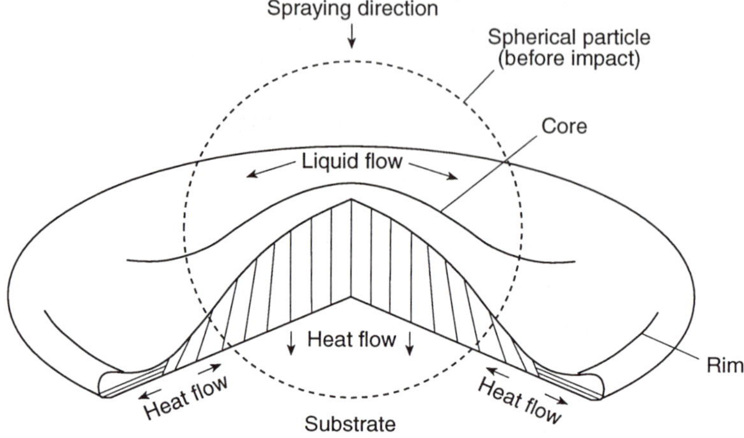

Figure 10.7 A model depicting the solidification and microstructure development of a single splat [29].

micrograph of a fractured plasma spray alumina deposit, illustrating the columnar grain morphology. Similar microstructures have been observed in plasma spray metals and other thermal spray materials. Further, it has been shown that texture is likely in these columnar solidified deposits, because under high solidification rates, it is expected that the fastest crystal growth direction will prevail. X-ray diffraction results from the back face of the

Figure 10.8 A fractured cross section of a plasma spray ceramic coating showing the columnar grain morphology.

atmospheric and vacuum plasma spray nickel, Ni–5 Al, and molybdenum alloy deposits show strong $\langle 100 \rangle$ type texture [30].

10.6 MATERIALS SCIENCE OF THERMAL SPRAY DEPOSITS

The model for microstructure development during thermal spray deposition suggests heterogeneous nucleation of a solid phase followed by columnar growth. The high cooling rates (large solidification rates) suggest massive or partitionless solidification via supercooling. This, associated with large undercooling, is expected to result in considerable solute trapping and the formation of metastable phases. Metastable phases are a frequent occurrence in thermal spray materials. Researchers have reported solid solubility extension in numerous plasma spray systems, namely Sampath in Ni–Cr [31], Krishnanand and Cahn in Al–Cu [32], Bhat *et al.* in Fe–C [33] and Sampath and Wayne in Mo–C [34]. From an earlier study, the present authors have shown complete supersaturation of chromium in nickel when an NiCr eutectic composition is plasma sprayed [35]. Metastable and amorphous phases are observed commonly in plasma spray ceramics, namely γ-alumina, amorphous cordierite, amorphous hydroxyapatite, tetragonal zirconia, etc. In addition to formation of metastable phases, thermal spray deposits typically show nanometer to submicron grain size distributions

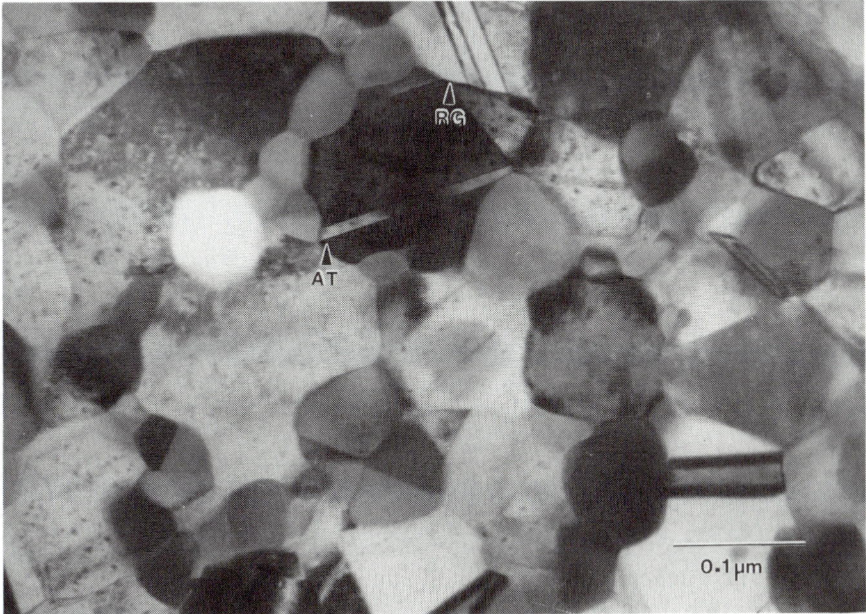

Figure 10.9 Transmission electron micrograph showing submicron grain size in a vacuum plasma spray Ni-based alloy [36].

associated with rapid solidification, Fig. 10.9 [36]. All of the above features have interesting implications for physical and mechanical properties of the deposits.

10.6.1 Coating testing and characterization

As indicated above, thermal spray deposition involves the quasi-continuous rapid deposition and solidification of molten droplets. This process results in microstructures with many defects, including porosity and, in the case of metals plasma sprayed in air, oxidation. There are industrial standards which specify limits for these defects, but it is difficult to control and to characterize them in sufficient detail to allow the establishment of realistic standards and specifications. These imperfections are generally determined by optical metallography, although SEM is occasionally employed for more detailed analysis. In addition, since the buildup of the coating involves rapid solidification processing there will result a significant percentage of metastable phases, which will commonly revert to more equilibrium structures at elevated temperatures. Thus, phase analysis is of the greatest importance, using X-ray diffraction methods, for example. Listed below are the characterization methods currently employed and those with potential for incorporation in the technology.

Microstructure

Metallographic analysis is usually carried out on coating cross sections in order to evaluate imperfection structure and to achieve a detailed examination of the coating–substrate interface. Both optical and scanning electron microscope (SEM) studies are common. For optical studies, one must be careful during polishing not to apply too great a pressure to avoid pulling out particles, otherwise the porosity value would be too large (see below). Several strategies help to avoid pullout particles in ceramic coatings, which are most susceptible to this weak interbonding problem. One approach is to infiltrate the coating with an epoxy (vacuum impregnation); this effectively strengthens the coatings and maintains interparticle integrity during grinding and polishing. These methods are discussed in the literature [37].

Porosity

Voids, cracks and coating decohesion are usually detected by metallographic techniques. However, two-dimensional cuts of a complex, defect-ridden sprayed microstructure are extremely difficult to analyze and the true three-dimensional character of the microstructure is complex to ascertain. Porosity can be viewed generically as the absence of material within the coating. The pores that are observed are nominally disk-like and lie between

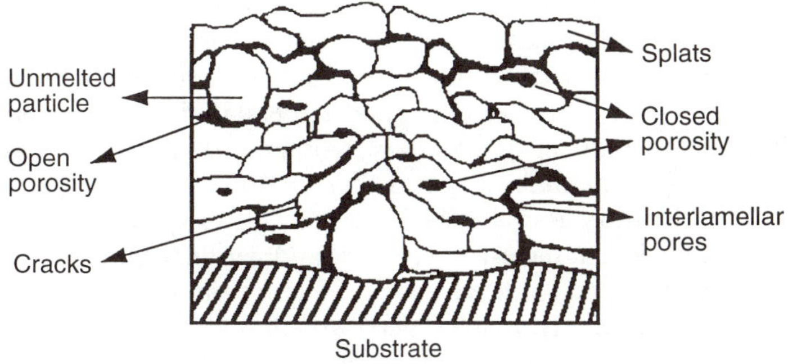

Figure 10.10 A model for porosity in plasma spray ceramic coatings [38].

the splats, forming during rapid solidification, Fig. 10.10 [38]. There are also vertical crack-like 'pores' that are both within the splats and positioned in torturous orientations among them. A reasonably full characterization has been accomplished through a combination of metallography and small-angle neutron scattering (SANS) [38], an important technique for elucidating porosity. In this method, long wavelength (cold) neutrons transit a thick specimen (millimetric or centimetric) and the scattering from internal pores is detected by a sensitive area detector. The scattered neutron profile can be analyzed to yield information on internal specific surface (area/volume), which is, in fact, related to the total surface of the pores sampled by the neutron beam. Preliminary SANS experiments on plasma spray alumina ceramics, performed at the High Flux Beam Reactor (HFBR) at Brookhaven National Laboratory and at the National Institute for Standards and Technology, are very promising [39]. Using various profile analysis methods, it has been possible to demonstrate that SANS can be used in a relatively practical manner to give information on pore size distribution and orientation.

The porosity in question is found to be either surface connected or totally enclosed. Surface-connected porosity can be examined through techniques such as mercury or helium porosimetry, but enclosed pores are not greatly accessible by traditional means. SANS, however, can be used to 'image' these pores. Plasma spray ceramics, for example, are comprised of flat, pancake-shaped particles (splats) which are 'welded' together around their peripheries [40]. Consistent with thermal conductivity measurements, the splats are essentially separated by lenticularly shaped pores with these tenuous connection points [41]. Both the splats and the pores are thought to lie normal to the direction of particle impact. SANS is ideally suited to study this pore structure. Thus, the pore structure, which may be so voluminous (perhaps 10 vol% and greater) as to be beyond the limit of percolation, presents a highly anisotropic array.

More commonly, porosity is determined by water displacement and mercury intrusion techniques. Water displacement measures all of the pores but mercury intrusion detects only those that are surface connected [42]. Both techniques allow a reasonably good understanding of how pores and cracks form in thermal spray materials. It is important to note that metal coatings generally contain fewer pores than ceramic coatings.

A key consideration is the dependency of porosity on spray ambient environment, powder characteristics (i.e. size and size distribution) and plasma spray parameters (e.g. power level, gas flow features and spray distance). The atmosphere in which spraying is carried out will have a significant influence on, for example, oxidation of metals, leading to greater porosity. Therefore, an inert atmosphere will decrease the porosity level in most metal coatings. As the spray ambient pressure decreases, the flame velocity goes up, along with the particle velocity, leading to a higher impact velocity of the molten particles onto the workpiece. This will result in flattening of the splats to a far greater extent than for atmospheric spraying, achieving a denser, lower porosity deposit. Relative to the torch spray parameters, it can generally be said that when the parameters are such, as the particle velocity increases, the porosity decreases. The same kind of generalization can be made for feedstock particle size: the larger the particle size, the greater the average porosity.

A common feature of plasma spray ceramic deposits is porosity and sometimes microcracks. These imperfections can be both beneficial and deleterious to the performance of ceramic deposits. The pores can act as crack stoppers and thus enhance fracture toughness, whereas the cracks can give rise to either transformation toughening (e.g. for partially stabilized zirconia) or to fracture initiators. Whatever the case, it is imperative to be able to characterize the imperfections in plasma spray ceramic bodies.

The usual ceramographic means of characterizing porosity and microcrack depend on proper polishing of cross sections, with subsequent examination using an inverted-stage reflection microscope. Image analysis techniques are applied which, when properly executed, can yield important information. This, in conjunction with intrusion methods (e.g. mercury intrusion porosimetry) and possibly permeability methods and SANS, can yield a reasonably good picture of the defect structure. Unfortunately, there are numerous well-documented ambiguities to contend with in the interpretation of these measurements. For example, porosimetry only detects surface-connected pores and cracks, but a large portion of the defects are within the sprayed body, not connected to the surface. Interior defects would presumably be detected by ceramographic techniques, but as discussed above, they very much depend on how the specimen is cut, mounted and polished; particle pullouts are an all too frequent occurrence.

An advanced method of evaluating the defect structures of thermal spray ceramics (HVOF and plasma) has been exploited by workers at Stony Brook working jointly with the Department of Applied Science at Brookhaven

National Laboratory. In a program sponsored by the General Electric Gas Turbine Division, high intensity X-ray computed microtomography was employed to image voids in plasma spray ceramic deposits [43]. These measurements, obtained at the X-21 beam line of the National Synchrotron Light Source, allowed an accurate determination of the overall deviation from theoretical density and the detection of voids and cracks in the deposit. Future studies should be aimed at relating these images to thermal spray processing parameters.

Phase analysis

As in any materials analysis study, thermal spray coatings are commonly subjected to phase determination using X-ray diffraction. In the case of spray coatings, however, the deposits have undergone various levels of rapid solidification and thus metastable phases are common. In addition, as is commonly observed in studies of rapidly solidified materials, amorphous phases are readily observed. Both of these considerations will add complexities to interpretations of diffraction patterns.

Adhesion

From an applications point of view, coating–substrate adhesion will be one of the most important properties to be considered. A recent review concerning this question has been written by Berndt and Lin [44]. The American Society for Testing and Materials (ASTM) defines adhesion as 'the state in which two surfaces may consist of valence forces or interlocking forces or both' [45]. However, the term *adhesion* requires special definition for the purposes of thermal spray coatings because they can be considered as 'composites' at the microstructural level. The basic bonding mechanisms which have been defined for thermal spray coatings onto metal substrates can be categorized into three major groups: (i) mechanical interlocking and anchoring; (ii) metal–metal or ceramic–metal bonding; and (iii) chemical bonding (formation of intermetallic compounds with the substrate).

The specific thermal spray will influence the microstructure of the coating and, therefore, it can be inferred that the adhesion strength of the deposit will vary with the process. Factors affected by the spray parameters, such as the size and distribution of porosity, oxide content, residual stresses and macro- or micro-cracks, influence the performance and failure of the coating system. Substrate preparation (e.g. grit blasted, cleaned, preheated) also plays an important role in determining the properties of the coating.

One difficulty with any mechanical property assessment of coatings is how to attach a loading device to the coating without influencing the property that is being measured. The adhesion measurement can be taken as a control parameter that can be used as a guide to optimization of the many process parameters which are associated with thermal spray. Usually, tensile adhesion tests are widely performed as quality control inspections [46]. A fracture mechanics approach is based on defining adhesion in terms of a stress intensity factor K or strain energy release rate G. Methods of measurement include a double cantilever beam (DCB) test, a double torsion test, a bending test and a hardness indentation test. The scratch test is being considered within the biomedical industry to examine the adhesion of plasma spray hydroxyapatite to orthopedic prostheses. Acoustic emission has also been studied to asses the 'crack density function', i.e. a product of the number of cracks and crack size. And various means of ultrasonics testing are potentially nondestructive methods for examining the adhesion strength of coatings. There are many other qualitative and quantitative methods that can be employed to evaluate adhesion. These techniques may be used for both fundamental understanding of coating performance or as tests for quality control. However, there is still no ideal adhesion test which can satisfy all requirements, so modifications to existing techniques and newly designed methods can further improve adhesion tests.

10.7 THERMAL SPRAY COATING APPLICATIONS

10.7.1 Thermal spraying for wear control

Thermal spray coatings find extensive applications in wear and corrosion resistance for industrial machinery. Some of the materials include metals such as Mo-based alloys, aluminum bronze and stainless steels; cermets, such as WC–Co and Cr_3C_2–NiCr; and ceramics, such as Al_2O_3–TiO_2 and Cr_2O_3. The coatings are subjected to a variety of abrasive, adhesive and erosive wear conditions and, therefore, an understanding of the tribological behavior of the deposits in relation to properties is critical in evaluating their performance. The lamellar nature of the deposit results in nonuniform wear with little or no incubative processes. Thus, the tribological behavior of the coatings is closely linked to the microstructure. Inhomogeneities in mechanical properties, such as hardness, anisotropy in toughness and varying surface profiles, all lead to wide variations in wear behavior. However, these variations can be included to develop predictive relationships.

Alloys based on refractory metals, especially Mo-based composites, have been used as thermal spray coatings for numerous adhesive wear environments. Flame spray Mo wire and plasma spray Mo pseudoalloys are used in the automotive, pulp and paper, and aerospace industries. For example, Mo-based coatings are used in automotive piston rings to provide scuff resistance and reduce adhesive sliding wear [47]. The beneficial frictional behavior of Mo-based coatings has been identified by Overs *et al.*, who attribute the lowered friction to the formation of a surface film on molybdenum [48]. A detailed review of the wear mechanisms of thermal spray Mo-based coatings is provided by Wayne, Sampath and Anand [49].

WC–Co and Cr_3C_2–NiCr based cermets are used extensively as thermal spray coatings for protection against abrasive, fretting and erosive wear environments. The single largest user of such coatings is the aircraft engine industry, where turbine compressors, midspan stiffeners and other components are coated to reduce wear and increase component life. Carbide coatings have been applied to aircraft engine components since the early 1950s using the detonation gun (D-gun) and plasma spray processes. The D-gun process has been proven in its ability to produce dense, well-bonded, hard deposits which enhance the component life by reduced wear. These coatings are applied in numerous sectors, such as printing, petrochemicals and machinery construction. The advent of the HVOF process has resulted in coatings which are equivalent, or sometimes superior, to those obtained by D-gun coatings. Furthermore, the HVOF process offers considerable versatility in mobility and in handling and, therefore, for applications. This has significantly enhanced the commercial potential for wear-resistant thermal spray coatings.

In a recent structure–wear property relationship study by Wayne *et al.*, several interesting observations were reported [50]. WC–Co thermal spray coatings processed by plasma and HVOF techniques show large anisotropy in indentation fracture toughness. Typically, the toughness was 10:1 between directions perpendicular and parallel to the lamellae. This has important implications on wear behavior. It was also found that abrasive and erosive wear can be empirically related to hardness, indentation fracture toughness and a microstructural parameter as follows:

$$\text{Wear resistance } \alpha K_{Ic}^{3/8} H^{1/2} \left(\frac{V_f^{Co}}{1 - V_f^{Co}} \right)$$

where K_{Ic} is the indentation fracture toughness, H is the hardness and V_f^{Co} is the volume fraction of cobalt.

Figure 10.11 compares the abrasion and erosion wear resistance for the WC–Co coatings from various thermal spray processes [50]. For reference, the wear resistance of sintered compositions are also included. It is clear that the HVOF coatings have excellent wear resistance, approaching that of sintered WC–Co materials.

Recent developments in the use of thermal spray coatings for wear control include coatings on aluminum alloy cylinders for internal combustion engines [51, 52]. Although this technology is still in the development stage, all of the preliminary results indicate that thermal spray coatings do perform well even under severe conditions. Such technologies will not only increase the utilization of thermal spray coatings, but will also positively reflect on the role of thermal spray technology in advanced manufacturing.

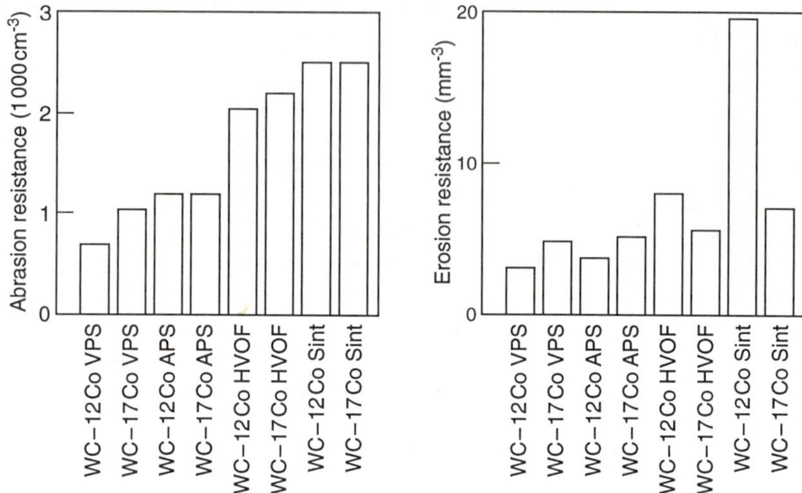

Figure 10.11 Abrasion and erosion resistance of thermally sprayed WC–Co coatings in comparison to their sintered counterparts [50].

10.7.2 Thermal spraying for corrosion protection and infrastructure maintenance

Thermal spraying a metal onto the strucure is used to prevent corrosion on steel. Active metals, such as zinc, aluminum and zinc–aluminum alloys, are normally chosen for this purpose. These inexpensive metals and alloys are readily in wire and powder form, which makes them suitable for spraying large structures. When sprayed to thicknesses between 5 and 20 mils (0.13–0.51 mm), they provide cathodic protection (CP) to the steel. CP is achieved when an external source is providing electrons to a region in which corrosion would normally take place. As in the case of corrosion protection of steel bridges, the coatings of Zn, Al and Zn–Al alloys sacrificially corrode, providing the electrons needed to protect the steel; this type of protection is specifically called *galvanic protection*. For concrete bridges CP is also used to protect the reinforcement steel, but the electrons come from a different source; instead of obtaining the electrons from the sacrificial metal, the electrons are provided by an external DC power source. The Zn, Al and Zn–Al alloy coatings are used as a conductive layer; this type of protection is specifically called *impressed current cathodic protection*. By using cathodic protection, the corrosion protection extends into regions of the coating which have been damaged or regions which could not be sprayed.

Other parts of the infrastructure which could greatly benefit from thermal spraying include underwater tunnels, concrete and steel gates on a river lock and gate system, river dams, offshore oil rigs, steel piping which runs underground or above ground, water towers, railroad cars and ships. Thermal spraying can be used in harsh environments to protect many engineering components made of ferrous alloys.

Barrier protection is another type of corrosion protection which can be achieved through thermal spraying. The thermal spray coating acts as a barrier between the outside environment and the steel. Compared to cathodic protection, thermal spray barrier coatings are better because they do not act sacrificially, but on the other hand, they do not give protection to uncoated regions or regions where the coating becomes damaged. Stainless steel has a naturally formed protective outer layer of chromium oxide. Other materials which produce a stable oxide layer are alloys containing molybdenum, chromium, nickel, ceramics and polymers.

10.7.3 Thermal spraying of polymers

Thermal spraying of polymers is gaining increased attention because of its ability to apply polymer coatings having a range of thicknesses, 0.002–0.25 in. (0.05–6.35 mm), onto a wide variety of substrate materials. Intrinsic polymer properties of interest include high chemical resistance, high impact resistance and high abrasion resistance, enabling applications to include plow blades, pump impellers, tank linings, external pipe coatings, structural steel coatings, light poles, transfer chutes and vacuum systems.

Fabrication of compound coatings with interesting chemistries is made possible by the incorporation of a second phase of polymer, metal or ceramic. Polymer matrix composites, (PMCs) are used as high temperature adhesives, in fuselage skins to reduce fatigue cracking, in engine components for increased propulsion, and in any structure requiring higher strength-to-weight ratios and lower manufacturing costs.

Thermal spray polymer coatings are generally achieved by the melting and high velocity deposition (in some cases atomization deposition) of polymeric feedstock material. The polymer powders can be prepared by freezing the feedstock followed by a grinding operation. One can dissolve a polymer in a suitable solvent to obtain a suspended solution containing the dispersed polymer chains. When the polymer powder is dried, UV stabilizers, special additives and pigments are melt compounded with the base polymer and subsequently propelled through a flame. Upon melting, they become molten droplets and are transported to the substrate. As the molten particles impinge on the substrate, they coalesce and flow to form a homogeneous coating. Polymers that have been successfully sprayed to date include polyethylene, polypropylene, polyester, polyamides, polyvinylidene fluoride, polytetrafluoroethylene and ethylene methacrylic acid copolymer [47].

The properties of the polymer coating depend upon the chemical nature of the polymer and the influence of process conditions. Variations in particle size and the distribution of molecular weights will affect the melting points of the polymer powders, which in turn will affect the coating's surface characteristics. For instance, fine particles will pyrolize whereas larger particles may remain unmelted. Substrate preheating can be an important consideration because it will enable the flow of impacting polymer droplets. However, excessive heating can cause polymer outgassing, resulting in increased porosity.

The performance of thermal spray coatings also depends upon the heat output applied to the polymer as well as the polymer's inherent properties. Overheating the polymer will lead to a reduction in coating performance before any visual signs of thermal degradation are evident. The optimum thermal window is polymer specific. The upper temperature limit is controlled by the thermal degradation sensitivity, and the lower limit is a function of the melting characteristics. Incorporating polar groups into a polymer increases the adhesion to polar substrates due to strong hydrogen and ionomeric bonding. However, excessive thermal input to such a material and to thermoplastics in general will increase the degree of cross-linking to the point of delamination from the substrate.

One must optimize thermal spray process parameters coupled with a knowledge of polymer chemistry and process dynamics in order to spray successful coatings that encompass an engineering spectrum of applications.

10.7.4 Thermal spraying of biomaterials

The alloys used for implantation in the human body were adopted from the aircraft industry where the performance of materials such as 316L stainless steel, Co–Cr and Ti–6 Al–4 V alloys are well documented. Various types of porous surfaces were manufactured to aid bone ingrowth, known as osseointegration. Among other things, the functionality of the implant was limited by its ability to bond with bone.

Thermal spray technology was the obvious choice for coating implants with Ti–6 Al–4 V and pure titanium. Hip prostheses were plasma sprayed with titanium to provide a larger surface area for bonding. Dental implants with threaded sections could interdigitate with bone, further increasing the attachment strength. Other techniques used to roughen the surface did not prove to be as good. The bone grew up to the implant, but was separated by a thin nonbonding layer. Bioactive glasses were developed to enable a direct bond with the surrounding bone. Due to their brittle nature, they were limited to percutaneous devices or used in the middle ear, and also for bacterial seals. Calcium phosphates, in particlar hydroxyapatite, were synthesized and applied as coatings. Hydroxyapatite, already found in the body as the mineral phase of bone, was believed to be intrinsically biocompatible, hence the interest in this material. Thermal spraying of hydroxyapatite, which has a rich thermal chemistry, results in a coating with by-products such as tricalcium phosphate of the α- and β-phases, calcium oxide and an amorphous phase. Nevertheless, hydroxyapatite deposited to a thickness of 50 μm is used widely in hip, knee, tooth and other prostheses where an intimate bond with bone leads to good fixation of the implant.

Hydroxyapatite-coated prostheses on the market differ in phase composition and microstructure. The behavior of the coated object depends upon the composition, porosity, thickness, residual stresses, roughness, size of lamellae and mechanical properties of the coating. Hydroxyapatite is not completely understood, but its ability to hasten the growth of bone up to the impant is recognized as being an advantageous factor. The capacity of hydroxyapatite to encourage faster fixation is defined as osteoconduction. The dissolution gives rise to a supersaturation of the ionic species at the implant site, which causes epitaxial growth of crystals from the crystalline areas of the coating into the bone. Implantation of a hydroxyapatite-coated implant in porous bone will lead to a higher success rate than using an uncoated implant because of earlier and possibly better fixation with bone, however, the location of the amorphous phase, being more resorbable, can lead to coating delamination if it is found adjacent to the substrate. In the presence of bone, a gap between the coating and the bone can be bridged. Studies are still being performed to discover if hydroxyapatite has a more powerful capacity, known as osteoinduction. This is the capacity of bone to grow upon a hydroxyapatite material in an ectopic site of the body (i.e. where there is no bone in the vicinity of the implant).

10.7.5 Thermal barrier coatings (TBCs)

Thermal barrier coatings (TBCs), such as partially stabilized zirconia, are applied by plasma spray to thicknesses of 0.005 in. (125 μm) and up, depending on the application. This class of ceramic has a very low thermal conductivity, effectively insulating the underlying superalloy substrate from the high temperature environment. For gas turbine applications, the thickness of the TBC will be 0.005–0.010 in. (125–250 μm). At this thickness range, for hot sections operating at temperatures around 1000 °C, the surface of the insulated superalloy component can be reduced by approximately 100 °C, enabling extended lifetime at the turbine's operating temperature or allowing the turbine to function at a higher, more efficient temperature. The sprayed TBC has a porosity of a few vol% which contributes further to a decrease in thermal conductivity. The remarkable feature of the TBC is its survivability, both in thermal shock and in the difficult environmental conditions within the gas turbine. The tenacity of the TBC originates from porosity, microcracks and inherent toughness of the particular ceramic.

It is widely recognized that TBCs can have process-induced and application-related stresses which can limit the utility of the ceramic under thermal shock conditions, for example. Using two or more powder feeders, it is possible to spray or to grade from one feeder to the next, producing a graded deposit, that is, from all metal at the substrate to all ceramic facing the outside environment. This approach is highly effective for reducing interfacial stresses and ensuring enhanced adhesive bonding in high performance applications. The grading can be extended to larger distances (e.g. millimeters), yielding functionally graded materials. Another approach to composite deposits is to employ a composite powder, perhaps produced by standard powder agglomeration methods.

10.8 THERMAL SPRAY AND ADVANCED MANUFACTURING

Thermal spray technology is poised to experience a new recognition on the part of the general engineering community. Although active for some 30 years, principally in vital repair and maintenance for heavy industry, and in servicing aerospace with a diversity of protective coatings, the field has up until now played a limited role in manufacturing. But this is changing rapidly. Numerous industries, in recognition of thermal spray's versatility and inherent economics, have introduced the technology into the manufacturing environment. Automotive companies plasma spray alumina onto millions of alternators each year, and there is a strong thrust towards thermal spraying of wear resistant cylinder coatings on aluminum engine blocks. Insulated metal substrates are being formed with plasma spraying for use in the microelectronics industry. And diesel engine manufacturers, in producing fuel efficient, low heat rejection engines, are using plasma spraying to form thick thermal barrier coatings onto piston crowns. Manufacturers the world over are employing plasma spray to produce solid oxide fuel cells so they can be economically competitive for producing power.

There are numerous applications of thermal spray which are now evolving and which will represent a sea change for the technology, with many concomitant opportunities and challenges. But although manufacturing appears ready for thermal spray, it may be that thermal spray is not yet prepared to enter a high production environment, in which performance, reproducibility and costs are the controlling factors. Until recently, thermal spray has developed mainly by trial and error, nevertheless, yielding some astounding results and helping it to become the mainstay of any number of industries such as aerospace and paper manufacturing. Now its versatility (virtual freedom in materials selection) and its inherent process complexity (large numbers of interacting process parameters) are starting to limit an Edisonian approach; they require that a concerted effort be made to place the technology on a sound scientific basis.

10.9 SUMMARY

Thermal spray science and technology is now beginning to enter the vocabulary of materials science. This was not the case one or two decades ago when the field was a very applied activity, well behind the neighboring technology of welding, already widely appreciated by the materials community. Things have now changed, and there is a worldwide excitement about the future possibilities of thermal spray. But much needs to be done to gain a fundamental understanding of the phenomena which govern its complex processes.

REFERENCES

1. Thorpe, M.L. (1993) *Adv. Mater. Process.*, **143**(5), 50.
2. Herman, H. (1988) Plasma spray deposition processes *MRS Bull.*, **13**, 60–7.
3. Kreye, H. (1991) in *Proceedings of the 2nd Plasma-Technik Symposium* (eds P. Huber and H. Eschnauer), Lucerne, Switzerland, Plasma-Technik AG, Vol. 1, p. 39.
4. Greenlaw, R. (1995) M.S. Thesis, SUNY, Stony Brook, New York.
5. Herman, H. (1988) *Sci. Amer.*, **256**(9), 112.
6. Meyer, P.J. (1991) in *Proceedings of the 2nd Plasma-Technik Symposium* (eds P. Huber and H. Eschnauer), Lucerne, Switzerland, Plasma-Technik AG, Vol. 1, p. 229.
7. Jackson, M.R., Rairden, J.R., Smith, J.S. and Smith, R.W. (1981) Production of metallurgical structures by rapid solidification plasma deposition. *J. Metals*, **33**, 23–7.
8. Sampath, S. and Herman, H. (1993) *J. Metals*, **45**(7), 42.
9. Tiwari, R., Herman, H., Sampath, S. and Gudmundsson, B. (1991) Plasma spray consolidation of high temperature composites. *Mat. Sci. Engng*, **A144**, 127–31.
10. Chraska, P. and Hrabovsky, M. (1992) in *Proceedings of International Thermal Spray Conference* (ed. C.C. Berndt), Orlando FL, ASM International, Metals Park OH, p. 81.
11. Marantz, D.R. (1991) Electromagnetically coalesced multi-arc plasma torch with true axial powder feed, in *Proceedings of the 1990 National Thermal Spray Conference*, Long Beach CA, ASM.
12. Kowlaksy, K.A., Marantz, D.R. and Herman, H. (1992) Characterization of coatings produced by the wire-arc-plasma process, in *Proceedings of the 4th Thermal Spray Conference*, Pittsburgh PA, ASM International, Metals Park OH.
13. Herman, H. (1991) *Powder Sci. Technol.*, **9**, 187.
14. Berndt, C.C. *et al.* (1992) *J. Thermal Spray Technol.*, **1**(4), 341.
15. Houck, D.L. (1980) *Modern Dev. Powder Metall.*, **12**, 485.
16. Fincke, J., Swank, W.D. and Haggard, D.C. (1993) *J. Thermal Spray Technol.*, **2**(4), 345.
17. Pfender, E. (1988) *Surf. Coat. Technol.*, **34**(1), 1.
18. Hackett, C.M. and Settles, G.S. (1994) Turbulent mixing of the HVOF thermal spray and coating oxidation, in *Proceedings of 7th National Thermal Spray Conference*, ASM International, Metals Park OH, p. 307.
19. ASM International (1990–1994) *Proceedings of National Thermal Spray Conference*, ASM International, Metals Park, OH.
20. Pfender, E. (1989) *Particle behavior in thermal plasma. Plasma Chem. Plasma Process.*, **9**, 167S–194S.
21. Vardelle, M.A., Vardelle, A, Fauchais, P. and Boulos, M. (1983) Plasma–particle momentum and heat transfer in modeling and measurements. *AIChEJ*, **29**, 286.
22. Agapakis, J. and Hoffman, T. (1992) Real time imaging for thermal spray process development and control. *J. Thermal Spray Technol.*, **1**(1), 19.
23. Houben, J.M. (1988) Ph.D. thesis, Eindhoven University of Technology, Holland.
24. Montavon, G., Sampath, S., Berndt, C.C., Herman, H. and Coddet, C. (1995) Effect of vacuum plasma spray processing parameters on splat morphology. *J. Thermal Spray Technol.*, **4**, 67–74.

25. Moss, M. (1968) *Acta. Metall.*, **16**, 321.
26. Wilms, V. (1978) Ph.D. thesis, SUNY, Stony Brook, New York.
27. Moreau, C., Cielo, P., Lamontagne, M. *et al.* (1991) *Surf. Coat. Technol.*, **46**, 173.
28. Moreau, C., Cielo, P. and Lamontagne, M. (1992) *J. Thermal Spray Technol.*, **1**(4), 317.
29. Sampath, S. and Herman, H. (1995) (in preparation).
30. Sampath, S. and Herman, H. (1989) Microstructure development in plasma sprayed coatings, in *Proceedings of International Spray Conference*, Welding Institute, London, Vol. 1, p. 53.
31. Sampath, S. (1989) Ph.D. thesis, SUNY, Stony Brook, New York.
32. Krishnanand, K.D. and Cahn, R.W. (1976) in *Rapidly Quenched Metals*, (eds N.J. Grant and B.C. Giessen), MIT Press, Cambridge, MA, pp. 67–75.
33. Herman, H. and Bhat, H. (1980) in *Synthesis and Properties of Metastable Phases* (eds E.S. Machlin and T.J. Rowland), TMS, AIME, Pittsburgh PA, pp. 115–37.
34. Sampath, S. and Wayne, S.F. (1994) Microstructure and properties of plasma sprayed $Mo-Mo_2C$ composites. *J. Thermal Spray Technol.*, **3**(3), 282.
35. Sampath, S. (1989) Ph.D. thesis, SUNY, Stony Brook, New York.
36. Sampath, S., Neiser, R.A., Herman, H. *et al.* (1993) A structural investigation of a plasma sprayed Ni–Cr based alloy coating. *J. Mater. Res.*, **8**(1), 78.
37. Diaz, D.J. and Blann, G.A. (1991) Thermal sprayed coatings. *ASM Standardization News*, May 1991, p. 48.
38. Ilavsky, J. (1994) Ph.D. thesis, SUNY, Stony Brook, New York.
39. Ilavsky, J., Herman, H., Berndt, C.C. *et al.* (1994) Porosity in plasma sprayed alumina deposits, in *Thermal Spray Industrial Applications*. Proceedings of NTSC (eds C.C. Berndt and S. Sampath), ASM International, Metals Park OH, 1994, pp. 709–14.
40. Ilavsky, J., Allen, A.J., Long, G.G. *et al.* (1995) Anisotropy of the surfaces of pores in plasma sprayed alumina deposits, in *Proceedings of the 1995 ITSC*, Kobe, Japan, May 1995 (in press).
41. Hollis, K.J. (1994) Unpublished work, Sandia National Laboratories.
42. Ilavsky, J., Berndt, C.C., Herman, H. *et al.* (1993) Analysis of porosity of free-standing ceramic manufactured by plasma spraying, in *Thermal Spray Coatings: Research, Design and Applications*. Proceedings of NTSC (eds C.C. Berndt and T. Bernecki), ASM International, Metals Park OH, pp. 505–11.
43. Spanne, P., Jones, K.W., Herman, H. and Riggs, W.L. (1993) Measurement of imperfections in thermal spray coatings using synchrotron radiation. *J. Thermal Spray Technol.*, **2**(2), 121.
44. Berndt, C.C. and Lin, C.K. (1993) Measurement of adhesion for thermally sprayed materials. *J. Adhesion Sci. Technol.*, **7**(12), 1235.
45. ASTM (1991) *D 907-91b, Terminology of Adhesives*, ASTM, Philadelphia PA.
46. ASTM (1982) Designation C633-79, standard test methodology for adhesion or cohesive strength of flame sprayed coatings, in *The Annual Book of ASTM Standards*, ASTM, Philadelphia PA.
47. Taylor, B.J. and Eyre, T.S. (1979) A review of piston ring and cylinder materials. *Tribol. Int.*, **12**, 78.
48. Overs, M.P., Harris, S.J. and Waterhouse, R.B. (1979) The fretting wear of sprayed molybdenum coatings at temperatures up to 300 °C, in *Wear of Materials*, ASME, pp. 379–87.
49. Wayne, S.F., Sampath, S. and Anand, V. Wear mechanisms on Thermally-Sprayed Mo-Based Coatings. *Tribol. Trans.*, **37**(3), 636.
50. Wayne, S.F. and Sampath, S. (1992) Structure–Property relationship in sintered and thermally sprayed WC–Co. *J. Thermal Spray Technol.*, **1**(4), 307.
51. Byrnes, L. and Kramer, M. (1994) Method and apparatus for application of

thermal spray coatings onto aluminum engine cylinder bores, in *Thermal Spray Industrial Applications*. Proceedings of NTSC (eds C.C. Berndt and S. Sampath), ASM International, Metals Park OH, p. 39.

52. Zaluzec, M.J., McCune, R.C., Reatherford, L.V. and Cartwright, E.L. (1994) Thermal spray self-lubricating composites from cored wires, in *Thermal Spray Industrial Applications*. Proceedings of NTSC (eds C.C. Berndt and S. Sampath), ASM International, Metals Park OH, p. 33.

Degradation of coatings by high temperature atmospheric corrosion and molten salt deposits

N. Birks, G.H. Meier and F.S. Pettit

11.1 INTRODUCTION

In discussing the degradation of coatings, it is necessary to consider the environments to be examined and the particular sources of data to be utilized. The questions of generation and use of test data are important because the degradation of coatings is important to many industrial applications. A wide variety of tests are available to examine coatings degradation. Some tests attempt to simulate the conditions of use. Others are more concerned with the definition of conditions causing the degradation, such as temperature and gas composition. An important parameter in coatings degradation is the microstructure developed during use in the application of interest. Relevant test data then relate to those tests that duplicate this degradation microstructure.

In this chapter the important types of environmentally induced degradation of coatings will be described. Both metallic and ceramic coatings will be considered. Oxygen is the principal reactant in many industrial environments, and the effects of oxidation on coatings will be examined first. The influence of other reactants in gases, such as sulfur and carbon, will then be described, considering the effect of very low oxygen pressures as can arise in some petrochemical processes. Next, degradation caused by deposits such as

Metallurgical and Ceramic Protective Coatings. Edited by Kurt H. Stern.
Published in 1996 by Chapman & Hall, London. ISBN 0 412 54440 7

Na_2SO_4 and NaCl will be described. Finally the effects of erosion on the various types of corrosion will be described.

11.2 METALLIC COATINGS

11.2.1 Oxidation

Coatings that confer oxidation resistance do so by forming protective reaction product barriers such as Al_2O_3, SiO_2 or Cr_2O_3 when exposed to the oxidizing environment. Such coatings are formed on alloys by aluminizing, siliconizing or chromizing, respectively, or by applying MCrAlY coatings. Degradation of these coatings occurs by cracking and spalling of the oxide scales. Cracking of the coating also can occur in some systems. The coatings consequently become depleted of the elements necessary to form the protective oxide barriers, as shown in Fig. 11.1 for the case of an aluminide coating, and eventually the protective barrier can no longer be formed upon the coating. The coating has failed.

There are numerous factors which affect the lives of coatings in addition to the use or exposure conditions. Some factors are the composition and

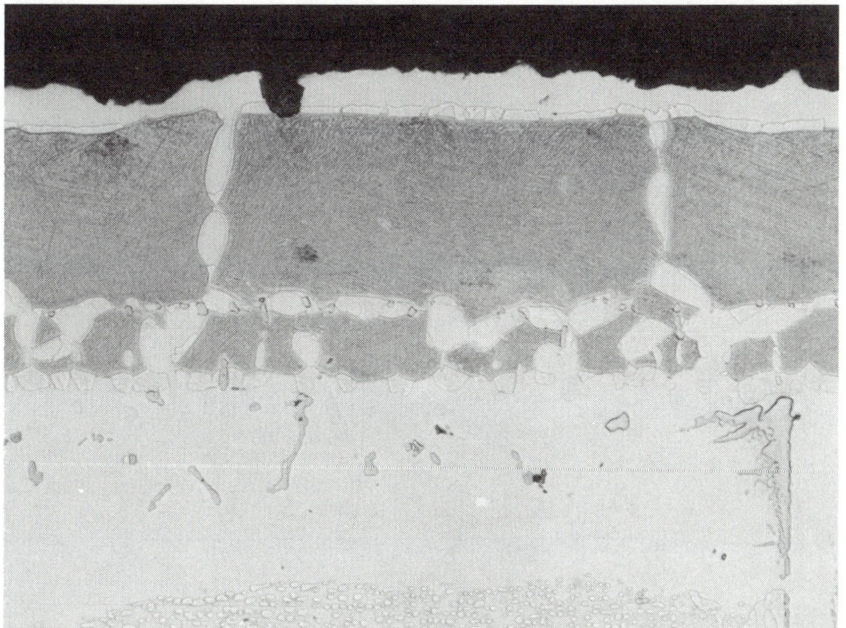

Figure 11.1 Aluminide coating on nickel-base superalloy after 30 h of oxidation at 1200 °C in air. Proceeding from the gas interface, zones of γ(nickel solid solution), $\gamma'(Ni_3Al)$ and $\beta(NiAl)$ are evident. The as-processed coating contained 100% β-phase.

thickness of the coating, the composition of the substrate upon which the coating is placed, and the mechanical properties of the coating. Higher concentrations of the element used to form the protective oxide barrier leads to longer lives, but there are limits that cannot be exceeded due to lack of ductility or melting, for example, which depend upon the systems under consideration. Other elemental additions can also be important. The addition of chromium permits an MCrAlY coating to remain an Al_2O_3 former down to lower aluminum concentrations [1, 2], and yttrium improves the adherence of Al_2O_3 to the coating [3, 4]. It is also well established that platinum in diffusion aluminides extends the lives of such coatings [5] as shown in Fig. 11.2. The mechanism by which platinum affects the coating lives is not completely understood; it may improve Al_2O_3 adherence, causing an Al_2O_3 scale to develop with slower transport properties due to higher purity, and it may inhibit interdiffusion between the coating and the substrates [6, 7]. Other precious elements such as rhodium produce similar effects [8], and more recent results show that palladium also may produce beneficial effects [9].

An example of substrate compositional effects is shown in Fig. 11.3, where the life of an aluminide coating on the superalloy B-1900 is significantly greater than on Mar M-200. This figure also shows that the oxidation resistance of B-1900 is better than that of Mar M-200. The alloy B-1900 has

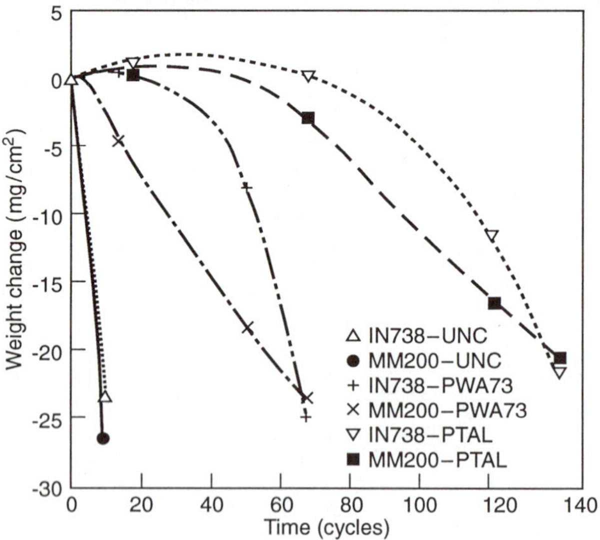

Figure 11.2 Cyclic oxidation at 1200 °C. Data for uncoated superalloys and for the superalloys with aluminide coatings and aluminides containing platinum. Large weight losses indicate that protective alumina scales are no longer formed on the specimens.

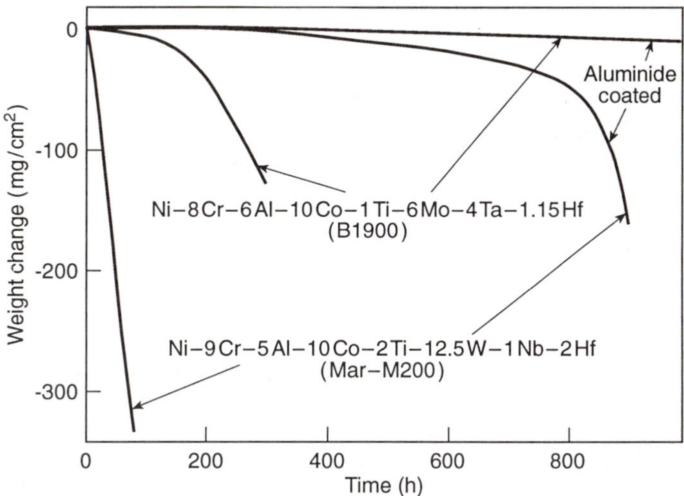

Figure 11.3 Weight change versus time data for the cyclic oxidation of two uncoated nickel-base superalloys and the same two alloys coated using a diffusion aluminizing treatment. The degradation of these alloys consists of two stages. An initial stage of less severe attack (not evident on one of the uncoated alloys for the timescale used) and a subsequent stage of more rapid attack.

a higher aluminum concentration than Mar M-200, partially responsible for the observed results, but other factors may also be important. For example, recent results show that sulfur in alloys (Fig. 11.4), can significantly affect oxidation resistance. It appears that small concentrations of sulfur (e.g. >1 ppm) adversely affect the adherence of Al_2O_3 scales [10, 11]. It is quite possible that sulfur in alloy substrates can adversely affect coating performances. It has also been observed that small boron concentrations ($\sim$0.1 wt%) in superalloys can cause oxide pit formation during oxidation of MCrAlY coatings, but the reasons for this condition are not understood.

Cracking in coatings can occur as a result of thermally induced stresses, arising from differences in thermal expansion coefficients, or stresses imposed by the load carried in the application. Silicon carbide and silicon nitride coatings will be discussed in more detail in the section on ceramic coatings. When these coatings are used on carbon–carbon composites, cracking occurs due to differences between thermal expansion coefficients, Fig. 11.5 [12, 13]. Silicide coatings can present similar problems. At high temperatures these cracks may seal upon subsequent oxidation by SiO_2 formation in the cracks, but at temperatures below about 1200 °C, SiO_2 formation is not sufficient and elements which result in the formation of liquid oxides, such as boron, are added to coatings to attempt to seal cracks. Cracking can occur in aluminide coatings as a result of the imposed load (Fig. 11.6), then the required

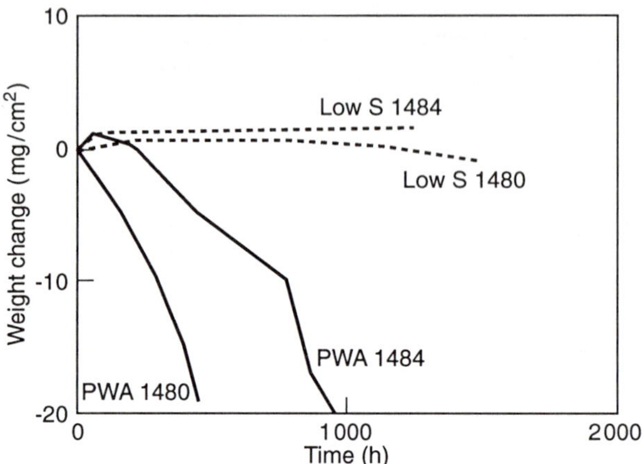

Figure 11.4 Weight change versus time measurements obtained for the cyclic oxidation at 1100 °C in air of as-received and hydrogen-annealed specimens of PWA 1480 and PWA 1484. The annealed specimens had sulfur concentrations of less than 1 ppm whereas the as-received alloys contained about 8 ppm sulfur.

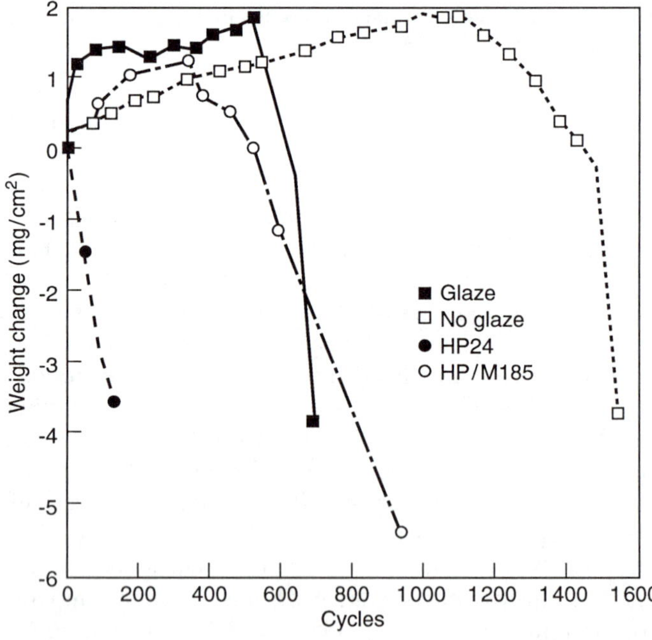

Figure 11.5 Cyclic oxidation at 900 °C . Data for the oxidation of an SiC-coated carbon–carbon composite. Large weight losses show that the coating has cracked and that carbon is being oxidized. (HP24–SiC coating, HP/M185–SiC + vitreous silica overlayer, glaze–SiC + B_4C + vitreous silica overlayer, no glaze–SiC + B_4C).

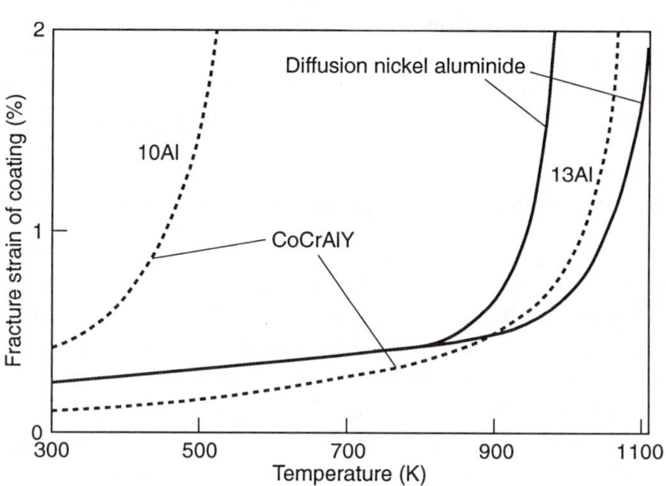

Figure 11.6 Ductility of CoCrAlY overlay and diffusion nickel aluminide coatings.

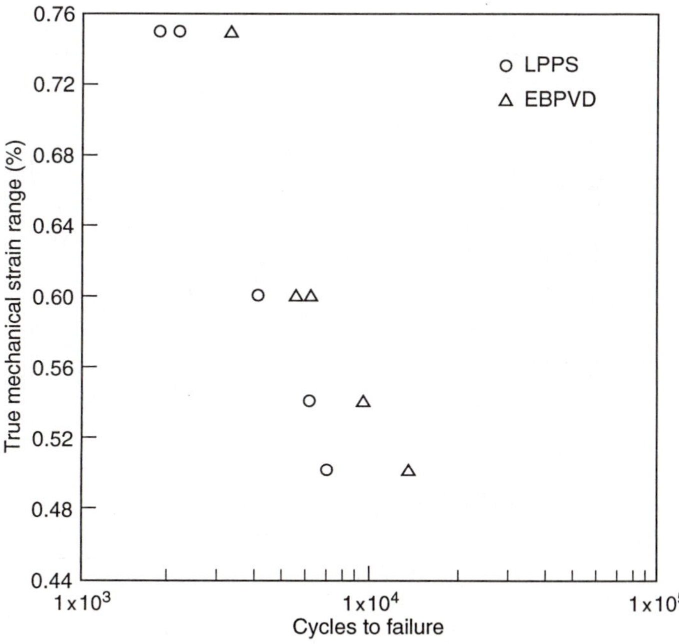

Figure 11.7 Cycles to failure versus strain range for EB-PVD and LPPS NiCoCrAlY coatings (Ni–23 Co–18 Cr–12 Al–0.5 Y) on alloy single crystal superalloy substrates.

increase in ductility can be obtained by reducing the aluminum concentration of the coating at the expense of coating oxidation life. Some MCrAlY coatings are susceptible to cracking caused by thermal fatigue (Fig. 11.7); and this problem can be overcome by changes in coating type or composition [14, 15].

11.2.2 Degradation of coatings in gases with multireactants including oxygen

Most metallic elements have greater affinities for oxygen than other reactive components in gases, such as sulfur, carbon or chlorine. Coatings in mixed gases with oxygen therefore degrade by oxide formation, but the other reactants generally cause the degradation to proceed more rapidly. The increase in degradation rate can arise from a number of causes. Figure 11.8 presents results obtained for the cyclic oxidation of various alloys in an O_2–SO_2 gas mixture [16]. Initially all of the alloys degrade at rates comparable to those in simple oxidation but eventually the degradation rates become higher. Onset of the more rapid attack indicates the scale is no longer pure oxide but rather a mixture of oxides and sulfides with degradation rate controlled by transport through the sulfide. Similar effects can be observed with other reactants such as carbon [16], chlorine [17] and nitrogen [18]. The mechanism by which any of these second reactants affect degradation

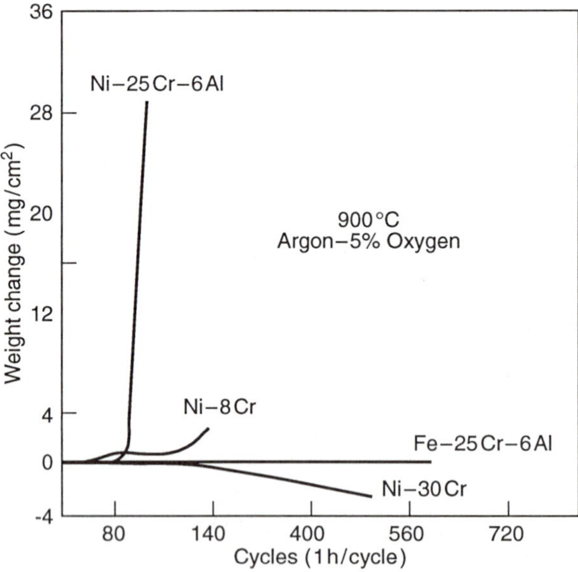

Figure 11.8 Cyclic oxidation of alloys in SO_2, initially protective oxides were formed on all of the alloys but eventually much less protective scales with sulfides developed at times where rapid weight increases or losses became evident.

of the coating certainly involves the formation of corrosion products through which transport is more rapid compared to oxides. But there are other important effects. The second reactant, especially sulfur, may adversely affect oxide adherence. Also, products of the additional reactants are formed in the coating just beneath the oxide scale, and when the oxide spalls, selective oxidation of aluminum and chromium is more difficult because of their presence. This is illustrated in Fig. 11.9 where an Fe–15 Cr alloy oxidizes much more rapidly in CO_2 than oxygen because of carbide formation.

In gases where the oxygen partial pressure is very low or negligible, oxide formation is not as important and the other phases, such as sulfides or chlorides, play dominant roles in the degradation processes. Since transport through these phases is more rapid than oxides, degradation of the coatings is generally more rapid compared to oxidation. It is also possible to have some volatile products form which can result in very accelerated degradation rates. When environments are encountered which contain very little oxygen, an entirely different coating may be required in order to develop corrosion resistance. For example, in gases very rich in chlorine, aluminide coatings degrade extremely rapidly due to formation of volatile corrosion products, and coatings which form solid corrosion products will degrade at slower rates.

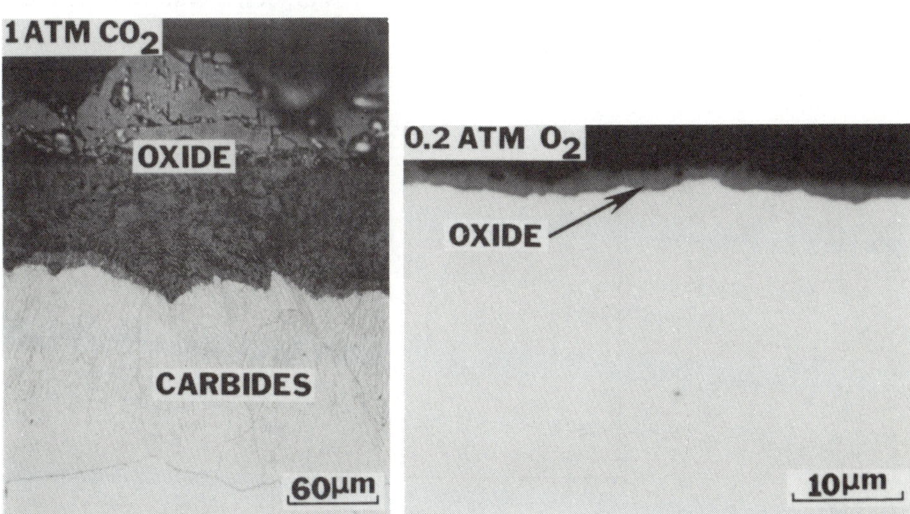

Figure 11.9 Photomicrographs showing products formed on an Fe–15 Cr alloy after exposure at 900 °C to (a) 0.2 atm (20 kPa) oxygen (16.2 h) and (b) pure flowing CO_2 (4 h). The chromium is selectively oxidized in oxygen but carbide formation in CO_2 prevents the selective oxidation of chromium and much less protective iron oxides are formed.

11.2.3 Degradation of coatings induced by molten salt deposits

Degradation of alloys is usually more severe when molten deposits accumulate on the surface. The degradation sequence usually consists of an initiation stage, during which the attack is virtually the same as for the alloy in the absence of the deposit, and a propagation stage, during which the attack is substantially increased. The length of the initiation stage and the type of degradation during the propagation stage depend upon the type of deposit. A great variety of deposits are relevant to industrial processes. Some typical deposits are Na_2SO_4 and solutions of Na_2SO_4 and $NaVO_3$. In considering the degradation of coatings induced by deposits, it is first necessary to describe the composition ranges over which the deposits exist. These compositional ranges depend upon the gas environments. Furthermore, as the coated alloys become separated from the gas environments by the deposits, compositional gradients are developed across the deposits, extending from the coating–deposit interface to the deposit–gas interface. Stability diagrams, such as Fig. 11.10, can be used to describe compositional gradients across deposits on coatings. In Fig. 11.10, various compositions of Na_2SO_4 are defined by the SO_3 and oxygen pressures. The deposits may become acidic or basic, depending upon the nature of the reactions that occur between the deposit and the coating.

The thickness of the deposit also has a significant effect on the hot corrosion process. When the deposit is thicker, or materials are immersed in a melt, the gas environment is less important. Hot corrosion of materials during the

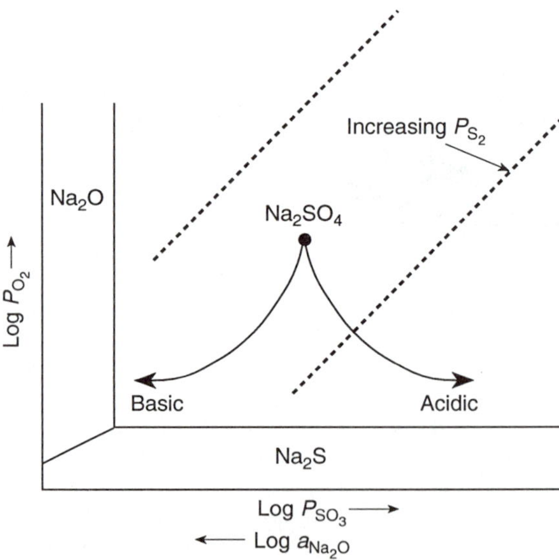

Figure 11.10 Thermodynamic stability diagram for the Na–O–S system showing how the composition of Na_2SO_4 may change due to reaction of the alloy with the deposit.

propagation stage can be viewed as consisting of three sequential processes:

- Reactions between the metallic or ceramic coating and the deposit,
- Diffusion through the molten deposit,
- Reactions between the gas and the molten deposit.

In order to understand the overall hot corrosion process, it is necessary to describe in thermodynamic and kinetic terms the essential features of these three sequential processes. A great variety of possibilities exist, depending upon the coating composition, the composition and thickness of the deposit, the temperature and the composition of the gas. A number of references describe some of these processes in more detail [19–22]. The point to emphasize is that the microstructural features can be used to help identify the forms of the hot corrosion degradation. For example, some microstructures of degraded coatings are presented in Figs 11.11 and 11.12. The microstructure in Fig. 11.11 typifies low temperature hot corrosion (type II) of a CoCrAlY overlay coating after 4200 h of marine service. The coating microstructure in Fig. 11.12 is after service in an aircraft gas turbine and the coating has degraded via high temperature (type I) hot corrosion. Figure 11.12 also shows the attack of the alloy substrate at a point where the coating has been penetrated. This microstructure is typical for hot corrosion of superalloys.

11.3 CERAMIC COATINGS

11.3.1 Gaseous attack

In discussing the attack and degradation of ceramic coatings, few data are available on mixed gas attack and it is reasonable to group all types together.

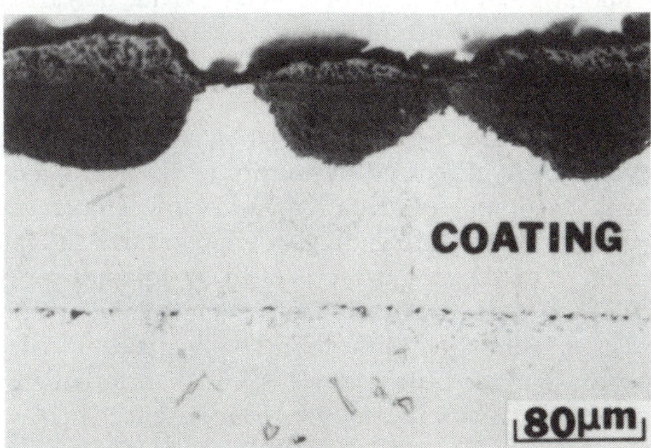

Figure 11.11 Photomicrograph showing features developed during the hot corrosion of CoCrAlY coatings in marine service (coated vane after 4200 h service).

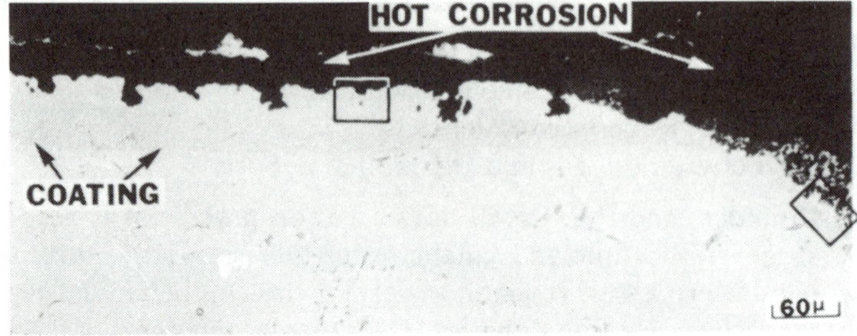

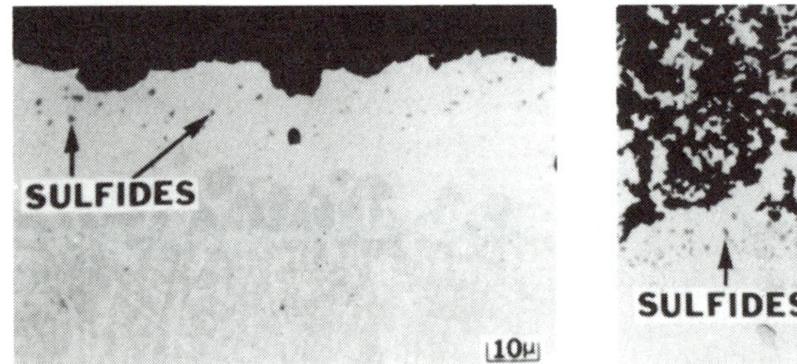

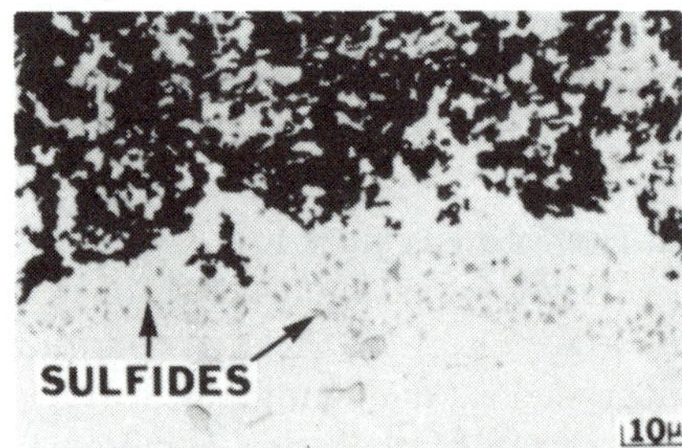

Figure 11.12 Photomicrographs of aluminide-coated nickel-base superalloy after service in an aircraft gas turbine. Photomicrographs (b) and (c) are typical of high temperature hot corrosion: (b) shows hot corrosion features developed in the coating, whereas (c) shows features developed in the superalloy substrate after the coating was penetrated.

There are essentially two types of ceramic coatings, oxide and nonoxide ceramics. Examples of nonoxide ceramics consist of SiC and Si_3N_4. These ceramics rely upon the development of protective oxide barriers for resistance to attack similar to that of metallic coatings. Oxide ceramics do not react substantially with oxygen-rich gases. A typical oxide ceramic coating is yttria stabilized zirconia.

When SiC or Si_3N_4 are exposed to oxygen at elevated temperatures, silica scales are developed. As discussed by Luthra [23], these scales are usually amorphous at the beginning of oxidation and at low temperatures, and they tend to crystallize at longer times, at higher temperatures and in the presence of impurities. The growth rates are controlled by complex processes, but as proposed by Luthra, probably involve mixed control, being influenced both by an interface reaction and diffusion. The temperature dependence of parabolic rate constants for silica scales is less than for Al_2O_3 and Cr_2O_3 (Fig. 11.13), consequently silica scales may be more effective reaction product barriers at temperatures above about 1400 °C.

Cracking is the principal problem encountered with ceramic coatings. As mentioned previously, this problem is most severe in the case of SiC and

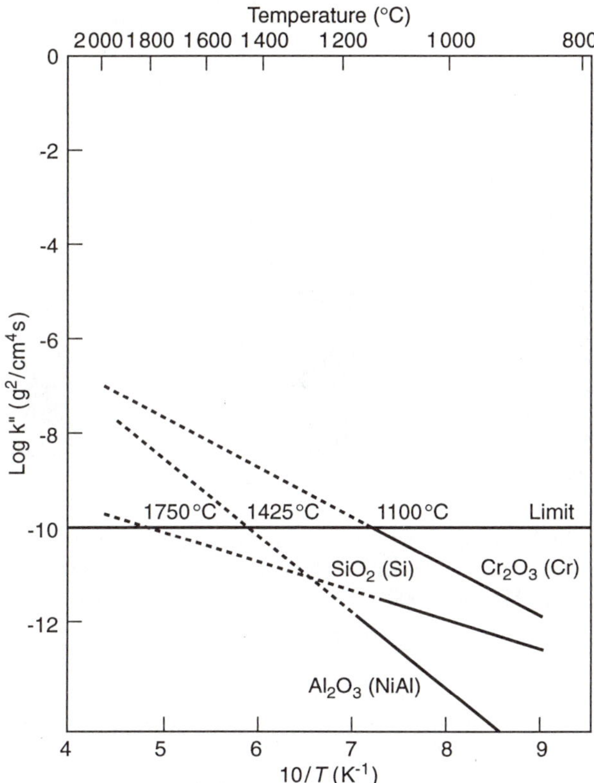

Figure 11.13 Parabolic rate constants versus temperature for growth of oxides that provide adequate protection at temperatures above 1200 °C. A rate constant of 10^{-10} (g^2/cm^4 s) is taken as a limiting rate constant for service. Weight gains at 10^{-10} g^2/cm^4 s: 6 mg/cm² for 100 h, 19 mg/cm² for 1000 h and 60 mg/cm² for 10 000 h.

Si_3N_4 at temperatures below about 1200 °C for which the oxidation rates are too low to permit sealing of cracks via silica formation. However, even at temperatures above 1200 °C cracking is a problem for long exposures with thermal excursions. In the case of oxide ceramics, sealing of cracks can only be attempted using B_2O_3 or similar sealants, not very effective with extensive times at temperature. Consequently, oxide ceramic coatings are currently used mainly as thermal barrier coatings.

11.3.2 Hot corrosion

Very little data is available on the hot corrosion of ceramic coatings. Generally, nonoxide ceramics are more susceptible to hot corrosion attack

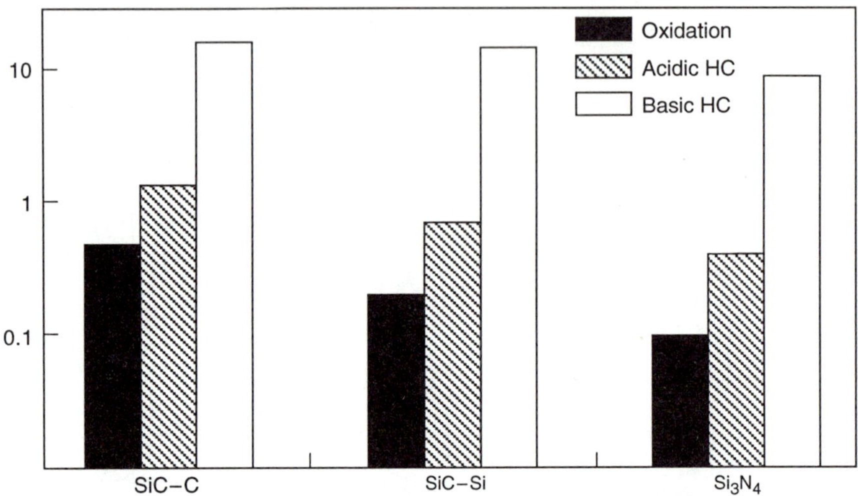

Figure 11.14 Thickness of layers formed for the oxidation, acidic hot corrosion and basic hot corrosion of C-side and Si-side single-crystal silicon carbide and CVD silicon nitride after 168 h at 1000°C. Note that oxide thicknesses follow O < A < B.

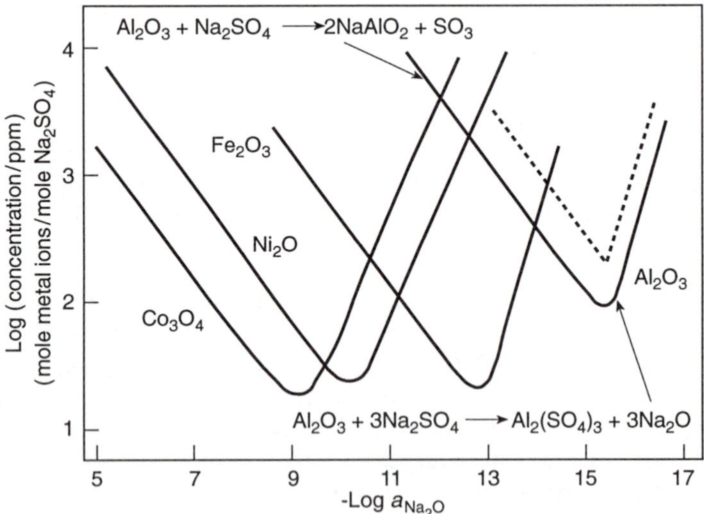

Figure 11.15 Solubilities of some oxides in Na_2SO_4 at 1200 K as determined by Rapp and coworkers. Dissolution reactions consistent with the slopes of the curves are presented for Al_2O_3. The dashed lines give calculated solubilities using Temkin's model for ionic melts and the acidic or basic dissolution reactions discussed in the text.

than oxide ceramics, but results can be significantly affected by impurities in the ceramics as well as the compositions of the deposits. Figure 11.14 compares the hot corrosion of high purity SiC and Si_3N_4. Hot corrosion attack is more severe than oxidation. And basic hot corrosion conditions are more aggresive than acidic; this is due to the silica scales that are formed, which are more reactive in basic deposits [24, 25].

In the case of the hot corrosion of oxide ceramics, solubility data generated by Rapp and coworkers are especially valuable, Fig. 11.15 [26]. As illustrated in Fig. 11.15, the reaction of Al_2O_3 with Na_2SO_4 can occur via either basic or acidic reactions, depending upon the composition of the Na_2SO_4. For such reactions the deposits can become saturated with the dissolved oxide, and reactions will stop until additional deposits accumulate.

In some cases the molten deposit can preferentially remove one component in the oxide. For example, deposits of $NaVO_3$ preferentially remove yttria from stabilized zirconia [27]. It is also possible for the reaction of oxides with molten deposits to occur in ways such that synergistic fluxing occurs. Such synergism arises when one oxide component dissolves basically and the other acidically [28, 29].

11.4 EROSION

Erosion and oxidation, or erosion and hot corrosion, can occur concomitantly in certain systems. Erosive conditions in combination with corrosion are especially undesirable because the reaction product barriers used to develop resistance against corrosion can be destroyed by the erosion process. A number of investigations have examined these interactions [30–32]. Although these studies have not necessarily been concerned specifically with coatings degradation, the results are applicable to coatings. In the case of metallic coatings, when either the erosive or corrosive component is small, the degradation occurs by erosion or corrosion; however, when both erosion and corrosion rates are significant, the interactions can be divided into two different regimes – erosion enhanced corrosion, and corrosion affected erosion. There are at least three types of erosion enhanced corrosion. The first consists of thinning of protective scales by the erosion process to a point where eventually a constant scale thickness is reached, representative of equal rates of scale growth and erosive removal. The second also involves thinning of reaction products, but the erosion process also alters transport processes in the scale. The third is usually observed when the size of the eroding particles is about equal to the scale thickness and spalling of the scales can occur.

As the intensity of the erosive component becomes large with respect to the still significant corrosive component, the corrosion affected erosion regime becomes important. Very often in this regime it is difficult to discern corrosion products because of their rapid removal by erosion, but the erosion process is affected by the corrosive environment. The initiation stage may be substantially shortened in the case of hot corrosion conditions.

The effects of erosive components on ceramic coatings at elevated temperatures have not been investigated to any great extent. For sufficiently large erosive components, cracking and spalling of these coatings can occur. It is necessary to determine the magnitudes of such erosive components for specific coatings, and to describe the damage characteristics.

REFERENCES

1. Giggins, C.S. and Pettit, F.S. (1971) *J. Electrochem. Soc.*, **118**, 1782–90.
2. Wood, G.C. and Chattopadhyay, B. (1970) *Corr. Sci.*, **10**, 471–80.
3. Tien, J.K. and Pettit, F.S. (1972) *Metall. Trans.*, **3**, 1587–99.
4. Ramanarayanan, R.A., Ayer, R., Petkovic-Luton, R. and Leta, D.P. (1988) *Oxid. Metals*, 445–72.
5. Bungardt, K. *et al.* (1972) U.S. Patent 33,677,789.
6. Schaeffer, J., Kim, G.M., Meier, G.H. and Pettit, F.S. (1989) The effects of precious metals on the oxidation and hot corrosion of coatings, in *The Role of Active Elements in the Oxidation Behavior of High Temperature Metals and Alloys* (ed. E. Lang), Elsevier Applied Science, New York, pp. 231–67.
7. Streiff, R. and Boone, D.H. (1988) *J. Mater. Engng*, **10**, 15–26.
8. Felten, E.J. (1976) *Oxid. Metals*, **10**(1), 23–8.
9. Alperine, S., Stienmetz, P., Josso, P. and Constantani, A. (1989) *Mater. Sci. Engng*, **A121**, 367–72.
10. Fuenkenbusch, A.W., Smeggil, J.G. and Bornstein, N.S. (1985) *Metall. Trans.*, **16A**, 1164.
11. Smialek, J.L. (1987) *Metall. Trans.*, **18A**, 164.
12. Schaeffer, J.C. (1989) The oxidation behavior of carbon–carbon composites and their coatings. Ph.D. dissertation, University of Pittsburgh, Pittsburgh PA.
13. Strife, J.R. and Sheehan, J.E. (1988) *Ceram. Bull.*, **67**, 369–74.
14. Goward, G.W. and Boone, D.H. (1967) *Trans. ASM*, **60**, 228–41.
15. Strangman, T.E. (1978) Thermal fatigue of oxidation resistant overlay coatings for superalloys. Ph.D. dissertation, University of Connecticut, Storrs CT.
16. Giggins, C.S. and Pettit, F.S. (1980) *Oxid. Metals*, **14**, 363.
17. Prescott, R., Stott, F.H. and Elliot, P. (1989) *Oxid. Metals*, **31**, 145–66.
18. Chen, I.C. and Douglass, D.L. (1980) *Oxid. Metals*, **34**, 473–96.
19. Rapp, R.A. (1987) *Mater. Engng*, **87**, 319–27.
20. Stringer, J. (1977) Hot corrosion of high temperature alloys, in *Annual Review of Materials Science* (ed. R.A. Huggins), Annual Reviews, Palo Alto CA, p. 477.
21. Shores, D.A. and Luthra, K.L. (1995) Hot corrosion of metals and alloys, in *Fundamentals of High Temperature Corrosion*, Academic Press, New York (in press).
22. Pettit, F.S. and Giggins, C.S. (1987) Hot corrosion, in *Superalloys II* (eds C.T. Sims, N.S. Stoloff and W.C. Hagel), Wiley, New York, Ch. 12.
23. Luthra, K.L. (1991) *J. Amer. Ceram. Soc.*, **74**, 1095–1103.
24. Jacobson, N.S. (1989) *Oxid. Metals*, **31**, 91–103.
25. Lawson, M.G., Kim, H.R., Pettit, F.S. and Blachere, J.R. (1990) *J. Amer. Ceram. Soc.*, **73**, 989–95.
26. Rapp, R.A. (1989) Hot corrosion of materials, in *Selected Topics in High Temperature Chemistry* (eds O. Johannesen and A. Anderson), Elsevier, New York, pp. 291–329.

27. Warnes, B.M. (1990) The influence of vanadium on the sodium sulfate induced hot corrosion of thermal barrier coating materials. Ph.D. dissertation, University of Pittsburgh PA.
28. Hwang, Y.S. and Rapp, R.A. (1988) Synergistic dissolution of oxides in molten salts, in *Oxidation of High-Temperature Intermetallics* (eds T. Grobstein and J. Doychak), TMS, Warrendale PA.
29. Lawson, M.G., Pettit, F.S. and Blachere, J.R. (1993) *J. Mater. Res.*, **8**, 1964–71.
30. Levy, A.V. and Man, Y.-F. (1986) *Wear*, **131**, 36.
31. Wright, I.G., Nagarajan, G. and Stringer, J. (1986) *Oxid. Metals*, **25**, 175.
32. Rishel, D., Pettit, F.S. and Birks, N. (1991) *Mater. Sci. Engng*, **A143**, 197–211.

12

Measurement of coating adhesion

David Rickerby

12.1 INTRODUCTION

The ability to change the surface characteristics of a component independent of the bulk material properties by use of surface coatings has opened up many new and diverse applications in a number of technology areas [1]. There are many definitions of what is meant by surface engineering, perhaps the most appropriate is – the application and design of surface treated material systems to achieve a cost-effective performance enhancement of which neither is capable on its own. In most applications the minimum criterion for acceptable performance is that the adhesion of the coating to the substrate should be sufficient so that the coating remains in place for the lifetime of the component in its operating environment. In such developments there is a requirement to measure the level of adhesion between substrate and coating in order to guarantee that only coatings 'fit for purpose' are released for evaluation. In many instances the slow take-up of surface coating technology can be traced, in part, to a lack of end-user confidence in the quality of supplied coatings and thus it is essential that a simple and reliable adhesion test is available, which should ideally be of relevance to the intended application.

The main theme of this chapter is coating adhesion, a property which is central to the functionality of the coating–substrate system. However, the nature of the interfacial region is linked to the choice of surfacing technology, so this chapter begins by briefly reviewing surfacing techniques and the failure

Metallurgical and Ceramic Protective Coatings. Edited by Kurt H. Stern.
Published in 1996 by Chapman & Hall, London. ISBN 0 412 54440 7

modes for thin films. It concludes by highlighting the future needs for adhesion test methods.

12.2 SURFACE ENGINEERING

The range of surfacing treatments available to modify the properties of a component can perhaps be best illustrated by making reference to the aerospace industry. Such was the pace of engine development in the 1960s and 1970s, driven by the commercial pressure to improve reliability and reduce the cost of ownership, that rapid progress was made in both the development and application of surface treatment techniques.

Figure 12.1 is a half-sectional view of a modern turbofan engine which serves to illustrate the extensive range of surfacing treatments and coating technologies specified; the number of surface treated components has increased from around 25% to cover 50% in the last 30 years. The surfacing technologies are categorized against the main reason for usage in Table 12.1, and it is immediately apparent that some (paint, electroplating, thermal spraying) are very versatile, whereas others have a more specific but often

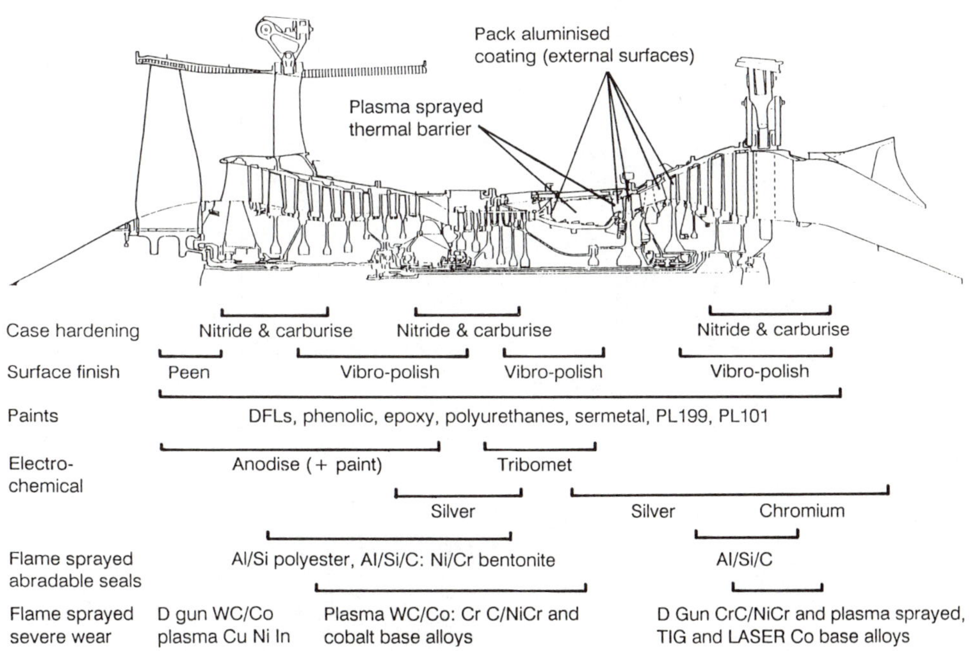

Figure 12.1 Half-sectional view of a large turbofan engine showing the wide range of surface treatment and coating technologies currently used in the aero-engine industry.

Table 12.1 Usage of surface treatments and coating technologies in the aero-engine industry

Main use	Paints	Electroplating anodizing	Thermal spray	Impact finishing	Case hardening	Weld deposition	Diffusion coatings (CVD)	Chemical conversion	PVD
Corrosion and wear	√	√	√	√	√	√	√	√	√
Cosmetic	√	√		√					
Mechanical properties				√	√				
Thermal insulation	√		√						√
Engine sealing			√						
Earthing	√	√	√						
Emissivity and infrared	√								
Salvage and repair		√	√			√			
Aerodynamic efficiency	√			√					√
Composite materials			√				√		√

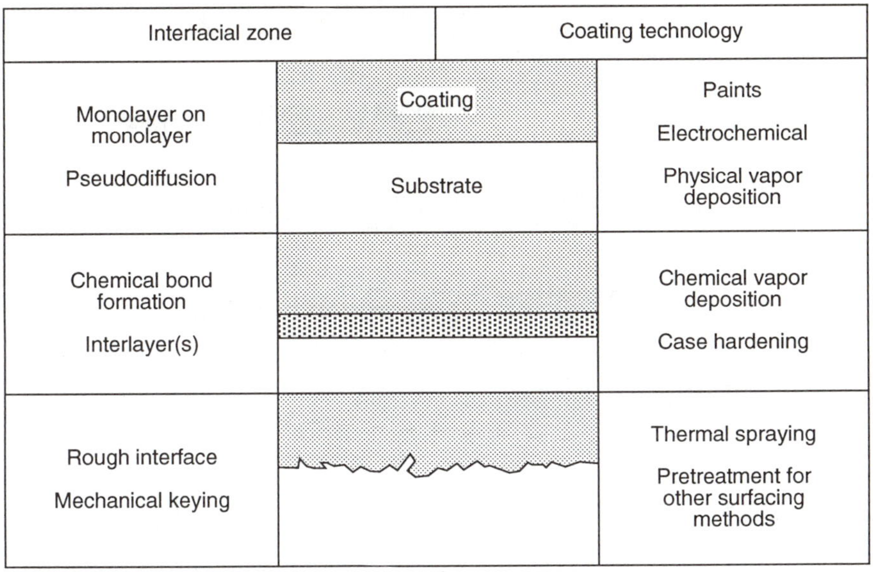

Interfacial zone		Coating technology
Monolayer on monolayer Pseudodiffusion	Coating Substrate	Paints Electrochemical Physical vapor deposition
Chemical bond formation Interlayer(s)		Chemical vapor deposition Case hardening
Rough interface Mechanical keying		Thermal spraying Pretreatment for other surfacing methods

Figure 12.2 Types of interfacial region formed between coating and substrate.

unique role to play (case hardening, PVD). Regardless of the treatment, an interface is formed between the coating and underlying substrate, and its nature depends to a great extent on the deposition technology employed; in many instances it is tailored to give the correct blend of material properties in this critical region (Fig. 12.2). One key property which will now be discussed further is the required level of coating to substrate adhesion. Details of the various surfacing techniques illustrated in Figure 12.1 can be obtained by reference to previous chapters and to a number of specialized texts [1–4].

12.3 THE REQUIREMENTS OF AN IDEAL ADHESION TEST

Before describing the range of adhesion tests that exist to determine practical adhesion, it is worth considering the fundamental property of adhesion and how experimentally measured quantities differ from the basic adhesion defined by the American Society for Testing and Materials (D907-70) as 'the state in which two surfaces are held together by interfacial forces which may consist of valence or interlocking forces or both.' The nature of these bonding forces may be van der Waals, electrostatic and/or chemical; they are effective across the coating–substrate interface. The approximate ranges of binding energies for each of these types of interaction are illustrated in Fig. 12.3 [5]. An important distinction is made between basic adhesion (BA), the maximum possible attainable value and experimental or practical adhesion (EA), which

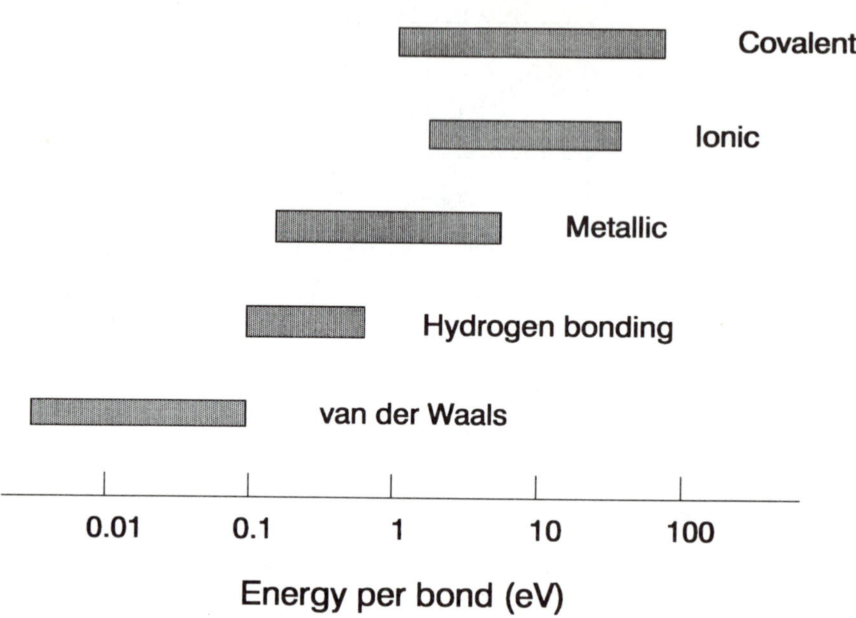

Figure 12.3 Typical bond strengths for a range of potential interfacial binding forces.

may be referred to as the bond or adhesion strength [6]. The relationship between EA and BA is given by [7]

$$EA = f(BA, \text{ other factors}) \tag{12.1}$$

such that EA ≪ BA due to the influence of other factors on the test, and these include internal stresses in the coating and the technique used for measuring the bond strength. Pulker *et al.* [8] further define the relationship between experimentally measured adhesion and basic adhesion as

$$EA = BA - IS \pm MSM \tag{12.2}$$

where IS is the internal stress factor and MSM is the method-specific error in measurement. From equation 12.2 it can be readily deduced that the true value of basic adhesion cannot usually be determined because the size of the measurement error inherent in the adhesion measurement technique can seldom be estimated.

The measurement of experimental adhesion can be achieved in two ways: (a) by defining the force of adhesion as the maximum force per unit area exerted when two materials are separated, and (b) by defining the work of adhesion as the work done in separating or detaching two materials from one another. If film A is detached from substrate B, and provided that the interface remains abrupt (sharp), then the energetic criterion for the system to remain adhered will be given by

$$W_{AB} = \gamma_A + \gamma_B - \gamma_{AB} \tag{12.3}$$

where W_{AB} is the reversible work of adhesion, γ_A and γ_B are the specific surface free energies of components A and B, and γ_{AB} is the interfacial specific free energy. In practice it is never possible to determine the true work of adhesion because some plasticity will occur during the coating removal process and an extra term γ_p should be added to equation 12.3 to account for this behaviour. As in the case of the fracture of bulk materials, γ_p will usually dominate over the surface energy terms. The total force of adhesion F_{AB} can be related to the work of adhesion W_{AB} by

$$W_{AB} = F_{AB}(x)\,\mathrm{d}x \tag{12.4}$$

This relationship relies on assumptions being made about the changes in force with distance of separation x, where the distance x is generally of atomic dimensions [6]. The use of equation 12.4 to determine the work of adhesion relies on a very simple planar model of the coating–substrate interface. This is never found in real systems, as illustrated previously in Fig. 12.2, in which interfacial reactions, surface roughness and bonding layers all play a part in defining coating–substrate adhesion. If the break occurs at the interface between A and B then it is termed adhesive failure; if it occurs either within A or B then it is referred to as cohesive failure. Depending on the ductile–brittle properties of the coating–substrate system, a number of possible failure modes can occur in any adhesion test, and though the failure may start at the interface, it can propagate into the substrate (Fig. 12.4). Alternatively, through-thickness cracking can lead to failure at the coating–substrate interface. For this reason it is important to characterize the failure modes associated with any adhesion test; these are summarized in Tables 12.2 and 12.3 for both tensile and compressive failures.

Ideally, the adhesion test should be nondestructive and easily adaptable to routine testing of geometrically complex shapes in which the degree of bonding may vary from point to point. Interpretation of the test should be relatively simple, and the method should be amenable to automation and standardization. The ideal adhesion test should also be reproducible, quantitative and directly related to coating reliability in specific applications. There are currently no tests for the measurement of practical adhesion which fulfill these requirements and all of the commonly used tests are destructive in nature. Indeed it is difficult to see how a nondestructive test can be developed, given the level of theoretical understanding which exists, and therefore it is necessary to make the best use of the available tests.

Tensile stress cracking

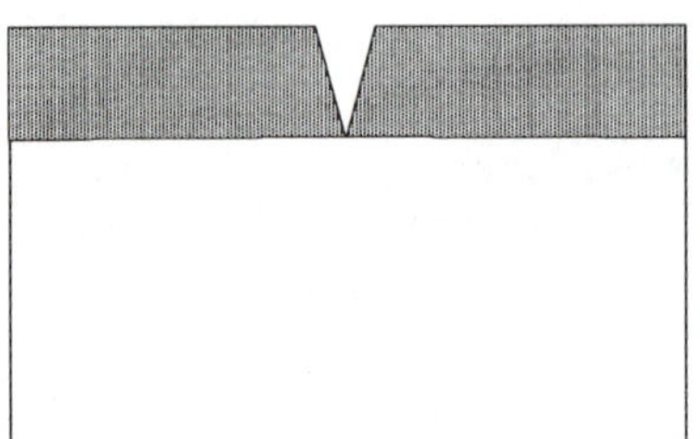

Tensile stress cracking
and delamination

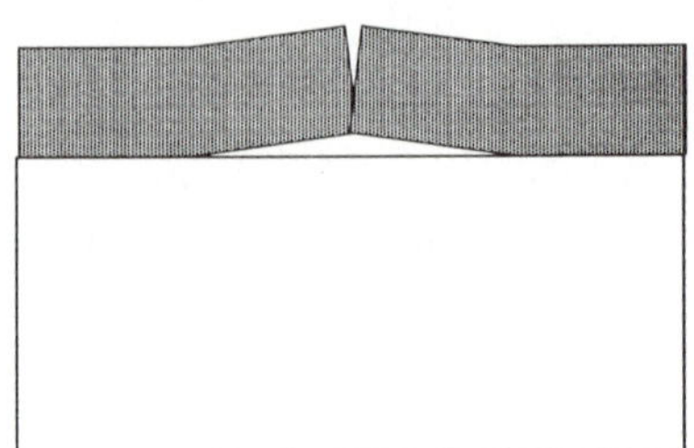

Tensile stress cracking and
failure of the substrate

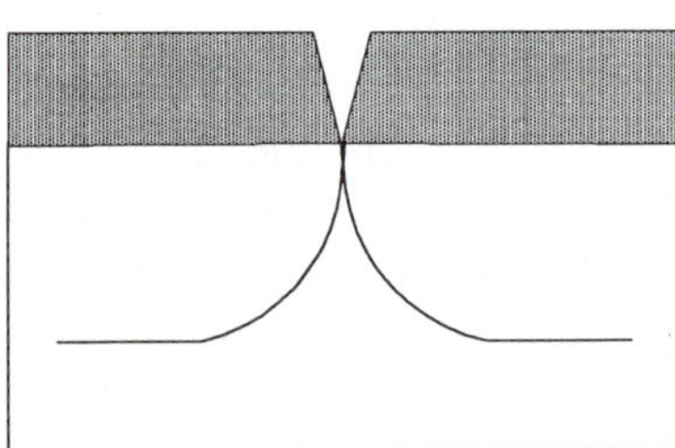

Compressive stress cracking
leading to delamination

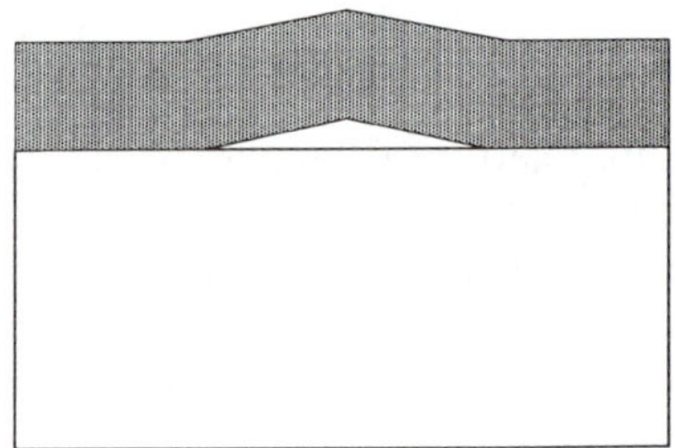

Figure 12.4 Schematic diagram of the surface cracking which can lead to coating detachment under compressive or tensile stresses.

Table 12.2 Tensile failure modes for thin films

Film	Substrate	Interface bonding	Decohesion mechanism(s)
Brittle	Ductile	Good	Film cracking – no decohesion
		Poor	Film cracking – interface decohesion
Ductile	Brittle	Good	Edge decohesion in substrate
		Poor	Edge decohesion at interface
Ductile	Ductile	Good	Film–substrate splitting – substrate decohesion
		Poor	Edge decohesion at interface
Brittle	Brittle	Poor	Edge decohesion at interface
			Film cracking – interface decohesion

Table 12.3 Compressive failure modes for thin films

Film	Substrate	Interface bonding	Decohesion mechanism(s)
Brittle	Ductile	Good	Buckle propagation in film
		Poor	Buckle propagation at interface
Ductile/brittle	Brittle	Good	Substrate splitting
		Poor	Buckle propagation at interface
Ductile	Ductile	Good	No decohesion
		Poor	Buckle propagation at interface

12.4 METHODS OF ADHESION EVALUATION

Adhesion test methods fall broadly into three categories: nucleation methods, mechanical methods and miscellaneous methods [7]. Table 12.4 shows some of the methods developed to measure the adhesion of thin films that have appeared in the literature. At the atomic level the nucleation methods probe adhesion via the breaking of individual coating–substrate atomic bonds, generating lamellar defects. The macroscopic experimental adhesion is simply a summation of the individual atomic forces which it should, in principle, be possible to relate to the adsorption energy of single adatoms on the atomically clean substrate. Although the nucleation methods are conceptually very simple, they are based on the measurement of (i) nucleation rate, (ii) island density, (iii) critical condensation and (iv) residence time of the depositing atoms comprising the film. Such measurements need access to an

Table 12.4 Methods that can be used to determine coating–substrate properties

Qualitative	Quantitative
Mechanical methods	
Scotch tape test [9]	Direct pull-off test [12–15]
Abrasion test [10]	Laser spallation test [18–20]
Bend and scratch test [11]	Indentation test [21–24]
	Ultracentrifuge test [7]
	Scratch test [25–43]
	Bend test [44–46]
	Linear elastic fracture mechanics [47–50]
Nonmechanical methods	
X-ray diffraction [4]	Thermal method [16]
	Nucleation test [9]
	Capacity test [7, 17]

electron microscope and consequently such methods are not applicable to test the adhesion of coatings on practical surfaces.

In addition to some of the miscellaneous methods, the mechanical methods are of more practical interest. In the mechanical methods adhesion is determined by applying a force to the coating–substrate system. The force is either directly applied normal or parallel to the coating–substrate interface, as in the pull or shearing stress tests respectively, or the force may be introduced indirectly by some stimulus, as in the indentation and laser spallation tests. These methods of adhesion evaluation are now considered in more detail.

12.4.1 Pull-off methods

Several techniques are based on forming a bond with the coating and applying a force normal to the coating–substrate interface in order to determine a value for practical adhesion. Perhaps the most widely used test for adhesion assessment is the tape test where a pressure-sensitive tape is applied to the coating surface and thereafter pulled off to determine the coating detachment. The test is essentially a qualitative success or failure method and can only be applied to rather weakly adhering coatings, the upper limit being about $20\,\mathrm{MN/m^2}$ [9]. For engineering coatings the direct pull-off method is employed (Fig. 12.5), although this suffers from the following deficiencies: (a) simple tensile tests frequently involve a complex mixture of tensile and shear forces, making interpretation difficult, (b) alignment must be perfect to ensure uniform loading across the interface, (c) such tests are limited by the strengths of available adhesives and the maximum adhesion that can be measured must be less than the strength of the bonding material itself, which is usually an epoxy (between 6.5 and $90\,\mathrm{MN/m^2}$), (d) it is possible that the adhesive or solvent penetrates the coating, affecting the coating–substrate interface. Despite the above difficulties, Jankowski [15] has presented results of fracture modeling in the test, and theoretical results are in excellent agreement with the experimental values of bond fracture stress for coatings of PVD titanium onto a beryllium substrate.

Another form of the pull-off method is the tangential shear or lap shear technique in which the load for detachment is applied parallel to the coating–substrate interface and the measured shear stress is the tangential force per unit area required to break the bond between the film and substrate. The advantages that the lap shear test has over the tensile test [51] are (i) it avoids severe deformation of the substrate; (ii) the film is gripped over a relatively large area and thus the stress is less concentrated; (iii) it approximates to a nominally pure shear measurement. As a variant on the above tests, Argawal and Raj [52] proposed a method of measuring the shear strength of metal–ceramic interfaces by depositing a ceramic coating on a ductile substrate and pulling the sample in tension. During the test, cracking occurs

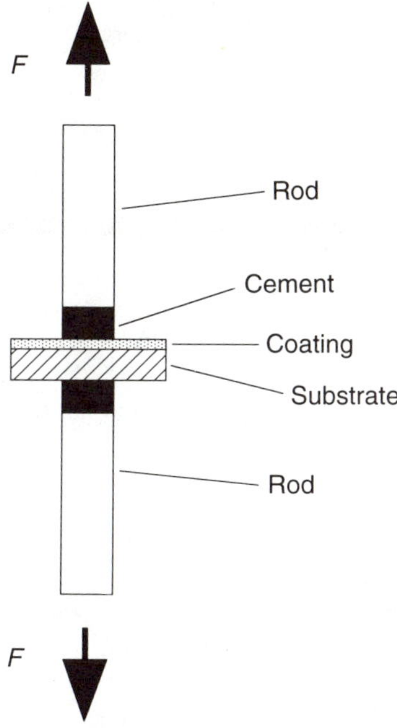

Figure 12.5 Schematic diagram of the direct pull-off test.

perpendicular to the loading direction, with the maximum crack density
being dependent on the interfacial shear strength τ according to

$$\tau = \pi t \sigma/\alpha \qquad (12.5)$$

where t is the crack thickness, α the maximum spacing between cracks and
σ the tensile strength of the coating (Fig. 12.6). The work of Agrawal and
Raj [52], performed with 60 nm films of silica bonded to pure copper
substrates, showed that the ultimate shear strength of the interface to lie in
the range 0.56–1.67 GPa. For thick glass ceramic films bonded to metal
substrates, Ashcroft and Derby [53] have shown that crack spacing is
proportional to coating thickness as predicted by equation 12.5; these
preliminary findings indicate that this particular method for assessment of
interfacial properties warrants further experimental work.

12.4.2 Ultracentrifugal and ultrasonic methods

The direct pull-off method suffers from the need to attach to the
coating–substrate system a device which will allow transmission of forces
either parallel or normal to the interface from which a measurement of

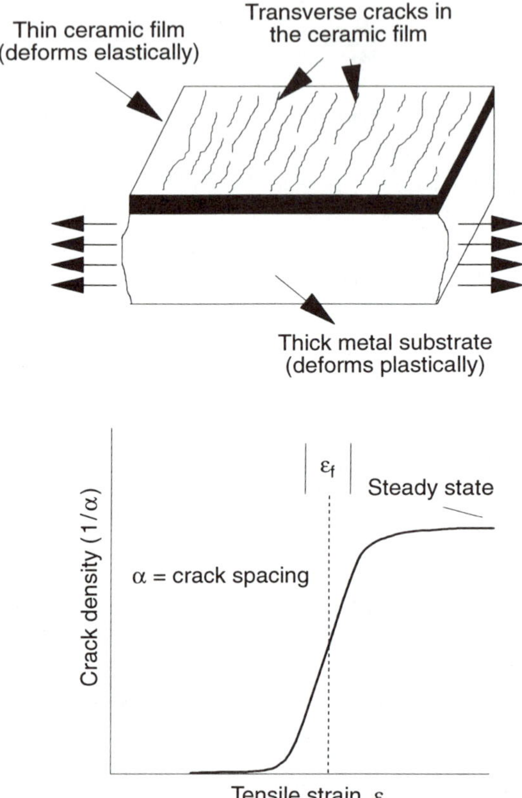

Figure 12.6 Schematic diagram of the tensile test method for measuring coating adhesion. After Agrawal and Raj [52].

practical adhesion can be gained. In the ultracentrifugal [54, 55] and ultrasonic methods [56] the adhesion of thin films can be assessed by subjecting the coating to acceleration or deceleration forces without using adhesives. In the ultracentrifugal method the specimen in the form of a rotor is levitated magnetically and rotated in vacuum at extremely high speeds to provide the centrifugal force. The instant of coating detachment can be detected by a pickup coil and the forces on the film can be calculated from a knowledge of the thickness and density of the coating, the radius of the rotor and the critical value of rotor speed [9]. Although data have been published for a number of coating–substrate systems, no systematic work has been carried out on the variation of adhesion with substrate for a single coating combination, and the method itself is not applicable to the testing of engineering components.

12.4.3 Bend and tensile test methods

As an alternative to determining the adhesive strength of the interfacial region formed in ceramic–metal joints, a number of tests based on fracture mechanics have been developed. These provide data such as fracture energy, which is closely interrelated with the bond or adhesive strength of the interfacial region [57, 58]. Of these, bend test methods require a stress concentration at the interface to initiate crack growth; this is usually provided by introducing a notch at the desired position, the notch length and the radius at the notch tip affecting the measured data [44]. As with the pull test, in addition to forming a suitable notch exactly at the coating–substrate interface, attachment of the test sample is usually by adhesives, which can give rise to practical problems in sample preparation and precludes tests on very strongly bonded coatings. Because of the scatter in fracture mechanics data, a set of identical specimens has to be tested and quantitative data evaluated with the aid of Weibull statistics. Figure 12.7 shows the experimental arrangement used by Muller *et al.* [45] in measuring the interface fracture toughness of physical vapor deposited (PVD) titanium nitride coatings ($<5\,\mu$m thick) deposited onto a steel substrate using a three-point bending test. As shown by Fig. 12.7, Muller *et al.* used two types of notching techniques, mechanical and a low adhesion film of copper, which resulted in differing degrees of scatter (Weibull modulus) in the fracture toughness measurements. The maximum values that could be measured using this method were in the range 200–$250\,$J/m^2 ($K_c < 8\,$N/m$^{3/2}$), representing the limiting strengths of the adhesives employed in sample fabrication. The best results were obtained with copper films approximately 100 nm thick.

For thicker coatings ($>100\,\mu$m), Clyne and coworkers [46] used a four-point bending test (Fig. 12.8) and made a through-thickness notch to the interface using a wire saw. Figure 12.8b shows a schematic of a typical load–displacement trace and indicates the events responsible for the shape of the plot. The critical energy release rate is calculated by comparing the total strain energy beam before and after a propagation event then dividing the difference by the area of contact.

As an alternative to simple bend testing, the critical strain energy release rate for ceramic–metal interfaces can be determined from a wide range of linear elastic fracture mechanics (LEFM) test specimens (Fig. 12.9); for further details see the literature [47–50]. In the case of the double cantilever beam (DCB) technique a typical sample arrangement is shown shematically in Fig. 12.9a. The interfacial critical strain energy release rate G_{Ic} is represented by

$$G_{Ic} = (P_c{}^2/2W)\,\mathrm{d}C_l/\mathrm{d}l \tag{12.6}$$

where P_c is the critical load for crack extension, W is the width of the specimen, l is the crack length measured from the end of the sample and C_l is the compliance of the test piece at the position of the applied load. In

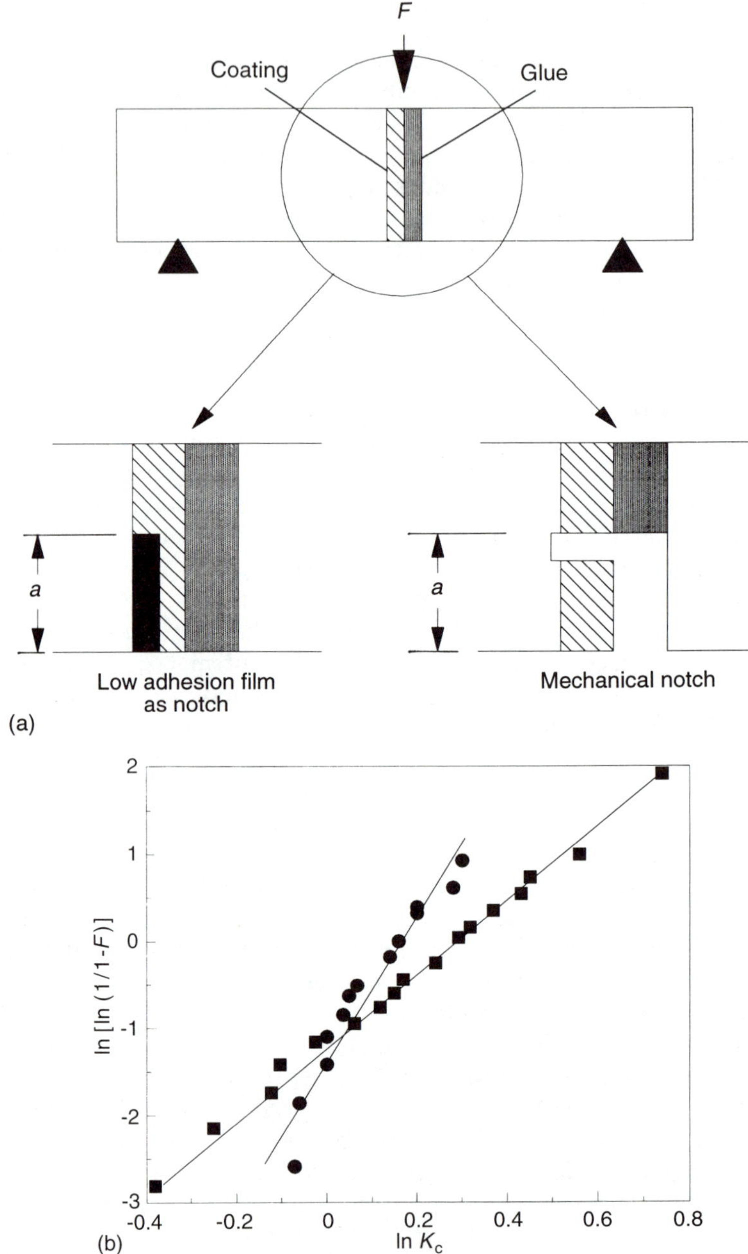

Figure 12.7 (a) Bend test sample used to measure the interfacial fracture energy of thin ceramic coatings. Note the use of a low adhesion film as a notch (left) and a cut perpendicular to the film as a mechanical notch (right). (b) Weibull diagram for bend test samples with a mechanical notch (■) or a copper film notch (●).

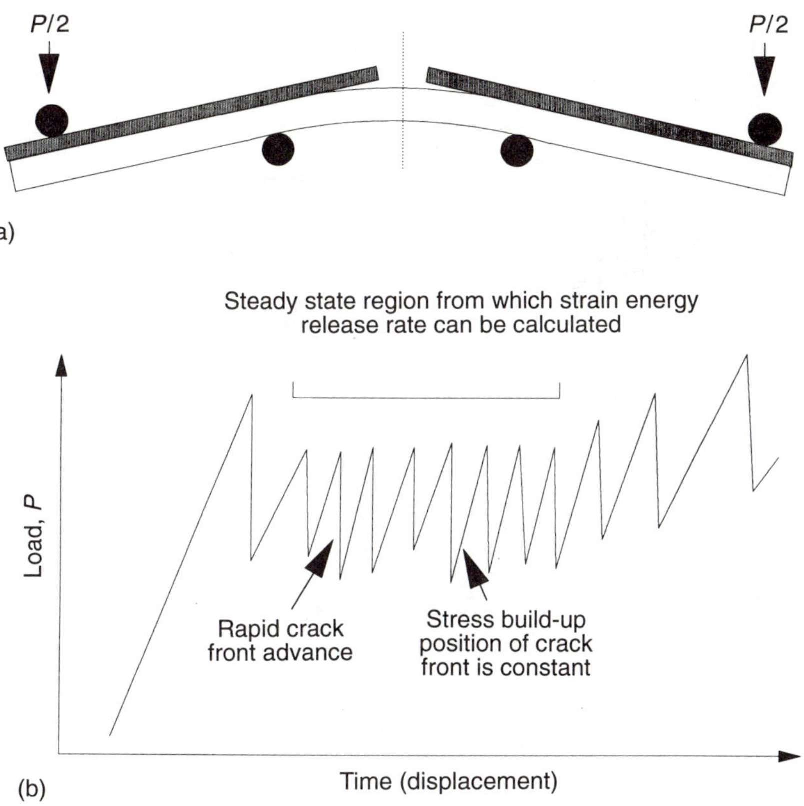

Figure 12.8 (a) Loading arrangement of the four-point bend test. (b) A schematic representation of the load–displacement trace employed in the testing of thick deposits.

equation 12.6 the precise determination of the crack length is critical but is difficult to establish by direct measurement, consequently it is usually obtained from a calibration curve of compliance versus crack length for a range of known crack lengths. By comparing a number of possible LEFM approaches for determining G_{Ic} of plasma-sprayed coatings, Guo and Wang [49] concluded that the DCB method appeared to offer a number of advantages: (i) a single specimen could produce several values of G_{Ic}, essential for any statistical evaluation of the data; (ii) the data reproducibility is good with reduced experimental scatter when compared with SEN (Single Edge Notched) and SB (Short Bar) methods; (iii) the method is theoretically more robust since the larger calilbration range for C_1 ensures a smaller relative calibration error.

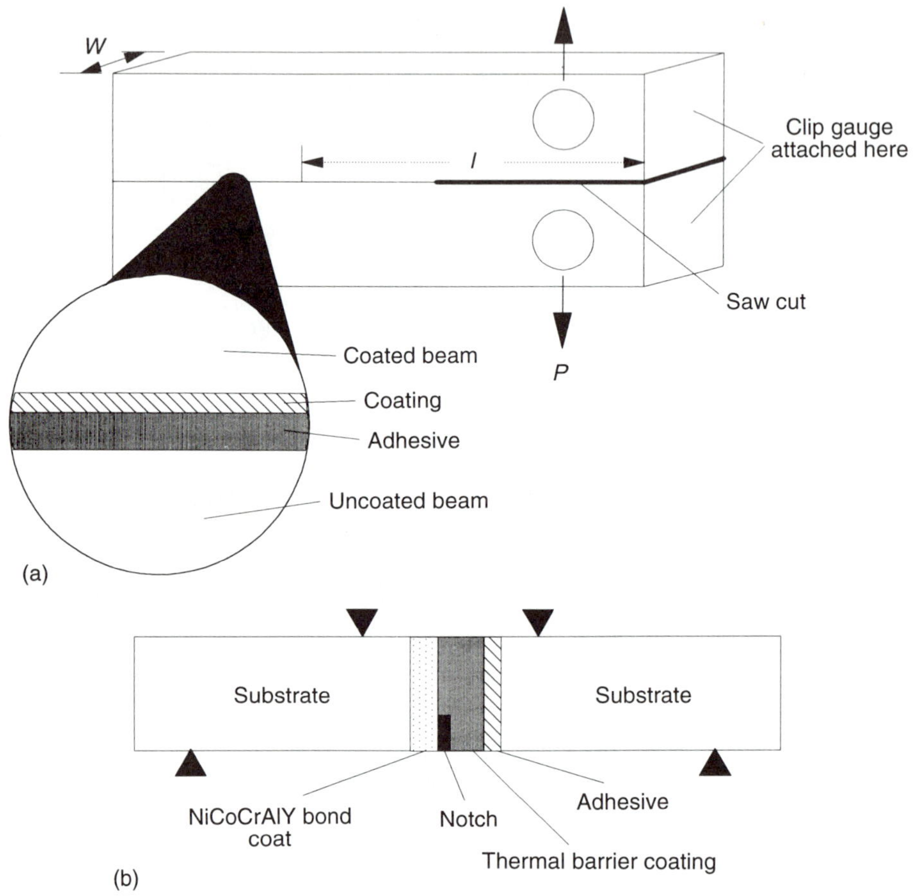

Figure 12.9 Examples of linear elastic fracture mechanics (LEFM) test specimens used in assessing ceramic–metal interfaces: (a) double cantilever beam (DCB) and (b) single edge notched bend (SENB) methods.

12.4.4 Indentation and shockwave-loading methods

The indentation adhesion test involves introducing a mechanically stable crack into the coating–substrate interface by the use of conventional indentation procedures using either Brale or Vickers indenters [21, 24]. The resistance to propagation of the crack along the interface is then used as a measure of adhesion and, by analogy with the fracture of homogeneous brittle solids [59, 60], this may be characterized by both a fracture resistance parameter and a strength parameter. The fracture resistance parameter relates uniquely to the bonding across the interface, and is a more fundamental measure of adhesion, whereas the strength is determined by the combined influences of the fracture resistance, the strength-controlling defects and residual stresses within the coating. The test assumes that the interface within the vicinity of the plastic zone created during indentation has a lower toughness than either the coating or substrate material, so it will be a site of preferential lateral crack formation. If fracture occurs in the coating or substrate rather than at the interface, it may be concluded that interface toughness is at least as large as that of the weaker component.

A schematic representation of the indentation test used by Jindal *et al.* [24] is given in Fig. 12.10 and shows the results of a series of indents made at different loads. The average change in lateral cracking is monitored as a function of load and the interfacial fracture toughness K_{Ii} is derived from the linear portion of the indentation load versus lateral crack length plot according to

$$K_{\mathrm{Ii}} = (G_{\mathrm{Ii}} E_{\mathrm{c}}/1 - v_{\mathrm{c}}{}^2)^{1/2} \tag{12.7}$$

where, in Fig. 12.10, A is a constant and E_{c} and v_{c} are Young's modulus and Poisson's ratio of the coating, respectively. An advantage of this technique is that the indentation adhesion parameters P_{c} and K_{Ii} are relatively

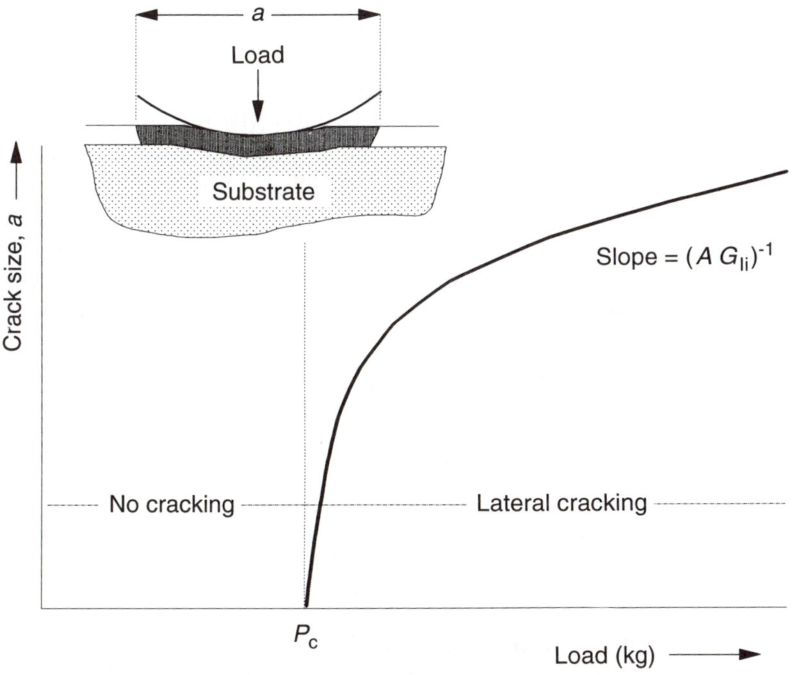

Figure 12.10 Schematic representation of the indentation adhesion test.

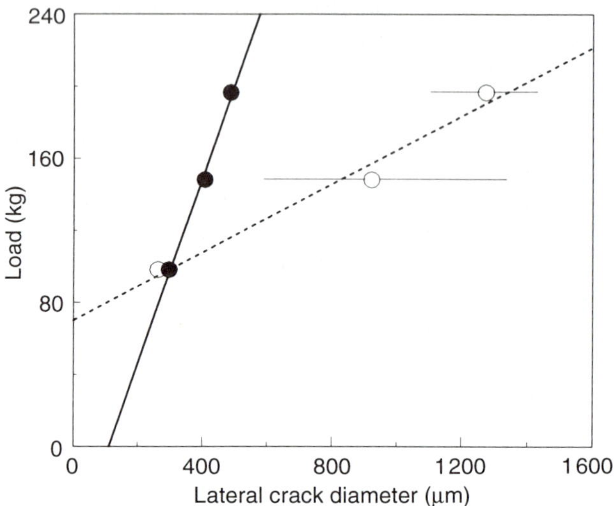

Figure 12.11 Indentation load versus lateral crack diameter for multilayer TiC–Al$_2$O$_3$ coated samples with thin discontinuous η-phase (●) and thick continuous η-phase (○).

insensitive to the substrate hardness, which is a problem with the scratch adhesion test (discussed later in the chapter). To further illustrate the differences between these two test methods, Fig. 12.11 shows indentation test results for carbide inserts CVD-coated with TiC–Al$_2$O$_3$ layers differentiated in terms of the η-phase occurrence at the coating–substrate interface. On the basis of the slope of the load/lateral crack diameter function, the coatings deposited onto a continuous layer of the brittle η-phase have poorer adhesion (toughness) compared with similar coatings formed with a discontinuous η-phase layer. But in both cases the scratch adhesion test indicated essentially identical L_c values for adhesive failure at the coating–substrate interface [24].

The approach of Evans and coworkers [21, 23] is based on the observation that, in the absence of buckling and for planar interfaces, there is no driving force for growth of a delamination already in existence at the coating–substrate interface; this initial delamination may arise due to interfacial contamination or by void formation and coalescence. Consequently, for such interfaces, buckling becomes a prerequisite for fracture propagation and eventual spalling. The critical stress for buckling of a circular delamination is given by [61]

$$\sigma_c = [KE_c/12(1 - v_c{}^2)](t/a)^2 \tag{12.8}$$

where t is the coating thickness, a is the delamination radius and $K \cong 14.7$. Once buckling occurs, a crack driving force G develops given by [62]

$$G = (1 - v_c)(1 - \kappa)t(\sigma^2 - \sigma_c{}^2)/E_c \tag{12.9}$$

where σ is the net compressive stress in the coating and $\kappa = 0.38$. Further growth of a delamination occurs if $G > G_{c}$, either for the interface or for the coating, and the delamination radius for coating spallation a_{s} is given by [62]

$$a_{s}/t = 1.9\,(E_{c}/\sigma)^{1/2} \tag{12.10}$$

For indentation-induced spalling [21–23]

$$\sigma = \sigma_{R} + E_{c}V/2\pi(1 - v_{c})\,ta^{2} \tag{12.11}$$

where σ_{R} is the initial residual stress and V is the indentation volume, with

$$V = 0.24\,(P/H)^{3}\cot\psi \text{ and } a = \lambda P^{3/4} \tag{12.12}$$

where P is the indentation load, ψ is the indenter half-angle, H is the hardness and λ is an experimentally determined coefficient. The critical indentation load for spalling P_{s} is given by

$$P_{s}^{3/2} = 3.7\,t^{2}E_{c}\,[\lambda^{2}\sigma_{R} + 0.24\cot\psi/2\pi(1 - v_{c})\,tH^{3/2}]^{-1} \tag{12.13}$$

Figure 12.12 shows some results obtained from indentation testing of ZrO_{2}–$Y_{2}O_{3}$ coatings [23]. The data exhibit the expected trend from equation (12.12) in that $a \sim P^{3/4}$, and fracture toughness G_{c} along the delamination path was found to be approximately $40\,\text{J/m}^{2}$, comparable with literature values for cubic ZrO_{2}. Similar results to those presented in Fig. 12.12 were

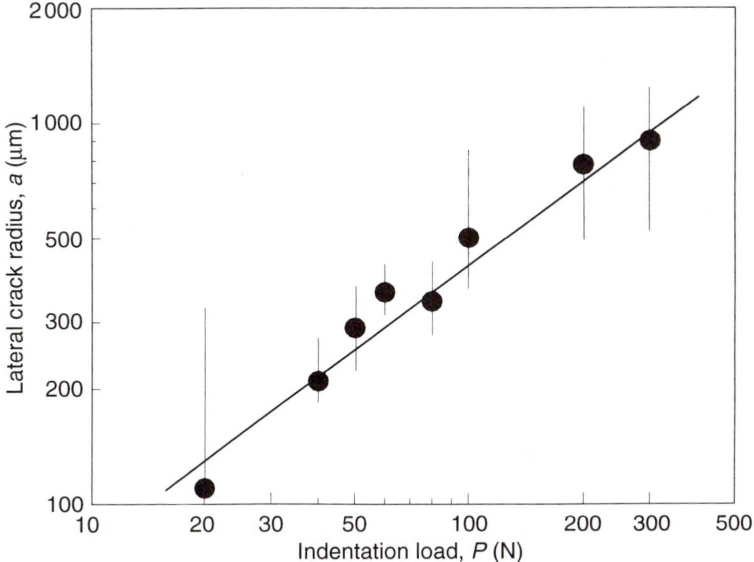

Figure 12.12 Plot of trend in delamination radius with indentation load for ZrO_{2}–$Y_{2}O_{3}$ coatings.

obtained for indentation spalling tests on a series of ZnO/Si samples, and excellent agreement was obtained between theoretical values of P_s (equation 12.13) and experimental measurements, indicating that the methods outlined above have some merit when it comes to assessing the adhesion of thin films. Recent work by Ashcroft and Derby [53] has confirmed the proportionality predicted by equation 12.12 for glass ceramic films bonded to metal substrates (Fig. 12.13). However, the values derived from this work for interfacial fracture energies, $1.5-2.7\,kJ/m^2$, are greatly in excess of those obtained from bend tests performed on similar samples, $3-5\,J/m^2$. The problem lies in the fact that, for other than ceramic–ceramic systems, it is impossible to propagate an indentation-induced interfacial crack without deforming the substrate and extensively damaging the coating. Until it is possible to accommodate these factors into the theoretical models of the indentation method, in the case of ceramic–metal systems, the test is only capable of ranking and not quantifying interfacial adhesion.

Alternatively, mechanical stimuli can be introduced into the coating–substrate system in order to produce delamination by shockwave-loading methods [20, 63, 64]. These techniques involve the absorption of energy either from the impact of erosive particles [64] or from an impinging laser beam [20, 63]. In the laser method the absorbed energy may induce a stress wave, and the acoustic waveform from the rear side of the substrate can be monitored while the laser probes the coating surface, as shown in Fig. 12.14. If the incident

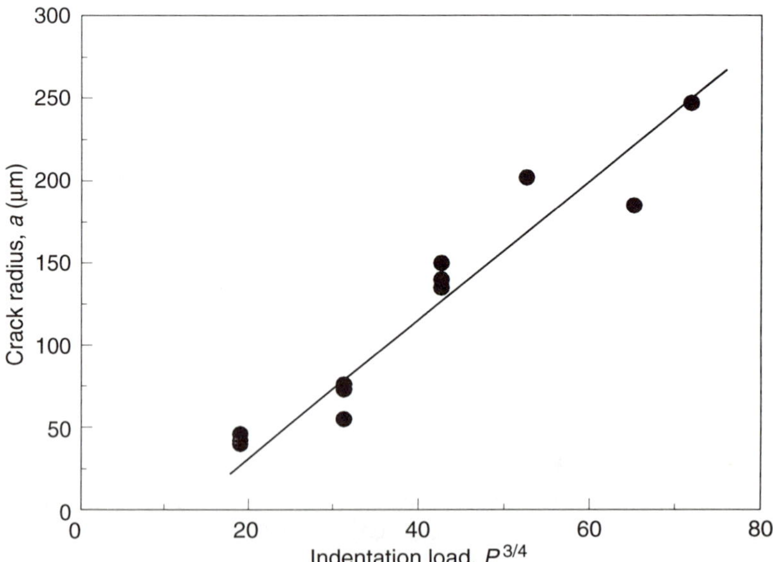

Figure 12.13 Indentation load against crack radius for a glass ceramic film.

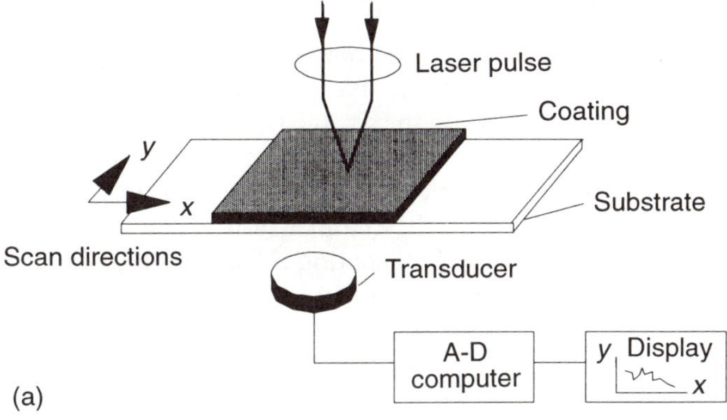

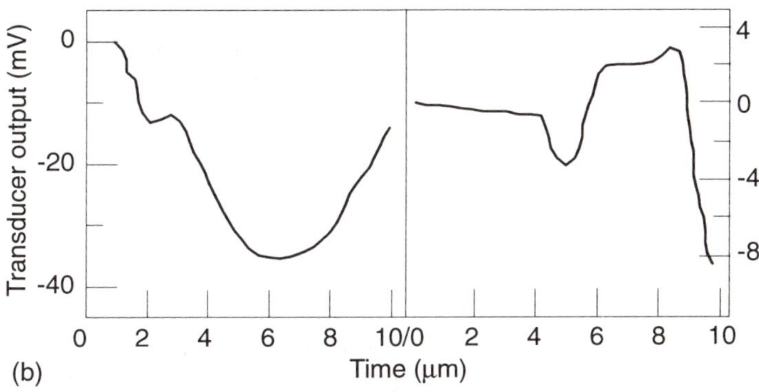

Figure 12.14 (a) Schematic diagram of the laser acoustic adhesion test. (b) Transducer outputs for plasma-sprayed coatings tested under the laser acoustic adhesion test. The main difference between the two curves is the signal propagation time, which is shorter when adhesion is good.

probe is pulsed, it is also possible to image the propagation of thermal wave within the coating; this can be used to detect the presence of cracks or debonded regions. Laser techniques have the advantage that they stimulate the spall problems which occur during thermal cycling and impact damage, as well as being relatively nondestructive and applicable to complex components.

Loh *et al.* [20] have examined the quasi-heating due to laser impingement on a coating system. This has the effect of introducing localized compression which, in conjunction with the existing residual stresses, was sufficient to cause spallation. By varying the power and duration of the laser pulses, the spall resistance of brittle coatings could be effectively measured.

12.4.5 The scratch adhesion test method

Of all the tests available for the measurement of coating adhesion, only the scratch test has achieved widespread use in assessing the adhesion of hard coating–substrate systems [65]. In the scratch test a stylus (usually a Rockwell C diamond) is drawn over the sample surface under a stepwise or continuously increasing normal force until the coating detaches [25]. In practice the film is seldom removed entirely from the channel so it is convenient to define a critical load (L_c) which is related to coating adhesion; this is usually taken as the load at which the coating is removed in a regular way along the whole track length [26], but because there is a distribution of flaws at the coating–substrate interface, isolated areas of coating removal do occur at lower loads, so the critical load transition can be somewhat subjective [39]. Coating detachment can be monitored using reflected light or scanning electron microscopy, acoustic emission [27] or frictional force measurement [28, 29]. Acoustic emission and frictional force measurement provide traceable signals which can be used to compare results from different samples; this may avoid some of the subjectivity of measurements made by eye. The sensitivity of the friction measurement procedure is particularly enhanced when the scratch test is applied to hard coatings less than 1 μm thick.

The initial analysis of the mechanics of the scratch test by Benjamin and Weaver [30] used the theories developed for fully plastic indentation to give an expression for the critical shearing force for coating removal in terms of the scratch geometry, the substrate properties and the friction force on the stylus.

$$F = KAH/(R^2 - A^2)^{1/2} \tag{12.14}$$

where $A = (L_c/\pi H)^{1/2}$, L_c is the critical load, R is the indenter tip radius, F is the shear strength of the interface per unit area, A is the radius of the circle of contact, H is the indentation hardness of the substrate material and K is a constant varying between 0.2 and 1.0 [30, 31]. More recent treatments consider the case of elastic–plastic indentation [31] and use a Griffith energy balance approach to relate the local stress σ responsible for coating detachment to the work of adhesion [32–34]:

$$W = \sigma^2 t/2E_c \tag{12.15}$$

where t is the coating thickness and E_c is the Young's modulus of the coating. Attempts have been made to calculate σ expressed as a combination of the applied stresses due to the indenter and the internal stresses within the coating. Laugier's proposed analysis of the applied stresses, based on the elastic equations of Hamilton and Goodman [35], is not generally applicable. Rickerby and coworkers [36, 41] have identified three contributions to the stresses for coating detachment (Figure 12.15):

- an elastic–plastic indentation stress
- an internal stress
- a tangential friction stress

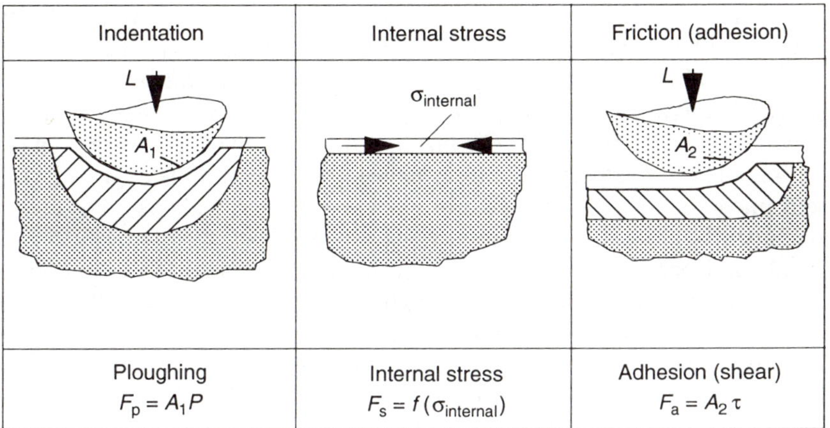

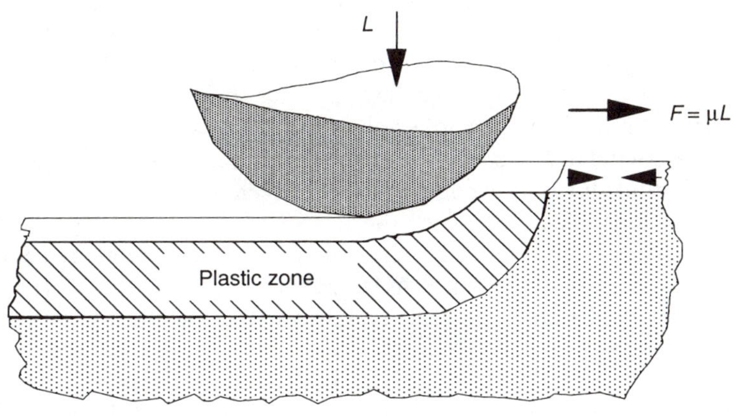

$$F = F_p + F_s + F_a$$

Figure 12.15 The scratch adhesion test represented as the sum of three contributions: an indentation term, an internal stress term and a frictional term. These may be represented as three frictional contributions; a ploughing component, an internal stress component and an adhesive component.

This more recent analysis leads to an equation relating critical load for coating detachment to the work of adhesion, which may be expressed as

$$L_c = (\pi d_c^2/8)(2EW/t)^{1/2} \tag{12.16}$$

where d_c is the width of the scratch track at the critical load, E is Young's modulus of the coating and t its thickness. The model of Bull and coworkers allows the work of adhesion to be calculated and their results showed that crack tip plasticity was important in the fracture process. However, at present, these theoretical models are not universally applicable and, given the complex stress state which exists under a moving indenter, it is difficult to see how the scratch test can ever be more than a qualitative measure of coating adhesion.

There are a number of extrinsic parameters (i.e. properties of the coating–substrate system) that influence the value of the critical load determined in the scratch test. For instance, the critical load for coating detachment increases with coating thickness or substrate hardness; but the critical load generally decreases as the levels of residual stress within the coating increase. In addition to the extrinsic parameters that influence the test, intrinsic parameters, such as indenter tip radius or the loading system of the tester, have an important bearing on the critical load determination. The more important of these are listed in Table 12.5 and are discussed in detail elsewhere [43]. Commercial scratch testers are basically of two types. In the automatic scratch test the normal load is continuously increased along the length of the scratch track by a spring-loading mechanism. This test is very quick to perform but has the disadvantage that catastrophic failure occurs at the first sufficiently large flaw and thus the critical load may be an underestimate of the practical adhesion in any application. Also, the critical load has been found to be a sensitive function of the machine loading geometry so it is quite difficult to make comparisons between laboratories with various types of machine [28, 41]. The alternative manual scratch tester uses deadweight loading and hence requires the performance of many scratches to assess the critical load, a much more time-consuming process. However, it has the advantage that the interfacial flaw distribution can also be assessed

Table 12.5 Factors affecting the critical load for coating detachment in the scratch test method

Intrinsic parameters	Extrinsic parameters
Loading rate [28, 41, 43]	Substrate properties [43, 67]
	Hardness
Scratching speed [28, 41, 43]	Modulus
	Thermal expansion coefficient
Indenter tip radius [28, 43, 66]	
	Coating properties [43, 68]
Indenter wear [43]	Hardness
	Modulus
Machine factors	Stress and interfacial properties
	Thickness
	Friction force and friction coefficient [28, 37, 43]
	Surface condition and testing environment

by counting the number of failures which occur at each load [39]. The number of failures is found to saturate at a certain load using reflected light microscopy or acoustic emission, where each failure generates a small burst of acoustic emission. Thus a cumulative failure probability $P(L)$ can be defined as

$$P(L) = N(L)/N_{\text{sat}} - N_0 \qquad (12.17)$$

where $N(L)$ is the number of failures at load L, N_0 is the number of failures at low loads (usually $N_0 = 0$) and N_{sat} is the saturation failure number (Fig. 12.16). In Fig. 12.16, L_a is a constant equal to the load at which there

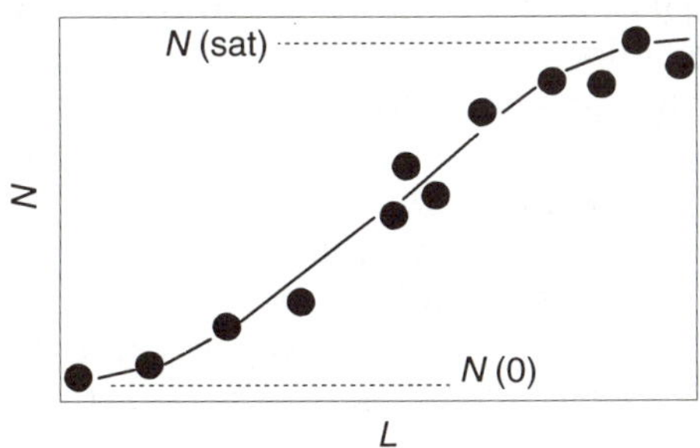

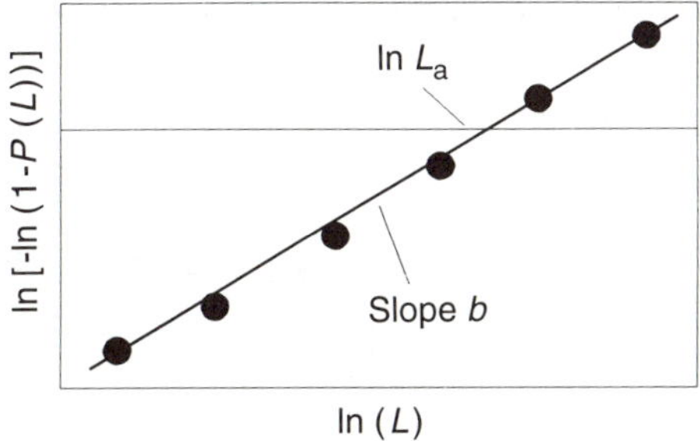

Figure 12.16 Weibull statistical analysis of the scratch test. Starting from the equation:

$$P(L) = 1 - \exp[-(L - L_0)/L_a]^b \qquad (*)$$

where L_a = characteristic value of L equivalent to 63.2% failure probability
L_0 = minimum value of L (generally $L = 0$)
b = Weibull parameter
$P(L)$ = cumulative distribution function

Can rewrite (*) as

$$\ln\ln\left[\frac{1}{1 - P(L)}\right] = b\ln(L = L_0) - b\ln L_a$$

or

$$\ln[-\ln(1 - P(L)] = b\ln(L - L_0) - b\ln L_a$$

is a 63.2% cumulative failure probability (which can be used as a critical load criterion related to the flaw distribution) and b is the Weibull parameter. The advantage of this method is that it offers the possibility of providing a completely automated way of determining the critical load for coatings which provide well-defined pulses of acoustic emission associated with coating detachment failure events. The acoustic emission trace at each load can be recorded and examined by computer, the number of failures counted and stored, and the variation with load determined. From this a value of L_a can be calculated which involves no subjective decisions by the operator and hence is a much more reliable adhesion criterion.

The scratch test has also been used to determine both interfacial and cohesive failure strengths for thick films [69–71]. In this case the material to be tested is in the form of a polished cross section and scratches are performed across the coating by starting from the substrate using stepwise-increasing loads (Fig. 12.17). A cone of damage is produced as the scratch diamond exits from the face of the coating, the size of which can be used to determine the cohesive strength of the coating. Figure 12.18, taken from the work of Kingswell and coworkers [71, 72], shows how the test can be used to identify subtle differences in cohesive strength, resulting from variations in the spraying conditions, as well as the larger changes due to oxide entrapment (present in coatings plasma-sprayed under air but not in coatings sprayed under vacuum). A crack is also produced at the

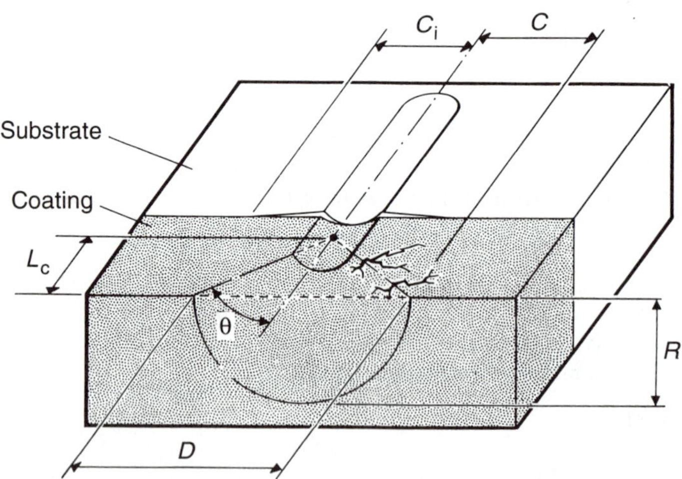

Figure 12.17 Schematic diagram of the damage produced during a scratch adhesion–cohesion test.

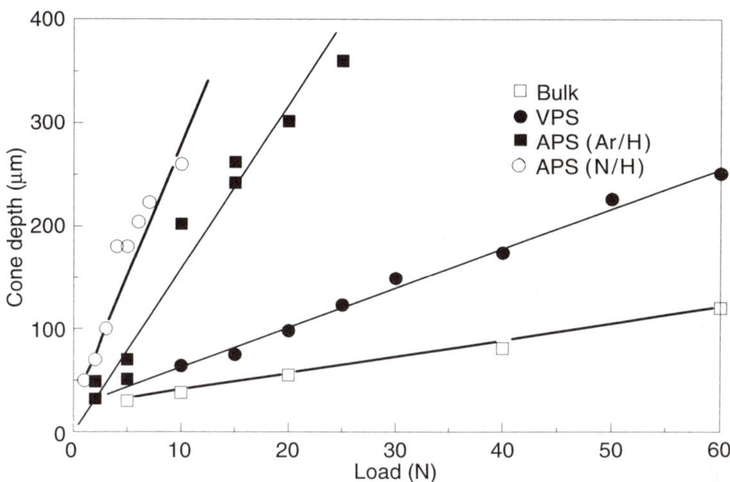

Figure 12.18 Variation in cone depth with indenter load for scratch tests performed on bulk and thermally sprayed tungsten.

coating–substrate interface once a critical load has been reached; this can be used to rank coating–substrate adhesion in these situations.

12.5 SUMMARY

Two quantifiable measures of adhesion are important in any application, namely adhesive strength (controlled by the interfacial flaw distribution as well as by interfacial bonding) and crack driving force (more closely related to bonding across the interface). Currently there is no ideal adhesion test, and none of the test methods available for the measurement of coating–substrate adhesion give a value for the true basic adhesion of the system; all measure what is termed *practical adhesion*. There is a genuine need to develop new test methods which are nondestructive in nature. Of the tests available today, indentation, bend, laser and scratch adhesion methods offer the best prospects of relating practical levels of adhesion to the performance of real engineering components.

REFERENCES

1. Rickerby, D.S. and Matthews, A. (1991) *Advanced Surface Coatings*, Blackie, London.
2. Grainger, S. (ed.) (1989) *Engineering Coatings – Design and Application*, Abingdon Publishing.
3. Meetham, G.W. (ed.) (1981) *The Development of Gas Turbine Materials*, Applied Science Publishers, London.
4. Chopra, K.L. (1969) *Thin Film Phenomena*, McGraw-Hill, New York.
5. Good, R.J. (1976) *J. Adhes.*, **8**, 1.

6. Mittal, K.L. (1974) *J. Adhes.*, **7**, 337.
7. Mittal, K.L. (1976) *Electrocomp. Sci. Technol.*, **3**, 21.
8. Pulker, H.K., Perry, A.J. and Berger, R. (1981) *Surf. Technol.*, **14**, 25.
9. Campbell, D.S. (1970) in *Handbook of Thin Film Technology* (eds L.I. Maissel and R. Glang), McGraw-Hill, New York, Ch. 12.3.
10. Ruggerio, E.M. (1965) in *Proceedings of the 14th Annual Microelectronics Symposium*, St Louis Section IEEE, 24–26 May, IEEE, New York, p. 68–71.
11. Davies, D. and Whittaker, B. (1976) *Metall. Rev.*, **12**, 15.
12. Jacobsson, R. (1976) *Thin Solid Films*, **34**, 191.
13. Katz, G. (1976) *Thin Solid Films*, **33**, 99.
14. Bodo, P. and Sundgren, J.E. (1986) *J. Appl. Phys.*, **60**, 1161.
15. Jankowski, A.F. (1987) *Thin Solid Films*, **154**, 183.
16. Chapman, B.N. (1974) *J. Vac. Sci. Technol.*, **11**, 106.
17. Bullet, T.R. and Prosser, J.L. (1972) *Prog. Org. Coat.*, **1**, 45.
18. Stephens, A.W. and Vossen, J.L. (1976) *J. Vac. Sci. Technol.*, **13**, 38.
19. Cooper, J.A., Dewhurst, R.J. and Palmer, S.B. (1985) in *Proceedings of the Ultrasonic International*, Brighton, 2–4 July, Butterworth, London.
20. Loh, R.L., Rossington, C. and Evans, A.G. (1986) *J. Am. Ceram. Soc.*, **69**, 139.
21. Chiang, S.S., Marshall, D.B. and Evans, A.G. (1981) in *Surfaces and Interfaces in Ceramic and Ceramic–Metal Systems* (eds J. Pask and A.G. Evans), Plenum Press, New York, p. 603.
22. Marshall, D.B. and Evans, A.G. (1984) *J. Appl. Phys.*, **56**, 2632.
23 Rossington, C., Marshall, D.B., Evans, A.G. and Khuri-Yakub, B.T. (1984) *J. Appl. Phys.*, **56**, 2639.
24. Jindal, P.C., Quinto, D.T and Wolfe, G.J. (1987) *Thin Solid Films*, **154**, 361.
25. Perry, A.J. (1986) *Surf. Engng*, **3**, 183.
26. Perry, A.J. (1983) *Thin Solid Films*, **107**, 167.
27. Perry, A.J. (1981) *Thin Solid Films*, **81**, 357.
28. Valli, J.J. (1986) *J. Vac. Sci. Technol. A*, **4**, 3007.
29. Jacobson, S., Jonsson, B. and Sundquist, B. (1983) *Thin Solid Films*, **107**, 89.
30. Benjamin, P. and Weaver, C. (1960) *Proc. R. Soc. Lond. A*, **254**, 177.
31. Weaver, C. (1975) *J. Vac. Sci. Technol.*, **12**, 18.
32. Laugier, M.J., (1984) *Thin Solid Films*, **117**, 243.
33. Laugier, M.J. (1986) *J. Mater. Sci.*, **21**, 2269.
34. Griffith, A.A. (1920) *Phil. Trans. R. Soc. Lond. A*, **221**, 163.
35. Hamilton, G.M. and Goodman, L.E. (1966) *J. Appl. Mech.*, **33**, 371.
36. Burnett, P.J. and Rickerby, D.S. (1987) *Thin Solid Films*, **154**, 403.
37. Bull, S.J., Rickerby, D.S., Matthews, A. *et al.* (1988) *Surf. Coat. Technol.*, **36**, 503.
38. Bull, S.J. and Rickerby, D.S. (1989) *Br. Ceram. Trans. J*, **88**, 177.
39. Bull, S.J. and Rickerby, D.S. (1990) *Surf. Coat. Technol.*, **42**, 149.
40. Bowden, F.P. and Tabor, D. (1954) *The Friction and Lubrication of Solids*, Clarendon, Oxford.
41. Bull, S.J., Rickerby, D.S., Matthews, A. *et al.* (1989) in *Proceedings of the 1st International Conference on Plasma Surface Engineering*, Garmisch-Partenkirchen, 19–23 Sept. 1988, DGM Informationsgesellschaft, Oberursel, p. 1227.
42. Wyatt, O.H. and Dew-Hughes, D. (1974) *Metals, Ceramics and Polymers*, Cambridge University Press, Cambridge, p. 411.
43. Steinmann, P.A., Tardy, Y. and Hintermann, H.E. (1987) *Thin Solid Films*, **154**, 333.
44. Jayatilaka, A.S. (1979) *Fracture Engineering Brittle Materials*, Applied Science Publishers, London.
45. Muller, D., Cho, Y.R., Berg, S. and Fromm, E. (1993) *Surf. Coat. Technol.*, **60**, 401.
46. Howard, S.J. and Clyne, T.W. (1991) *Surf. Coat. Technol.*, **45**, 333.

47. Kleer, G., Schonholz, R. and Doll, W. (1991) in *High Performance Ceramic Films and Coatings* (ed. P. Vincenzini), Elsevier, Amsterdam, p. 329.
48. Bernt, C.C. and McPherson, R. (1981) *Mater. Sci. Res.*, **14**, 619.
49. Guo, D.Z. and Wang, L.J. (1992) *Surf. Coat. Technol.*, **56**, 19.
50. Kingswell, R. (1992) Properties of vacuum plasma sprayed alumina and tungsten coatings. Ph.D. thesis, Brunel University, 144.
51. Lin, D.S. (1971) *J. Phys.*, **D4**, 1977.
52. Agrawal, D.C. and Raj, R. (1989) *Acta Metall.*, **37**, 1265.
53. Ashcroft, I.A. and Derby, B. (1993) *J. Mater. Sci.*, **28**, 2989.
54. Beams, J.W. (1954) *Science*, **120**, 169.
55. Beams, J.W., Breazeale, J.B. and Bart, W.L. (1955) *Phys. Rev.*, **100**, 1657.
56. Moses, S.T. and Witt, R.W. (1949) *Ind. Eng. Chem.*, **41**, 2334.
57. Suga, T. and Elssner, G.Z. (1985) *Werkstofftech.*, **16**, 75.
58. Elssner, G., Suga, T. and Turwitt, M. (1985) *J. Phys(Paris)*, *Colloq. C4*, **46**, 597.
59. Lawn, B.R. and Wilshaw, T.R. (1975) *Fracture of Brittle Solids*, Cambridge University Press, Cambridge, p. 53.
60. Evans, A.G. (1982) *J. Am. Ceram. Soc.*, **65**, 127.
61. Timoshenko, S. and Gere, J.M. (1961) *Theory of Elastic Stability*, 2nd edn, McGraw-Hill, New York, 390.
62. Evans, A.G. and Hutchinson, J.W. (1984) *Int. J. Solid Struct.*, **20**, 455.
63. Vossen, J.L. (1978) in *Adhesion Measurement of Thin Films, Thick Films and Bulk Coatings* (ed. K.L. Mittal), Special Tech. Pub. 640, ASTM, Philadelphia, p. 122.
64. Snowden, W.E. and Aksay, I.A. (1981) in *Surfaces and Interfaces in Ceramic and Ceramic–Metal Systems* (eds J. Pask and A.G. Evans), Plenum Press, New York, p. 651.
65. Rickerby, D.S. (1988) *Surf. Coat. Technol.*, **36**, 541.
66. Hamersky, J. (1969) *Thin Solid Films*, **3**, 263.
67. Hammer, B. and Perry, A.J. (1982) *Thin Solid Films*, **96**, 45.
68. Burnett, P.J. and Rickerby, D.S. (1988) *Thin Solid Films*, **157**, 233.
69. Lopez, E., Beltzung, F. and Zambeki, G. (1989) *J. Mater. Sci. Lett.*, **8**, 346.
70. Beltzung, F., Zambelli, G., Lopez, E. and Nicoll, A.R. (1989) *Thin Solid Films*, **181**, 407.
71. Kingswell, R., Rickerby, D.S., Bull, S.J. and Scott, K.T. (1991) *Thin Solid Films*, **198**, 139.
72. Scott, K.T. and Kingswell, R. (1991) in *Advanced Surface Coatings* (eds D.S. Rickerby and A. Matthews), Blackie, London, p. 217.

Index